World Health Organization
Regional Office for Europe
Copenhagen

Epidemiology of occupational health

Edited by

M. Karvonen

and

M.I. Mikheev

WHO Regional Publications, European Series No. 20

ISBN 92 890 1111 4
ISSN 0378-2255

614·8

Epidemiology
of occupational
health

LIBRARY, RO WESTCOTT
REGULATIONS FOR BORROWERS

1. Books are issued on loan for a period of <u>1 month</u> and must be returned to the library promptly.

2. Before books are taken from the Library receipts for them must be filled in, signed, and handed to a member of the Library Staff. Receipts for books received through the internal post must be signed and returned to the Library immediately.

3. Readers are responsible for books which they have borrowed, and are required to replace any such books which they lose or damage. In their own interest they are advised not to pass on to other readers books they have borrowed.

4. To enable the Library Staff to deal with urgent requests for books, borrowers who expect to be absent for more than a week are requested either to arrange for borrowed books to be made available to the PA or Clerk to the Section, or to return them to the Library for safekeeping during the period of absence.

Cover photo by courtesy of the Department of Occupational Medicine, Regional Hospital, Örebro, Sweden.

CONTENTS

Preface

Many definitions of epidemiology have emerged over the last few decades, in keeping with the rapid development and broadening of the science. According to the Dictionary of epidemiology edited by John M. Last in 1983, epidemiology is "the study of the distribution and determinants of health-related states and events in populations, and the application of this study to control of health problems". Epidemiological methods have been widely used in the field of occupational health to describe the health status of specific working populations, to study their morbidity in relation to the type of occupation, to identify specific occupational hazards, to generate and test hypotheses on cause–effect relationships, and to evaluate interventions. When the relationship has proved to be quite strong and specific, such studies have been very successful in increasing our knowledge of the effects of occupational hazards. In many cases, however, epidemiological studies have led to controversial and confusing results. This may have been due to the use of inappropriate methodology, but may also have been the result of overestimating the power of epidemiological and statistical methods. Those engaged in resource consuming studies must be well aware of the limitations of the tools they are using. In the particular field of occupational health there are limitations due to the characteristics of the population under study, which is usually rather small, selected, and subject to change over time with regard to exposure to occupational hazards. Other limitations arise from the type of hazards concerned, very often combined with other exposures that may or may not be related to occupation, and also with long latency periods for any outcome from the exposures. In addition, there are methodological limitations due to the lack of proper assessment of exposure and health effects. Retrospective studies may be subject to inappropriate information on exposure to the suspected hazard, as well as to other confounding environmental or behavioural factors. Prospective studies may overcome this type of problem but, besides their relatively high cost, they have their own methodological biases too, such as those linked to the quality of the reference groups, observer errors, and changes induced in the study population by the study itself.

Such limitations should not be seen as an obstacle to the use of epidemiology, but rather as justification of the use of sound and standardized methods. The present manual tries to respond to the specific needs of occupational health

vii

epidemiology. The WHO Regional Office for Europe, after a decade of work in this field, has secured the collaboration of leading experts to present and discuss in a practical way different approaches and methods and their application to specific problems in occupational health. Rather than a comprehensive review of the subject, the book presents a series of articles. The first four chapters deal with general principles and definitions in occupational epidemiology, and describe the work-related hazards and diseases. Chapters 5, 6 and 7 deal with information collection and the use of data in the assessment of health risks and in descriptive epidemiology. General methods for epidemiological studies are discussed in Chapters 8–10. The following chapters address specific aspects such as the study of combined effects (Chapter 11), the statistical analysis of epidemiological data (Chapter 13), the validity aspects of epidemiological studies, including consideration on the problems of "false positive" and "false negative" results and the basis for causality judgement (Chapter 14), or the particular interest of experimental epidemiology in occupational health (Chapter 15). Chapters 12 and 16 cover two special issues of importance to workers' health, namely occupational stress and the epidemiology of accidents. Chapter 17 gives an excellent overview of the uses of epidemiology in occupational health, and the last chapter presents a concrete case study, with an assessment of the use made of epidemiological methods. There are unavoidable repetitions and overlaps throughout this manual due to the way in which it was compiled, but it gives very useful and practical information to those interested in epidemiological research applied to occupational problems.

All this work was done in collaboration with the office of occupational health at WHO headquarters in Geneva.

It is hoped that this book will encourage and facilitate sound and reasonable application of epidemiology to the identification of hazards, assessment of risks and evaluation of control measures in the working environment, thus contributing to the achievement of the European target[a] calling for effective protection of people in the Region against work-related health risks. It is also hoped that it will have worldwide value in the development of epidemiological tools for the purposes of occupational health.

J.-P. Jardel

Director, Programme Management
WHO Regional Office for Europe

---

[a] *Targets for health for all.* Copenhagen, WHO Regional Office for Europe, 1985.

Independent reviewers

Professor M. Backett, Lidstone, South Town, Dartmouth, Devon, United Kingdom

Dr B. Bedrikow, Occupational Safety and Health Branch, International Labour Office, Geneva, Switzerland

Dr M. El Batawi, Chief, Occupational Health, World Health Organization, Geneva, Switzerland

Professor W.J. Eylenbosch, IEA Liaison Officer with WHO, International Epidemiological Association, University of Antwerp, Wilrijk, Belgium

Mr D. Hémon, Head of Research, INSERM, Villejuif Cédex, France

Dr V. Kodat, Director, Hygiene and Epidemiology Department, Ministry of Health, Prague, Czechoslovakia

Dr A. Lellouch, Ministry of Health, Paris, France

Professor R. Rothan, Chief, Department of Medical Inspection of Work, Ministry of Labour, Paris, France

Dr F. Varet, Chief, Office of Epidemiology, Prevention and Health Education, Ministry of Health, Paris, France

Epidemiology in the context of occupational health

M. Karvonen[a]

Epidemiology is a science concerned with morbidity and mortality: it studies the distribution of states of health and disease in the community as well as the distribution of health-related events and their determinants.

As applied to occupational health, epidemiology thus has the dual task of describing the distribution of deaths, accidents, illnesses, and their precursors in the various sections of the occupationally active population and of searching for the determinants of health, injury, and disease in the occupational environment.

The succinct definition requires some words of explanation. The preamble to the Constitution of the World Health Organization defines "health" as a state of complete wellbeing—physical, mental, and social—but "health" is also often used to encompass the entire continuum extending all the way from the ideal state of complete wellbeing to death. It should also be made quite clear at the outset that epidemiology, although it uses statistical methods, is not a mere application of biostatistics, but one of the two basic approaches in medical science, the other being concerned with disease mechanisms.

Epidemiological methods are increasingly used also for studying the functions of health services. Though obviously a sound and useful development, this aspect will not be discussed in the present volume.

Whereas clinical medicine is primarily concerned with sick individuals, epidemiology deals with communities. In an individual, the state of health can be described in terms of diagnosis and prognosis, but in a community rates are needed: e.g., the prevalence of ill subjects at a point—or short period—of time in a population, or the incidence of new cases in a population within a defined time, e.g., in a year. For mortality rates, the cases are deaths. The number of cases supplies the

---

[a] Pioppi, Salerno, Italy.

numerator for these rates, the size of the community in which the cases occur, the denominator.

The morbidity rates are community diagnoses. It is a further task of epidemiology to seek the determinants of these rates. Epidemiology is concerned with the causation of health and disease. The causes of ill health are to be sought (a) in the structural, functional, and behavioural characteristics of individuals, (b) in their physical, chemical, biological, and social environment, and (c) in the interactions between individuals and the environment. An epidemiologist searches for the causes by looking for individual and environmental variables that affect the morbidity and mortality rates. The mechanisms by which these causes exert their effects in the organism must be clarified by other means: by the study of pathogenesis. In clinical medicine, understanding of the nature of disease processes is being deepened essentially by biomedical pathogenetic studies, though epidemiological methods may sometimes also contribute to the analyses of mechanisms (see, for example, Ref. 1). Where the aim is to analyse the causation of diseases, however, epidemiology is the key science. In the pursuit of knowledge, the study of epidemiology and the study of pathogenesis are complementary and not in competition. The two approaches continuously provide each other with stimuli and challenges and thus have jointly become a potent accelerating force for medical progress.

In addition to its role as the science of causes of ill health, epidemiology also fulfils other functions. As the epidemiologist is concerned with rates, his work necessarily implies collecting numbers, both for the numerator and for the denominator. Both these figures have several uses. By providing quantitative descriptions of morbidity and analysing its determinants, epidemiology serves the health services and also the community at large. Health planning, several aspects of social policy, food and agricultural policy, and even education can derive guidance from epidemiological studies. Changing the ways people live may affect their health for better or worse. Epidemiology has, indeed, the necessary tools for measuring the health impact on populations of control measures and other changes (interventions), planned or unplanned, be they medical, economic, technical, social, or cultural. For community health, epidemiology is a basic science, necessary for meaningful planning and evaluation.

Today, epidemiology is a rapidly developing member in the family of medical sciences. In recent years, its scope has been expanded so that, besides its traditional role of studying epidemics of infectious diseases, it now also examines the causes of chronic noncommunicable diseases, including occupational ones; in addition, new vistas have been opened up in the study and control of mental diseases, and directives have been developed for curbing the epidemic of road and industrial accidents. Epidemiology adopts and adapts a wide variety of methods from clinical and laboratory medicine. Its statistical tools are partly specific to the field, partly common to demographic or biomedical research, or even to econometrics.

Like many other medical sciences, epidemiology also exists as a discipline, with personnel and facilities for teaching and research. In most centres of learning epidemiology is still a newcomer. A shortage of competent experts and teachers is felt in many parts of the world. This applies both to developed and to developing countries. In some developed countries with old, established institutions, the forces of inertia may retard any novel approaches. Lack of competence in epidemiology soon adversely affects other medical disciplines, however well established, and leads to their stagnation or sometimes to diversion of efforts into areas with little relevance to the major problems of health and disease.

Occupational health is one of the environmental health sciences, concerned broadly with the health effects of work and of working conditions. Physical, chemical, biological, organizational, and social variables associated with occupation may affect the physical or psychosocial wellbeing of the worker adversely or positively. Any environmental health research, when systematically conducted, must be concerned with

— the general characterization of the environment,
— the characterization of those exposed,
— the duration and intensity of exposures to various environmental factors,
— interactions between variables in the environment and those exposed to them, and
— health-related changes in the subjects exposed.

In applying this model to occupational health, work has to be looked upon as an exposure that needs detailed, many-sided characterization. The concept of interaction may need some clarification. Consider a straightforward example: the etiology of stress fractures. Such fractures occur in the leg and foot bones during training when long marches are performed. The fractures have been shown to increase in frequency with leg length asymmetry (2). Another type of interaction between the marcher and the road is physically mediated by the footwear: evidently, this also deserves epidemiological study.

Until recently, the concept of occupational disease denoted a specific clinical and pathological syndrome caused by a hazard specific to a particular type of work or the work environment. Epidemiological studies have, however, somewhat shaken the concept of specificity. On the one hand, the occurrence of occupational diseases may be affected by non-occupational factors, such as nutritional state. On the other, the prevalence and incidence of several common diseases may also be influenced by occupation. This is known to apply, for example, to some forms of cancer, which are not in the lists of occupational diseases, to a variety of common respiratory diseases, and to miscarriages, congenital malformations, and ischaemic heart disease. When work contributes to the causation, the term "work-related diseases" is being used.

3

The demographic concept of "social class" is often based on pooling together occupations considered similar in type, such a grouping of occupations being called a "social class". The breadwinner's family is included in his/her class. "Social class" differences in morbidity are thus often occupational differences, at least in regard to the breadwinner. Obviously, in comparing any two occupations—or groups of occupations like the "social classes"—there are generally differences also in education, income, housing standards, life habits, etc. The incidence of premature death from most major causes is connected directly or indirectly with the person's occupation (see Table 4.9 in Ref. 3). Occupational differences in morbidity may thus be ascribed to "social class" differences, as has been customary in traditionally stratified communities. However, marked occupational differences in total mortality, without a systematic "social class" pattern, are being observed in communities with a rather turbulent recent demographic and social history (4).

With modern technology, many hazardous exposures at work have been reduced. As a result, manifest occupational diseases are becoming rare, at least in the more economically advanced countries. For assessing the potential risk, new indicators are therefore needed. Clinically inapparent alterations in physiological variables, e.g., in lung function and in nerve conduction velocity, may be measured in groups of exposed workers and in suitably selected unexposed reference populations, the "controls". Sensitive indicators of incipient ill health have also been sought for in various subjective symptoms. The frequencies of headache or of complaints related to the musculoskeletal or gastrointestinal system have been found to vary according to the work situation. These differences deserve careful investigation.

Evidence of a health-related response to work or the work environment is much strengthened if no information gap remains regarding interactions between the environment and the worker. Sometimes the required information can be obtained with the aid of an experimental exposure test. With chemical hazards, the gap may often be narrowed by determining the substance or its metabolites in blood, urine, expired air, or even in hair. With physical hazards, bridging the gap by direct measurement is sometimes possible (e.g., in exposure to vibration), but with hazards that are psychosocial in nature such measurements have seldom been attempted. Often an exposure can be verified only by studying the work environment and organization. Occupational Exposure Limits, Threshold Limit Values (TLV), and Maximum Allowable Concentrations (MAC) offer empirical guidelines for controlling the work environment. When recommendations for such values are being made nationally or internationally, epidemiological data on exposure–response relations are essential.

Accidents are a major cause of health loss in many occupations. It is commonly believed that their incidence may be affected by such factors as organization of work, proper training, ergonomics, and safety campaigns. However, the amount of epidemiological research on

4

accidents at work and on their prevention has been meagre in relation to the importance of the problem.

Some studies of the unemployed have demonstrated that lack of work may also be a health hazard. Insufficient effort has yet been devoted to identifying those features of work that promote health. The present movement of "work enrichment" deserves as its companion proper epidemiological study.

What occupational health services offer to the epidemiologist

An occupational health service caters for a defined population. Thus, denominators for rates are fairly easy to define. The service usually records information on the state of health of the workers and hence is able to take care of some numerators. Complementary information on sickness absenteeism, pensions, and even mortality may be secured.

In some occupations and enterprises, the working population once established is remarkably stable. Longitudinal studies, historical or prospective, find a fertile field in such an environment, where both the subjects and their health records are easily available. Even a truly prospective cohort study can, with little extra effort, be organized within an occupational health service, but only if those leaving work can later be traced.

Records exist not only for working populations and their health but also for their exposures. Industrial hygiene measurements may have been performed routinely or occasionally. If not, it is often possible to reconstruct an approximate grading of exposures with the aid of the work record and skilled help. Whether a study is cross-sectional or prospective, an adequate sampling strategy of exposures should be built into the study plan from the very beginning. Epidemiological principles should not remain foreign to the industrial hygienist.

Occupational health epidemiology often faces problems similar to those in environmental health at the community level: air pollution, noise exposure, etc. The study strategy, however, may be rather different. The occupational health epidemiologist generally has to contend with smaller and selected populations, but this is amply compensated for by much higher exposures and sometimes even by their documentation in the past.

The interface of man and work supplies essential information on health hazards. Many occupational exposures in an enterprise vary from task to task. The possibility of becoming acquainted with the entire spectrum of tasks carried out in an enterprise—with their varied exposures—is an asset to the physician in occupational health that his more clinically oriented colleagues do not share. It would be both poor epidemiology and poor occupational health practice not to know the exact nature of the work that is being done.

Work processes and work places change. The changes are mostly dictated by organizational, technical, or economic considerations. Sometimes an ergonomic improvement may also be the target. Whatever the motivation of the changes, the doctor or the ergonomist is seldom consulted or informed. Some of these "natural experiments" may, however, provide unique possibilities for evaluative studies. The occupational health epidemiologist should be on the alert for such interventions. The ergonomist, concerned with the products or the production, can also use the skills of epidemiology and apply them to ergonomic problems.

Changes may also be made to achieve health objectives. One approach would be to plan from the beginning a controlled study in one or several enterprises. Some epidemiologists think only in terms of double-blind randomized controlled trials. Such approaches can seldom—or never—be realized in an occupational setting and the results also have limited generalizability. It is consoling to remember that of the sum total of human knowledge, most has been gained by using less "perfect" strategies. Information from "natural experiments" or "quasi-experiments" is not to be frowned upon: there is a wide and growing experience in their utilization. The definition of new problems is at least as important a function of research as the solution of old ones. Handbooks of epidemiology or statistics do not include any orthodox, codified standard methods for charting the unknown. There is still scope for the innovative mind.

The occupational health service is a link in the work organization. Its *raison d'être* is the health of the workers. It should be able to serve them by a wide spectrum of activities, all the way from health education to curative medicine and rehabilitation. The inputs of time and money to the various tasks by the health service can be measured. What is gained in terms of health is, however, far from self-evident. Carefully planned epidemiological studies might be able to give at least some answers. Epidemiological studies provide the means for the critical self-appraisal of any institution delivering health services: a valuable guideline for optimizing the always limited resources.

How epidemiology helps protect workers' health

An occupational health epidemiologist does not work only in the crystal-clear atmosphere of pure science, but rather for human welfare in the world of labour where interests clash. The situation calls for a simple, well defined code of ethics. A code of values is essential, with rules of conduct and formulation of standard practices.

For the health professions, human life is high in the hierarchy of values. The general public also considers health a major determinant of the quality of life. Societal values have great importance in the world of labour, with its complex informal and formal social structures. Cultural values, both those of the workers and those of the community at large, have to be considered. Not least, the privacy and individuality of each worker deserve respect. Other, often competing, values (e.g.,

6

economic and political ones) must be taken into account, but they are not a primary concern of the epidemiologist.

In searching for guidelines for the relation between the epidemiologist and the population studied, the code of ethics of the doctor–patient relationship offers a well established parallel observed all over the world. However, since the epidemiologist is making a *community* diagnosis, the moral obligations of the community studied on the one hand, and of the epidemiologist towards the community on the other, need to be discussed, weighed, and codified. When being asked to take part in an epidemiological study, a worker should make his decision not only as an individual, but also bearing in mind his obligations to his fellow workers and to society. Agreement to participate may help to improve the health of present or future fellow workers.

More problematic are the ethical issues connected with third parties. The news media assume that everyone has the right to know everything. Such a creed obviously serves the profits of the information industry, but it raises serious questions of discretion and responsibility. As a natural and well founded reaction against large-scale breaches of privacy, steps are taken to protect the individual. Unfortunately, this has resulted in some ill-advised and unfortunate legislation, which has gravely handicapped the study of health hazards, particularly those that are long-term. It is quite evident that good long-term records on the individual's work and health are absolutely necessary for any scientific attempt to improve workers' health. This must be stated categorically to prevent irresponsible mismanagement of data and—even worse—the enactment of legislation that retards innovation in health.

The epidemiologist's role in occupational health

Research as an aid to decision-making
Scientific research serves two functions: it helps to extend human knowledge and this, in turn, can be used as an aid to making practical and administrative decisions. Scientists are rightly concerned with the quality of their work and prefer to exclude information of questionable value.

The manager, or administrator, on the other hand, must make decisions based on the best information available to him, often within a time limit, and therefore sometimes on insufficient evidence. In real life, even the common decision to continue the *status quo* seldom rests on scientific evidence. In other situations, the decision-maker may have to deal with conflicting evidence, often made worse by competing interests, e.g., those of health, power, and profit. Even in the health field, it would be unrealistic, impertinent, and illogical to require the decision-makers to limit themselves to deciding issues only when backed by faultless epidemiological studies.

Furthermore, a decision has sometimes to be made in a situation where alternatives are supported by results falling far short of the arbitrary 5% level of significance. In practical life, industry, and trade, decisions are often made on evidence statistically "softer" than that required for acceptance in the realms of science. There would indeed be very little business life if new developments were launched only when the odds were 10 to 1 or higher in favour. Human lives have a high intrinsic value and, other things being equal, most of us would probably opt for any safe preventive or curative procedure that offered a reasonable chance of success—certainly far short of 95%. Measures aimed at improving work safety have until now seldom been based on hard epidemiological data; case histories of accidents, engineering skills, and cost-effectiveness considerations have been the main determinants.

Democratic communities and decision-makers need information. In deciding on health policies and programmes, they are best served by versatile *ad hoc* information systems combining various types of hard and soft health data. Pertinent epidemiological studies, if available, would rank high as providers of data for such systems. However, less certain data, often indirect in nature, can also contribute much, if intelligently used (5). No epidemiologist in occupational health can work in an ivory tower; he should have a wide grasp of the real world and its potential sources of information.

Epidemiology and planning

Goal setting
The first step in planning is to set goals. In the community at large, these are often codified through political decisions. This applies also to occupational health. The three partners in the labour market are workers, management, and government. In order to set feasible and reasonable goals for occupational health and safety, all three need information and advice. All the better if this is unbiased. A source serving all three is the International Labour Office, which has a quasi-legislative function, while the World Health Organization has contacts primarily with the ministries of health and tends to have a more advisory role. Sound data for goal setting in occupational health can be provided by epidemiologists; a country without them will not be able to develop its services independently.

Strategic planning
Once the goals have been set, decisions are to be made on means for attaining them. Further, information on hazards and on measures to control them (and on their effectiveness) will be needed. In other words, a national occupational health information system has to be developed. Both fixed and versatile elements may be included in such an information system. Some guidance may be obtained from a comprehensive occupational disease register (6) and from work accident statistics (see, for example, Ref. 7). The keepers of such registers should obviously have proper epidemiological training and expertise.

Tactical planning

The tactics used in occupational health consist of measures for health promotion, prevention of illness and accidents, and treatment and rehabilitative measures. In all these areas, pertinent epidemiological problems can be formulated. Innovations are frequently introduced. Their value can occasionally be assessed with the aid of controlled studies. Even when this is not possible, the process of change should be assessed as a "quasi-experiment"—a task for a skilful and experienced epidemiologist.

Implementation

The above outline of the steps in occupational health planning implies that there is also a need for epidemiological knowledge in the implementation phase. A WHO Working Group has recently recommended that the evaluation of occupational health services should be a regular activity, fully integrated into the planning and implementation of occupational health and safety programmes (*8*). Social, technological, and economic factors should all be taken into account. The appropriate information systems clearly need a spectrum of skills to which an epidemiologist can contribute.

The social partners—the trade unions, and the employers' federations—need epidemiological expertise in order to be adequately informed about current questions of occupational health and safety. Whether these organizations prefer to have experts among their staff or to rely on outside institutions will vary from country to country. Major epidemiological investigations on occupational health in an industry are most effective if supported, at least morally, by both social partners, but an enlightened trade union, employer, or government department may alone sponsor a study organized in a manner likely to produce valid results. To date, most large-scale epidemiological studies of good quality in occupational health have been undertaken by universities or governmental research agencies (such as the Medical Research Council in the United Kingdom, the National Institute of Occupational Safety and Health in the United States, or the various institutes of occupational health in the Soviet Union). One good solution is to have an adequately supported, reasonably independent, many-sided national body with appropriate competence in epidemiology as well as in other fields.

Questions of "bias" are often raised, particularly when results are unpopular with one of the interested parties. However, disinterested parties really do not exist (this applies even to government agencies). In the end, scientific quality depends on the competence, training, financial independence, and character of the investigators, and on having all sides of industry well informed as to the nature and purpose of epidemiology.

All enterprises have need of epidemiological studies. The Joint ILO/WHO Committee on Occupational Health has recently outlined the educational requirements for the tasks of an occupational health physician (*9*). He should be able, among other things:

9

— to assess the incidence and prevalence of ill health in relation to work conditions;

— to identify occupational health problems in the light of the general health of the working population;

— to prepare and evaluate statistical records of sickness absences, to use such records to identify causes, and to propose measures to eliminate those causes;

— to use epidemiological and other methods to investigate occupational risk factors, the possibility of their prevention, and the means by which they may be prevented.

Present day technology is a valuable asset if its limitations are understood. Computers and their software already provide thoroughly tested systems of data recording, processing, retrieval, and analysis; these can be on-line, obviating the need for other records of occupational health services. Many enterprises use computer technology extensively in production, storage, sales, and accounting. Few have yet considered its application to the health of its labour force. However, good epidemiology does not depend on computers, but on understanding logical principles of design and purpose.

An occupational health service keeps in close touch with the population it is designed to cover. It has the duty—and also a realistic possibility—to discover and monitor health hazards associated with work. Both actual and potential hazards should be detected, even ahead of the appearance of clinical cases of occupational disease, and brought under control. However, some health hazards at work cannot be totally eliminated. Continuous surveillance is then required. Epidemiological principles are essential in planning an effective and efficient system of surveillance, as well as in the detection and definition of health hazards.

Physical and chemical hazards to health in the working environment are monitored by using technical and analytical methods of environmental hygiene. The sampling strategies have often been designed to discover peak values rather than average exposures. Peak loads may be quite relevant to the safety of an individual worker, but are often of short duration and affect only a small part of the labour force. Should peak exposure values be incorrectly related to the health of all the workers of the enterprise—or even of those at the workplace—the hazard may easily be underestimated. The early occupational exposure limits were, in fact, often based on hygienic measurements carried out for immediate practical purposes in exceptional situations and were poorly suited to give guidelines for continuous exposures. This is one reason why many early occupational exposure limits have later been found unduly high. More frequently, the mistake can be attributed to poor epidemiology with too few workers observed for too short a time or observations limited to healthy survivors. Of course, another reason for lowering the limits has been the development of increasingly sensitive diagnostic methods for early health effects. Obviously, it would be desirable to obtain a representative picture of the distribution of exposures in the working population. For this purpose, epidemiologically appropriate strategies for industrial hygiene measurements have been developed and tested (*10*).

For exposures to many toxic chemicals, biological monitoring of the substance or of its metabolites in blood, urine, expired air, or hair is a particularly effective form of surveillance. Here, too, epidemiology provides guidelines for planning the sampling strategy (*11*).

Another approach to surveillance is through the observation of the state of health of the workers. Occupational physicians (and their staff) see patients, enquire about their complaints, use laboratory aids, and make diagnoses. The occupational health services are thus in an excellent position to observe health impairment caused by work and to recognize it as such. For example, a clinically apparent case of lead poisoning is less likely to be misdiagnosed in the occupational health clinic at the plant than would be the case in even a good general hospital. It should be appreciated, nevertheless, that normal clinical practice seldom provides data directly useful for epidemiological analyses; the diagnostic procedures are not usually sufficiently standardized nor are the findings systematically recorded. No doubt, both shortcomings can eventually be eliminated, but the transition will have to await radical changes in medical teaching and practice.

Special surveys of ill health among a working population generally prove more rewarding than a watch for manifest disease. Cases of occupational disease often remain "latent" among the labour force. The workers adapt to a slowly developing condition, and for social or financial reasons may be unwilling to report illness that may lose them their jobs. In a survey, more sensitive methods for discovering health impairment may be used. Hazards typical for an occupation can be brought under surveillance through periodic health control as, for example, in a hearing conservation programme. Epidemiological expertise is again essential both for planning health surveys and for their analysis.

The health of workers does not depend only on known occupational hazards. Heredity, past experience and illnesses, exposures outside work, and life habits all include potential determinants. Some of these have been identified, others are being sought. There are several examples of causes of disease outside work being first discovered among working populations. Further epidemiological studies may lead to new discoveries.

Present situation

The occupational health services vary much in structure and function, more so than primary health care or hospital services, even among the industrialized countries. In some countries, epidemiology has in recent years already been introduced into the training and practice of at least some categories of occupational health personnel. Mostly, however, epidemiology is not yet contributing its full potential to occupational health.

Research activities in occupational health epidemiology are best judged by studying the published literature. Recent handbooks are available in some major languages (*12–14*). For journal articles, the best way to obtain a comprehensive listing is to order an *ad hoc* computer-processed bibliography. The library of the University of Kuopio, Finland, kindly provided such a search for papers published in 1979 and 1980 on occupational health epidemiology. The off-line bibliographic citation list generated by MEDLARS II included 455 items. Half of the references were supplied with abstracts, half listed only. As usual, there were also "false positives", articles pertaining to occupational health or its border areas, but with emphasis mainly on clinical study instead of epidemiology. No doubt, also many of the "false positives" would have offered interesting and useful reading material. Most articles were in English, some in French, German, Polish, or Russian, while no other language was represented by more than 1 % of the articles listed.

An important and somewhat disconcerting observation was that publications on occupational health epidemiology appear in an amazing variety of medical journals. The journal most often listed had published 8 % of the articles. Other leading journals in occupational health approached at most half that figure. It clearly emerges from this survey that in order to keep up with progress in occupational health epidemiology, following just a few journals is not enough. Fortunately, several abstract journals and modern computer-aided library services provide adequate help, even to those outside university centres.

Many of the papers published in English came from other than English-speaking countries. Active interest in occupational health epidemiology is widespread. The topics covered an extensive field; many articles dealt with occupational toxicology, lung diseases, and cancerogenesis, but relatively few with musculoskeletal illness, ergonomics, or psychosocial problems. Altogether, the bibliography of 455 articles proved interesting and instructive—and stimulated the reader's curiosity for further reading.

Epidemiology still lacks competent professional representatives in many places. Many articles on occupational health epidemiology are published by "self-made" epidemiologists, who sometimes use peculiar methods and terminology. The reader should bear this in mind when examining the data and conclusions, and should use his judgement in regard to unimportant formal weaknesses. In fact, the reader would be much helped by some knowledge of epidemiology in judging the value of published studies in a variety of health fields.

Conferences and symposia bring the information more up to date than published work. For example, a Symposium on Epidemiology in Occupational Health, organized by the Permanent Commission and International Association on Occupational Health, was held in 1981 in Helsinki, Finland (*15*) and was followed by a similar symposium in 1982 in Montreal, Canada. Fifty papers were presented at the Helsinki

Symposium and 80 in Montreal. Symposia bring together investigators actively involved in epidemiological studies and thus provide a valuable forum for discussing methodological issues. Epidemiological studies in occupational health are, of course, also reported at several other international and national meetings.

An umbrella organization for occupational health is the Permanent Commission and International Association on Occupational Health. One of its subcommittees is devoted to epidemiology. The two main functions of the Permanent Commission are large international congresses and small working meetings of the subcommittees. At both, epidemiology has its deserved place.

Several institutions of occupational health and public health in Europe and North America offer regular long-term or short courses in epidemiology; widely known are those held at the London School of Hygiene and Tropical Medicine. Special short intensive courses afford a further chance of learning. Such courses have been arranged both by the World Health Organization (in English, French, and Russian) and by some national institutions and universities. Internationally well known are the Advanced Courses in Epidemiological Methods held (in English) since 1972 in summer at the Institute of Occupational Health, Helsinki.

In the wellkept records of many occupational health services valuable data are lying fallow, waiting for an epidemiologist to utilize them. The incidence of visits to the doctor for complaints, manifest illness, and absenteeism often differs in contrasting groups of workers. With the aid of such data, various hypotheses could be put to the test retrospectively — an approach that has occasionally proved of special value during a labour dispute.

Multifactorial causes

Achievements in the control of infectious diseases brought about general acceptance of the concept of one specific cause for each disease, replacing the earlier held views of multifactorial etiology. The search for the causation of diseases was strongly influenced by this philosophy: one chemical, physical, or microbial agent had to be discoverd for each specific disease or syndrome. The effect of this thinking is clearly evident in the development of industrial medicine and occupational health legislation the world over; for a long time it dictated the direction and limits of occupational health research.

The doctrine of a one-to-one relation between cause and disease has now been shown to be untenable, even for microbial diseases. This conclusion rests solidly on a quarter of a century of research on chronic noncommunicable diseases, notably on the causation of ischaemic (coronary) heart disease (see, for example, Ref. 16). Instead of being due to one cause, ischaemic heart disease is shown to be associated with exposure to several risk factors, i.e., predictors of incidence. To the question whether a predictor of risk is also a cause of the disease, an adequately planned experiment ought theoretically to give an unequivocal answer. However, it is very difficult if not impossible to make randomized, double-blind human experiments on the causation of chronic noncommunicable diseases. Therefore, less direct evidence of various kinds has to be collected, critically assessed and weighed, and practical conclusions reached accordingly. For ischaemic heart disease, there is a widespread consensus on the causal nature of the following risk factors: dietary intake of saturated fats, total serum cholesterol level, high blood pressure, and cigarette smoking. Psychosocial predictors of ischaemic heart disease have also been well established, the most widely known being Type A versus Type B behaviour (17, 18), but corresponding personality traits can also be identified by using factor analysis of responses to psychological questionnaires (19, 20).

A wide array of other predictors of risk have been found in prospective and case–referent studies. To deal with several causative factors at once, special statistical methods are required. Both multiple regression techniques (see Ref. 21) and multiple logistic analysis (22) are being used for this purpose. Such approaches are described in Chapter 11.

It is pertinent to ask, whether occupation and work also contribute to chronic noncommunicable diseases. In regard to ischaemic heart disease, physical activity at work, psychosocial attributes of occupation, and some toxic exposures are considered as potential modifiers of the risk. Chapter 13 reviews and critically assesses the strategy followed in a series of studies in which carbon disulfide exposure in the rayon industry was examined as a risk factor conducive to ischaemic heart disease.

Cancer, another group of chronic diseases for which there still is little firm evidence of communicability, has become a central topic in research on occupational exposures to chemicals and dusts. The combined effects of cigarette smoking and asbestos on the incidence of bronchial cancer (but not mesothelioma) are well known (23, 24). As with ischaemic heart disease in the rayon industry, one causative agent is occupational and the other a part of the worker's lifestyle. Statistical methods applicable to the epidemiology of cancers — and of many other diseases—with a special emphasis on case–control studies have recently been reviewed (25, 26).

The concept of multifactorial causation is a stimulus for widening both the horizon and the scope of activities in occupational health, from the more or less limited statutory requirements of the past to a comprehensive assessment of the effects of work on physical, mental, and social wellbeing.

Further research

The future holds promise of increasing control of the communicable diseases and means for the prevention of premature deaths and disability as caused by such major killers as ischaemic heart disease. Medical sciences and health services have to shift their focus to problems yet unsolved, in particular to those due to changes in the age structure of the population, social conditions, and lifestyles. In occupational health, many causes of occupational diseases could be practically eliminated even today by appropriate technology. Of the other work-related diseases, less is known. The associations of different occupations with risk factors and lifestyles adverse to health deserve much additional epidemiological research. Possibilities for preventing and treating musculoskeletal illness and injury among working populations and for rehabilitating the victims have been insufficiently studied. The entire field of psychosocial causation of ill health deserves a concerted attack: the psychosocial climate in which work is performed is indeed one of the dominant lifetime exposures. Not only should risk factors specific for this or that disease be looked for, but attention should equally be devoted to elements in work that promote health.

A novice may gain the impression that epidemiology is a complete collection of canonized methods. It is sobering to point out that epidemiology still has tasks to face for which it is not yet adequately prepared. One of the tasks for the future will be to reorient so-called health statistics from what they now mainly are, death and disease statistics, so that they pertain to the lifelong quality of life for the individual, to early predictors of longevity, and to determinants of independent and active old age (27). The populations of the future, when most major causes of premature death will have been controlled, need such epidemiological tools. They can be developed only through linking individual exposure, performance, and health records over the entire span of human life (28). Some occupational health epidemiologists may well decide to spearhead these and other new developments.

However, such projections into the future are likely to remain unrealistic and irrelevant for most of the world's working people for a long time to come. The emerging industrialization of developing countries exposes workers to risks almost forgotten in more fortunate parts of the world. It brings with it florid occupational diseases as well as frequent accidents, and their control is still a real challenge. Knowledge and skill are lacking more than material resources. The simple classical approaches of epidemiology and prevention are quite applicable—and are much needed.

References

1. **Nurminen, N. et al.** Quantitated effects of carbon disulfide exposure, elevated blood pressure and aging on coronary mortality. *American journal of epidemiology,* **115**: 107–118 (1982).
2. **Friberg, O.** Alaraajojen pituusasymmetria—eräs rasitusmurtuman etiologinen tekijä. [Length asymmetry of lower extremities—an etiological factor of stress fractures]. *Annales medicinae militaris Fenniae,* **55**: 149–154 (1980).
3. **Barker, D. J. P. & Rose, G.** *Epidemiology in medical practice,* 2nd ed. Edinburgh, Churchill Livingstone, 1979, p. 61.
4. **Sauli, H.** *Ammatti ja kuollleisuus 1971–75. [Occupational mortality in 1971–75].* Helsinki, Central Statistical Office of Finland, 1979.
5. **Härö, A. S., ed.** *Planning information services for health. Decision-stimulation-approach. Report of NOMESCO/ADAT working group.* Helsinki, Valtion painatuskeskus, 1981.

6. **Vaaranen, V. & Vasama, M.** *Occupational diseases in Finland in 1980–81. Occupational diseases reported to the Finnish occupational disease register in 1980–81.* Helsinki, Institute of Occupational Health, 1982.
7. **Abt, W.** *Unfallanalyse 1977. [Accident analysis 1977].* Bonn, Hauptverband der gewerblichen Berufsgenossenschaften e.V., 1979.
8. *Evaluation of occupational health and industrial hygiene services: Report on WHO Working Group.* WHO Regional Office for Europe, Copenhagen, 1982 (EURO Reports and Studies, No. 56).
9. WHO Technical Report Series, No. 663, 1981 (Eighth report of the Joint ILO/WHO Committee on Occupational Health).
10. **Corn, M.** Strategies of air sampling. In: McDonald, J. C., ed. *Recent advances in occupational health.* Edinburgh, Churchill Livingstone, 1981, pp. 199–209.
11. **Tola, S. & Hernberg, S.** Strategies of biological monitoring. In: McDonald, J. C., ed. *Recent advances in occupational health,* Edinburgh, Churchill Livingstone, 1981, pp. 185–198.
12. **Monson, R. M.** *Occupational epidemiology.* Boca Raton, FL, CRC Press, 1980.
13. **Lazar, O., ed.** *Pathologie industrielle. Approche épidémiologique.* Paris, Flammarion Médecine, 1979.
14. **Jedrychowski, W.** *Metody badan epidemiologicznych w medicynie przemyslowej. [Methods of epidemiological investigation in industrial medicine].* Warsaw, Panstwowy Zaklad Wydawnickow Lekarskich, 1978.
15. **Kurppa, K., ed.** Proceedings of the First Symposium on Epidemiology in Occupational Health, Helsinki, 10–12 June 1981. *Scandinavian journal of work, environment and health,* **8**: Suppl. 1, 1982.
16. **Keys, A.** *Seven countries. A multivariate analysis of death and coronary heart disease.* Cambridge, Mass., Harvard University Press, 1980.
17. **Dembroski, T. M. et al., ed.** *Coronary-prone behaviour.* New York, Springer, 1978.
18. **The Review Panel on Coronary-prone Behaviour and Coronary Heart Disease.** Coronary-prone behaviour and coronary heart disease: a critical review. *Circulation,* **63**: 1199–1215 (1981).
19. **Vanhala, K.** *[Psychological risk factors related to coronary heart disease. Prospective study on psychological factors predicting coronary heart disease morbidity of Helsinki policemen].* Helsinki, Monographs of Psychiatria Fennica, 1979 (in Finnish).
20. **Nirkko, O. et al.** Psychological risk factors related to coronary heart disease. Prospective studies among policemen in Helsinki. *Acta medica scandinavica,* Suppl. 660: 137–146 (1982).
21. **Draper, N. R. & Smith, H.** *Applied regression analysis.* New York, Wiley, 1966.
22. **Truett, J. et al.** Multivariate analysis of the risk of coronary heart disease. *Journal of chronic diseases,* **20**: 511–524 (1967).
23. **Saracci, R.** Asbestos and lung cancer. An analysis of the epidemiological evidence on the asbestos-smoking interaction. *International journal of cancer,* **20**: 323–331 (1977).
24. **Saracci, R.** Personal-environmental interactions in occupational epidemiology. In: McDonald, J. C., ed. *Recent advances in occupational health.* Edinburgh, Churchill Livingstone, 1981, pp. 119–128.
25. **Breslow, N. E. & Day, N. E.** *Statistical methods in cancer research. Volume 1. The analysis of case-control studies.* Lyon, International Agency for Research on Cancer, 1980 (IARC Scientific Publications, No. 32).
26. **Schlesselman, J. J. & Stolley, P. D.** *Case control studies: design, conduct, analysis.* New York, Oxford University Press, 1982.
27. **Rice, D. P.** Health statistics: past and present. *New England journal of medicine,* **305**: 219–220 (1981).
28. **Fox, A. J.** Linkage methods in occupational mortality. In: McDonald, J. C., ed. *Recent advances in occupational health.* Edinburgh, Churchill Livingstone, 1981, pp. 107–118.

2

Nature and health effects of occupational factors

N. F. Izmerov[a] & J. I. Kundiev[b]

Occupational factors may have both positive and negative effects on a worker's health, depending largely on the nature of a particular factor and the degree of exposure to it.

Occupational factors are classified as (1) chemical, (2) physical, (3) biological and (4) ergonomic and psychosocial. The combined effects of exposure to more than one occupational factor will also be considered briefly.

Chemical factors

Occupational poisons (chemical risk factors) are those chemical compounds to which a worker is exposed in the course of his work and that may produce harmful effects when taken into the organism in doses exceeding the capacity of the body to deal with them.

The number of chemical compounds used in industry today is vast and their biological effects are so diverse that it is unlikely that a common principle will be found that sufficiently reflects the relationship between the chemical properties of a substance and its biological effects. It is therefore virtually impossible to develop a universal classification of substances meeting all the requirements of occupational health, toxicology, and epidemiology. However, the identification of chemical factors in epidemiological studies may derive practical benefit from the use of some already available classifications that systemize the properties of individual chemical compounds.

First, it is possible to make use of the chemical classification of substances into organic and inorganic. Here, substances are classified or

[a] Institute of Industrial Hygiene and Occupational Diseases, Academy of Medical Sciences of the USSR, Moscow, USSR.

[b] Scientific Research Institute of Labour Hygiene and Occupational Diseases, Kiev, USSR.

grouped according to the customary chemical nomenclature. Moreover, depending on their state of aggregation in the atmosphere, all industrial substances may be classified as gases, vapours, or aerosols (liquid and solid).

Another principle consists in grouping substances according to their application. Such groups comprise substances different in chemical structure and biological effect. For instance, the group of solvents includes alcohols, ethers, ketones and other compounds. There are groups of pesticides, plasticizers, organic dyes, etc.

Several classifications of industrial substances are based on the quantitative characteristics of the toxicities.

The classification of industrial substances according to the character of their effect on the human organism is considered to be the most appropriate for use in epidemiological studies. Henderson & Haggard (1) used this principle as the basis of their classification of all volatile industrial substances into several large groups:

(1) chokedamps: (a) simple; (b) biochemically active;
(2) irritant gases;
(3) inhalation narcotics and related compounds: narcotic gases without any marked after-effect.

The same authors have also divided all substances into reactants and non-reactants. Reactants undergo chemical reactions and conversion inside the organism. The toxic effect may be determined by both the compound itself and its metabolites. Non-reactants undergo few changes in the organism and are excreted in the same form in which they were absorbed. Benzene may serve as a representative of the first group, since its effect on the haematopoietic system is produced by its metabolites (phenol, pyrocatechol, hydroquinone). The most typical compounds of the second group are the hydrocarbons of the aliphatic series (2).

Many classifications are based on the nature of the effect of poisons on enzyme systems. Depending on the particular organs and systems principally affected, the substances are divided into neurotoxic, hepatotoxic, nephrotoxic, cardiotoxic, etc. There are also classifications that depend on a specific adverse effect, e.g., sensitizing action, carcinogenic activity, or irritant action.

The character of the adverse effect on the organism depends on the level of exposure, i.e., the toxic effect of brief exposure to a high concentration is frequently different from that of prolonged exposure to a low concentration. In this regard, epidemiological analysis of the industrial environment is usually characterized by low intensity and long duration of exposure to the chemical factor. In these circumstances, the most suitable classification is one that takes account of effects due to chronic exposure at low levels and that is based on the criteria of long-term hazards associated with cumulation: (a) extremely hazardous; (b) very hazardous; (c) hazardous; (d) slightly hazardous (3).

I. *Substances producing a selective long-term effect*
    (1) carcinogenic;
    (2) mutagenic;

18

(3) atherogenic;
(4) sclerogenic;
(5) gonadotropic;
(6) embryotropic.

II. *Nerve poisons*
   (1) spasmodic and neuroparalytic;
   (2) narcotics affecting parenchymatous organs;
   (3) substances of purely narcotic action.

III. *Substances producing blood changes*
   (1) bone marrow depressants;
   (2) substances reacting with haemoglobin;
   (3) haemolytics.

IV. *Irritants and corrosive poisons*
   (1) irritants of mucous membranes of the eye and upper respiratory system;
   (2) skin irritants.

It should be kept in mind, however, that many poisons possess a variety of properties. Assigning a substance to a particular group indicates its dominant action under real conditions of production and exposure. To a certain degree, the last classification makes it possible to characterize the effect of a poison qualitatively in order to obtain a quantitative estimate of its hazard to the organism. The classification of hazards according to types of effect can be done on the basis of those properties, but it imposes great limitations in practice.

Thus, several investigations have enabled the chemical substances examined to be related to one or several groups according to their dominant effects. For instance, benzo[a]pyrene, asbestos, nickel carbonyl chromates, 2-naphthylamine, and 1-naphthylamine were related to the group of substances characterized by pronounced blastogenic activity; the group of substances possessing a selective gonadotropic effect comprised chloroprene, manganese, and some pesticides; substances manifesting mutagenic activity included ziram, chloroprene, and arsenic (4).

The epidemiological studies must be concerned with the raw materials and intermediates of the manufacturing process, their physical and chemical properties, and their possible transformations in the surrounding medium. The sources of hazardous substances escaping into the environment must also be studied, including their effect on a worker, particular features of the microclimate, and other environmental factors.

The working conditions are frequently characterized by considerable variability in the concentrations of dust, vapours, and gases. The proper quantitative estimation of exposure to a given chemical substance is the basis for the corresponding estimation of the response of the organism and determination of correlations between the quantitative indicators of the effects of chemical agents and the changes in the workers' health. The latter can be estimated on the basis of the following data:

(*a*) results of casual medical examinations of workers;

(*b*) results of occupation-oriented health inquiries;

(*c*) results of special medical examinations and clinical investigations;

(*d*) results of tests to determine the functional state of a worker's body;

(*e*) evaluation of records of morbidity with temporary invalidity; data on applications for medical attention;

(*f*) data on acute and chronic occupational poisonings, invalidity, mortality, etc.

The correlations between the quantitative indicators of the effects of a chemical agent and the changes in the workers' state of health are established with the help of the results of periodic medical examinations of workers compared with those of preliminary examinations at the time of entering the job.

In line with the aims of epidemiological studies, it is important to solve the problem of correlating the general (non-specific) and the specific aspects of the effects of chemical agents on the organism and the related problem of preclinical, prepathological conditions (i.e. those on the borderline between normal and pathological) resulting from the dominant effect of chemical agents, on the organism produced by low levels of exposure. Under modern industrial conditions, when workers are exposed to low concentrations of chemicals the distinction between the specific and nonspecific effects of industrial poisons is somewhat blurred. In these circumstances, the effect of these poisons on the organism undergoes a considerable (sometimes a complete) change in its intrinsic specificity. Consequently, it is often believed that a lengthy exposure to low concentrations of toxic substances causes the toxic effect to be manifest in a new form, with the appearance of nonspecific features. This situation is probably associated with the load on the homoeostatic mechanisms that maintain the stability of the internal environment (2).

One of the specific features of prolonged exposure to low concentrations of chemical substances is their ability to produce, long before pathological changes occur, a considerable modification in the interaction between the organism and the changing environmental conditions. This results in so-called premorbid conditions, characterized primarily by an unstable dynamic equilibrium between the organism and the environment, the functional basis for which lies in changes in the efficiency of the adaptation mechanisms. These changes lead to a succession of stable and unstable conditions, and may be accompanied by both a decrease and an increase in the adaptability of the organism. The decrease in adaptability caused by prolonged exposure to a chemical and the related functional instability determine, ultimately, the ability of premorbid conditions to become transformed into various pathological states. This statement may be corroborated by the effect of prolonged exposure to industrial chemicals on the general morbidity of workers. Epidemiological studies have made a considerable contribution to establishing a causal relationship between the appearance of a number of physical diseases and the presence in the environment of such industrial poisons as carbon disulfide, hydrogen sulfide, hydrocarbons, amino compounds, organic compounds of chlorine and phosphorus, and heavy metals.

Evaluation of numerous studies leads to the conclusion that prolonged exposure to different toxic substances, even at low concentrations, may cause a variety of adverse effects on health. In some cases these are expressed by changes in the whole morbidity pattern; in others they take the form of a chemical "depression of working functions", which, according to Pravdin, covers various functional disorders that lead to a reduction in working efficiency without any obvious manifestations. Functional changes in neurohumoral regulation can be among the very early indications of a prolonged exposure to toxic substances at low concentrations.

Analysis of data on the general morbidity of workers reveals a trend to higher susceptibility of those organs and systems that show marked tropism with regard to a given toxic substance. There is now a great deal of information available demonstrating the complexity of the interaction between the organism and harmful substances, including those of a chemical nature. The result of this process and the nature of the effect produced depend on characteristics of the agent and the peculiarities of the individual, and may be expressed either as adaptation (i.e., normal vital activity under new conditions) or as pathology.

It is of particular significance that one and the same agent at different intensities of exposure may cause different forms of occupational pathology while affecting one or different organ systems. For example, high concentrations of silica-containing dust cause silicosis, which, as a rule, is not associated with the clinical picture of bronchitis, while low concentrations lead to chronic dust bronchitis without radiographic evidence of silicosis. Different intensities of exposure may cause qualitatively unlike forms of occupational pathology, probably as a result of differences in the sensitivity of the various systems of the body to the effect of the given agent (5).

Data on the subsequent fate of toxicants in the organism are of no less importance. Information is now available on the ways in which different organic compounds are transformed in the organism, as well as on their metabolites, and data are also available on the behaviour of inorganic compounds. Of significance for the purposes of epidemiological studies is, in some cases, the quantitative investigation of biotransformation processes, i.e., the establishment of the quantitative dependence of the dynamic circulation of the poison in the organism on the different levels of its concentration in air. From the data on the dynamic circulation of the poison in the organism, it is possible to deduce information on intakes, which may be subject to considerable fluctuations depending on the length of exposure, interruptions during work, character and level of intermittent concentrations, intensity of work, and the effect of other concomitant factors (6).

In establishing correlations between levels of exposure to chemical substances and impairment of health, and especially in studies to check the validity of occupational exposure limits, particular importance attaches to the use, together with the common clinical methods (haematological, biochemical, etc.), of special methods for the early detection of the effects of chemical substances on a worker's health. Since the earliest and least obvious effects of the overwhelming majority of chemical substances are functional changes in the nervous system, methods that reveal such changes are of particular importance (EEG, chronaximetry, etc.). It has also been found that the cardiovascular system reacts sensitively to exposure to toxic substances; hence, tests for investigating the functional state of this system (ECG, graded exercise tests, etc.) are particularly informative (7).

The above-mentioned facts are gaining increasing importance in connection with epidemiological studies for the purpose of establishing hygienic standards. Here, the basic principle for the application of diagnostic tests for the detection of pathological and prepathological changes consists in selecting the most informative indicators, whether of a specific or nonspecific character. The results of a number of investigations have shown, for instance, that for some substances the threshold concentration for their allergic effect is much lower than that for their toxic action (8); the threshold concentration for the gonadotropic effect may also be drastically different from that for toxicity for one and the same substance (4). For example, epidemiological studies to establish hygienic standards for benzene and aniline in air in the working environment were undertaken largely by means of a thorough analysis of the specific action of these substances and their effects on the haemopoietic system. The exposure limit for formaldehyde was corrected with particular regard to its effect on the upper respiratory system. For mercury and carbon disulfide, which affect autonomic centres and limbic structures in the brain, the most informative studies were those concerned with functional changes in the nervous system and mechanisms for regulation of arterial blood pressure.

An extremely important and, at the same time, complex task is that of detecting hidden nonspecific changes in the organism and also specific changes: gonadotoxic, embryotoxic, etc. These studies are performed on workers who have undergone prolonged exposure to the substances under investigation. Recently, a number of cases of health impairment following prolonged exposure, which were not detected by experiment, have been revealed through the study of the effects on the organism produced by low concentrations of lead, antimony, mercury, some organomercuric compounds, and some pesticides (9).

Using the values for occupational exposure limits for a wide variety of hazardous substances determined by means of epidemiological investigations it was possible to formalize the information with the aid of mathematical simulation and to predict the toxic effects of different concentrations of chemical substances in the air of the workplace. The prediction of safe concentrations (in terms of the workers' health) based on mathematical simulation of health effects is feasible only for large groups of workers with sufficient length of service who have been exposed to different concentrations of the substance under study. By studying the "exposure–time response" dependence in exposed groups compared with a control group, and selecting the most informative indicators, it is possible to determine the threshold value at which the substance under investigation produces a detrimental effect. Thus, as a result of a long-term study of the state of health of workers handling dibutyl phthalate, it was discoverd that the early, most informative symptoms were those indicating disorders of the sensory system (hyperalgesia). The incidence of these disorders increased with the length of time a person had been working with dibutyl phthalate and with the level of exposure. By means of regression analysis, a linear model corresponding to empirical data was developed, which made it possible to validate the exposure limit for dibutyl phthalate (10). Among other factors that are important for the response of the organism is the physicochemical state in which a chemical compound enters the organism: coarse dispersion (dust, mist), colloidal dispersion (aerosol, lyosol), molecular or ion dispersion (solutions or vapours). For example, during the study of the effect of silica-containing dust on the organism, it was shown that the development of different forms of pneumofibrosis depends on the state (crystalline or amorphous) in which the dispersed phase is present.

It should be kept in mind that the health effects of a chemical substance may differ greatly depending on whether it enters the organism through inhalation or through the intact skin.

The toxic effect of a chemical substance may be intermittent—the type of effect most frequently encountered under actual working conditions — or it may be continuous at a constant level.

For some poisons—such as chloroform and gasoline—the intermittent toxic effect is more obvious than the continuous one. In this case, medium and high concentrations do not play a decisive role in determining the exposure level. On the contrary, for heavy metals exhibiting a very distinct cumulative action, the average weighted concentrations are of most significance.

Physical factors

Microclimate

The microclimate, as it affects the human body, is defined by a combination of such environmental factors as air temperature, humidity, air speed, and temperature of the surrounding surfaces (thermal radiation). These physical parameters, taken together, make up the thermal characteristics of workplaces. The effect of the microclimate on the organism is to create different conditions of heat exchange between the organism and the environment and to determine its thermal functional state (11–13). Heat exchange between the organism and the environment takes place in the form of convection ($\pm C$), radiation

($\pm R$), and evaporation resulting in heat loss ($-E$). Heat exchange in the human organism should be considered in terms of the character of the environmental parameters responsible for heat exchange and the processes that contribute to generation of heat inside the organism (heat production or metabolism, M). Schematically, heat exchange between the organism and the environment may be expressed in the following way:

$$M \pm C \pm R - E = 0$$

Depending on the thermal load and exposure time, the organism may react in different ways. The most severe thermal disorders are heat stroke, heat syncope, heat cramp, and heat exhaustion. Standardization of measurements of the heat load and the thermal state is based on the principle of stress analysis of thermal regulation (*12, 14*). Here, the following conditions should be defined.

Optimal microclimatic conditions providing thermal comfort, i.e., the optimal thermal state. In these conditions, the deep body temperature is maintained with little strain on thermoregulatory mechanisms and without the involvement of additional adaptive mechanisms. Such conditions are not limited in time. Optimal microclimatic conditions create prerequisites for a high level of work capacity.

Permissible microclimatic conditions failing to ensure thermal comfort. Prolonged exposure to such conditions causes strain on thermoregulatory mechanisms and creates transient changes in body temperature, which quickly returns to normal, without making excessive demands on physiological adaptive responses. This is associated with a perception of thermal discomfort, loss of the sense of wellbeing, and a decrease in working capacity without pathological changes or disorders.

Those microclimatic conditions that result in pathological changes or disorders are inadmissible. Clear and exact criteria have been developed for estimating the physiological state of health of workers under thermal loads and these mainly relate to the degree of thermoregulatory strain (*15, 16*).

The most informative indicators for estimating the thermal state of the organism (thermal comfort) are skin temperature, evaporative losses, and heat perception.

The basic criteria for determining the strain on thermoregulation due to cooling are increase in heat production, the temperature drop over the entire skin surface, especially over exposed areas, and a slowing of the heart rate (bradycardia). The basic criteria for determining the strain on thermoregulation due to high environmental temperatures, on the other hand, are increase in evaporative heat loss through perspiration, change in skin temperature and temperature topography, smoothing of the distal-proximal gradient, decrease in heat production, and increase in the heart rate (tachycardia). As for the skin temperature, this is stable under conditions of thermal comfort and even under permissible microclimatic conditions placing a considerable strain on thermoregulation. The skin temperature rises only when the thermal

24

load is very high and, despite the involvement of additional thermoregulatory mechanisms, constant deep body temperature cannot be maintained. Consequently three grades of thermoregulatory strain—weak, moderate and high—can be discerned under permissible microclimatic conditions (17).

A schematic classification of the thermal states of the organism based on recent findings is presented in Table 1 (11, 12, 15).

The exposure limits for permissible microclimatic conditions adopted in various countries are based on measurements of the thermal load in the occupational environment and the demonstration of a correlation between the intensity of the factors to which the organism is exposed and the thermal state of the organism with regard to the workload and individual characteristics.

In order to integrate all the factors affecting the thermal state of the organism and involved in the estimation of the microclimatic conditions of the environment, a number of indices have been suggested, including effective temperature, corrected effective temperature, heat stress index, and predicted four-hour sweat rate. Each of these indices should be correlated with the levels of heat strain on the organism (14, 18).

The thermal state of the organism is also affected by the optical radiation produced by industrial sources, which comprises electromagnetic radiation with wavelengths of 100 nm to 1 mm $(3 \times 10^{11} - 3 \times 10^{15}\,\text{Hz})$. An increase in the source temperature causes a shift in the radiation spectrum to the shorter wavelengths. At temperatures over 1500°C, the spectrum contains both infrared and visible radiation, together with ultraviolet radiation.

The International Commission on Radiological Protection (19) has suggested the shown in classification of optical radiation Table 2.

This classification is based on biological reactions that are most marked at particular wavelengths. The thermal effect is most pronounced with absorption of wavelengths from 315 nm to $30\,\mu$ m, i.e., with absorption of infrared and visible radiation as well as long-wave ultraviolet radiation directly adjacent to the visible spectrum region. The thermal effect also depends on the intensity and duration of the radiation, as well as on the area of the irradiated surface; these parameters determine the absorbed dose of radiation. The shorter the wavelength, the higher the energy of the radiation, and it produces a stronger biological effect. Physiological reactions to the absorption of thermal radiation are expressed in a rise in the temperature of the skin and body, increased pulse rate, change of vascular tension, etc. Radiation causes the appearance in the skin of such biologically active substances as histamine and choline. The content of phosphorus and sodium in the blood also increases. This may result in secretory hyperactivity of the stomach, pancreas, and salivary glands, excessive inhibition of the central nervous system, decrease in neuromuscular excitability, and changes in metabolic processes.

Ultraviolet radiation may cause photochemical reactions (erythema, skin pigmentation, burns). Prolonged and multiple irradiation of the

Table 1. Schematic classification of thermal states of the human organism by degree of thermoregulatory strain

Physiological indicator	Excessive heat loss to the environment[a]			Comfort	Excessive heat gain from the environment[a]		
	high	moderate	low		low	moderate	high
Rectal temperature (°C)	below 35.0	35.5±0.5	above 36.0	37.2±0.4 (drift, not more than 0.2/h)	37.2–37.6	37.6–37.8	progressive rise, more than 0.3/h
Skin temperature (mean weighted) (°C)	below 28.0	29.7	31.1±1.0	33.2±1.0	34.9±0.7	36.0±0.6	over 36.0
Evaporative losses (g/h)	L.S.[b]	L.S.[b]	below 40	50±10	60–25	250–500[c]	500–2000[c]
Evaporative losses (% of total heat emission)	10–15	10–15	15–20	20–30	30–50	50–80	80–100
External symptoms	cyanosis of skin and mucous membranes, shivering, chill	pallor, skin cyanosis (slight)	pallor	lack of perspiration	slight reddening of skin, slight perspiration	reddening of skin, swelling of veins in limbs, heavy perspiration	severe reddening of skin, swelling of veins in limbs and face, heavy perspiration
Decrease in working capacity (%)	50 or more for 30–60 min	up to 30–50 in 3–4 h	up to 10–20 in 6–8 h	none	up to 10–20 in 6–8 h	50 or more in 3–4 h	50 or more for 30 min
Heat perception	very cold	cold	chilly	comfortable	warm	hot	very hot
Permissible time required for skilled task	up to 30 min	4 h	12 h	unlimited	12 h	4 h	up to 30 min

[a] high, moderate, low refer to the degree of thermoregulatory strain.   [b] L.S. = little studied.   [c] dripping perspiration.

Table 2. Classification of optical radiation

Spectrum region	Designation	Wavelength (nm)
Ultraviolet	UV-C	100–280
	UV-B	280–315
	UV-A	315–400
Visible	Vis-A	400–500
	Vis-B	500–600
	Vis-C	600–700
Infrared	IR-A	700–1400
	IR-B	1400–3000
	IR-C	3000–10^6 (1 mm)

eyes with short-wave, infrared, and ultraviolet rays may induce acute disorders of the electric ophthalmia type, and possibly, lenticular opacity (occupational cataract), xerophthalmia, conjunctivitis, blepharitis, etc.

Noise

The occupational environment is frequently characterized by noise, i.e., any unwanted sound or combination of sounds. In characterizing the effect of noise on the organism the following physical parameters are used: (1) intensity, (2) frequency spectrum, (3) duration of exposure.

The intensity of noise is determined by the amplitude of the acoustic pressure, which represents the difference between alternating rises and falls in acoustic vibrations compared with atmospheric pressure. Because the sense of hearing does not respond to differences in acoustic pressure, but rather to the frequency of changes in pressure, it is assumed that the intensity of noise is judged not by absolute levels of acoustic pressure, but by their ratio to the threshold value. The levels of acoustic pressure within the range of hearing may vary millions of times. Therefore, in order to reduce the measurement scale, the noise levels are expressed in logarithmic units or decibels (dB). Zero decibel corresponds approximately to the hearing threshold at 1000 Hz, while the pain threshold exceeds 140 dB.

Noise levels may vary during the working day, as does the duration of certain noise effects. The character of these changes may be of considerable value for estimating the effect of occupational noise on the worker.

According to time characteristics, a noise may be considered constant if its level does not change by more than 5 dB(A) over the day; noises that, under the same conditions, change by a greater value are described as "fluctuating". Among fluctuating noises it is possible to differentiate "time-related" ones, noises that change continuously, and "intermittent" noises whose levels fall abruptly to the background value; there are also "pulsed" noises comprising one or several acoustic signals each less than 1 s long.

Wide-band noise covers a range of frequencies with a continuous spectrum, while tonal noise is characterized by audible discrete tones.

The response of the organism to noise having different physical characteristics exhibits some peculiarities. Thus, noise whose energy is concentrated mainly in the high-frequency band produces a much more adverse effect on the organism than low-frequency noise. Tonal noise produces a still more unfavourable effect than wide-band noise. Noise that periodically falls in level during the working day produces less effect than continuous noise of the same level.

In order to estimate the effect on health of fluctuating noise as a function of its levels and the duration of exposure, the International Organization for Standardization (ISO) recommends using the following method for calculating the equivalent noise level (20):

$$L_{eq} = K_{lg}\left(\frac{1}{T}\int_0^T 10^{L(t)/K}\, dt\right)$$

where

$L(t)$ = noise level in the time function;
T = observation period;
K = the constant of the "averaging law".

The value K depends on the averaging parameter (q) which shows the value of dB by which the noise level may increase when the duration of exposure is halved and the amount of energy remains unchanged:

$$K = \frac{q}{lg\, 2}.$$

Calculation of the equivalent noise level is now widely used in many countries, although the averaging parameter (q) differs from country to country (in international recommendations and in the USSR, $q = 3$ ($K = 10$); in the USA, $q = 5$ ($K = 16.6$); in the United Kingdom, $q = 4.5$ ($K = 15$); and in the Federal Republic of Germany, $q = 3.9$ ($K = 13.3$)).

This situation complicates the comparison of estimations of the health effects of noise undertaken in different countries.

In recent years, it has been suggested that efforts should be made to estimate the dose of noise energy to which a worker is exposed. The Brüel & Kjær Company (Denmark) is already manufacturing two types of portable personnel dosimeter, types 4424 ($q = 3$ dB) and 4425 ($q = 5$ dB).

Occupational noise produces diverse effects on the organism: it has adverse nonspecific effects on the functioning of the nervous and cardiovascular systems, the gastrointestinal tract, and other organs. The majority of countries have accepted the decrease in hearing ability, particularly the temporary lowering of the hearing threshold, as a criterion for assessing the effect of occupational noise on workers' health.

Hearing function in industrial conditions can be tested in one of the following ways:

28

(1) one-time examination of patients subjected to prolonged exposure to industrial noise by means of tonal threshold audiometry;

(2) one-time examination of patients subjected to a particular type of noise;

(3) lengthy observation (several years) of changes in the hearing function of workers who have shown progressive loss of hearing.

The major features of occupational hypoacusis are initial loss of the hearing function in the region of 4000–8000 Hz with further spreading into medium and low frequencies.

In addition to tonal threshold audiometry at frequencies of 125–8000 Hz, assessment of hearing function is carried out by means of vocal audiometry and whisper perception tests.

Occupational hearing disorders are divided into the following categories (Table 3). In category I hearing loss, disorders characteristic of cochlear neuritis are present in the region of the basal margin of the cochlea where perception of high-frequency sounds takes place. In category II, the mediobasal portion of the cochlea is affected, resulting in failure to perceive sounds in the vocal zone. In category III, the entire medium and apical levels of the cochlea are affected and patients start complaining of subjective noise and inability to perceive even loud speech.

In assessing hearing function, the hearing changes normal for the subject's age should be taken into account, as shown in Table 4 (*22*).

Individual sensitivity to noise varies in different people. Peyser (*23*) used the following test for revealing insensitivity and hypersensitivity to noise. For this purpose, the hearing thresholds are determined at a frequency of 1000 Hz by means of air and bone conduction, with additional noise load of the same pitch at a level of 100 dB for three minutes. After 15 seconds the hearing threshold is measured and the degree of sensitivity to noise determined from the data given in Table 5.

Besides its specific action on the hearing organ, noise has a nonspecific effect on a number of systems and organs. It is worth mentioning that, according to several studies, the specific effect of noise on the hearing organ becomes noticeable much later than the nonspecific effect on other organs, in particular, that on the nervous system. This may lead rapidly to the development of neurasthenic and asthenic syndromes, autonomic dysfunction, and neurocirculatory disturbances associated with hypertension, hypotension, and cardiac symptoms (*24*).

Vibration

Vibration may be defined as a motion of particles or a mechanical system characterized by intermittent cyclic variations in amplitude and intensity. The intensity of vibration is expressed in terms of its velocity or rate of acceleration. It is measured in dB, which reflect the magnitude of the oscillations relative to the decimal logarithm of the ratio between the estimated and the "reference" (zero) value. Spectrum analysis of vibrations is carried out in octave and 1/3-octave wavebands (Hz). A

Table 3. Assessment of hearing function in occupational disorders (21)

Degree of hearing loss	Hearing function	Tonal threshold audiometry (dB)		Vocal audiometry (dB)		Whisper perception (m)
		hearing deficit at 500, 1000 and 2000 Hz	hearing loss 4000 Hz	threshold, 50% perception of speech	threshold, 100% perception of speech	
0	Practically normal hearing	0–10	0–20	30–35	40–45	5.1–7
I	Cochlear neuritis with slight hearing loss	11–30	21–50	36–45	46–60	3.1–5
II	Cochlear neuritis with moderate hearing loss	31–40	51–70	46–55	61–75	1.1–3
III	Cochlear neuritis with severe hearing loss	41–50	over 70	over 56	over 75	below 1

Table 4. Decline in hearing function with age in males (M) and females (F), (dB)

Age (years)	Sex	Frequency (Hz)							
		250	500	1000	2000	3000	4000	6000	8000
25	M	0	0	0	0	0	0	0	0
	F	0	0	0	0	0	0	0	0
30	M	0	0	0	0	0	0	5	5
	F	0	0	0	0	0	0	0	0
35	M	0	0	0	0	5	5	5	5
	F	0	0	0	0	0	5	5	5
40	M	0	0	0	5	5	10	10	10
	F	0	0	0	5	5	5	10	5
45	M	5	5	5	5	10	15	15	15
	F	5	5	5	5	5	10	10	10
50	M	5	5	5	10	15	20	20	20
	F	5	5	5	5	10	10	15	15
55	M	5	5	5	10	20	25	25	30
	F	5	5	5	10	10	15	20	20
60	M	5	10	10	15	25	30	35	35
	F	5	10	10	10	15	20	25	25
65	M	10	10	10	20	30	35	40	45
	F	10	10	10	15	20	25	30	35

man can perceive vibrations from fractions of a hertz to 8000 Hz. Higher frequencies are perceived as a sense of warmth (caumaesthesia).

Conventionally, the spectrum of vibrations is split into high-frequency (over several tenths of a hertz) and low-frequency bands. In most cases, the vibrations are of wide-band character; less often, they have a sinusoidal form.

In measuring the exposure to vibrations it is admissible to estimate an integral frequency by means of special filters or a mathematical formula, e.g., that recommended in CMEA Standard No. 1932–79 (25):

$$\tilde{U} = \sqrt{\Sigma K_i^2 U_i^2},$$

where U_i = root-mean-square value of the parameter being standardized within the given i-frequency band;
K_i = weighting coefficient for frequency band obtained from a special table.

Table 5. Indices of sensitivity to noise

Level of sensitivity	Increase of hearing threshold (dB) after 15 s following exposure	
	air conduction	bone conduction
Insensitivity to noise	5	0
Sensitivity to noise	6–10	0
Hypersensitivity to noise	over 10	5

Vibrations may be stable or unstable (time-varying, intermittent); those lasting for less than 1 s are called pulse vibrations.

With regard to their effects on the human organism, vibrations are divided into whole-body and segmental (arm-transmitted). The character of the response of the organism to vibrations depends greatly on the way they are transmitted to it, their frequency spectrum, and their time-varying intensity.

Permissible vibration levels are closely associated with the duration of the exposure. According to the supplement to CMEA Standard No. 1932–79, the dependence of the permissible duration of exposure (if not more than 8 hours) on the vibration level can be determined by the formula:

$$U_T = U_{480} \sqrt{\frac{480}{T}},$$

where U_T = level for permissible time T;
$\quad U_{480}$ = level for 8-hour (480-min) exposure;
$\quad\quad T$ = length of actual exposure in minutes.

Whole-body vibrations may be vertical, i.e., spreading along the vertical axis of the orthogonal coordinate system (Z), or horizontal, i.e., effective along the horizontal axis X (from back to chest) or the horizontal axis Y (from one shoulder to another). Local vibrations are also classified according to the three mutually perpendicular directions in which they produce their effects. Determining the direction of effect is particularly important for any whole-body low-frequency vibration, since this may induce resonant phenomena in the organism owing to the fact that the body, its parts, and the elastically suspended organs possess their natural frequencies of vibration. Resonant phenomena character-ized by an abrupt increase in the amplitude of vibration may occur in response to total vertical vibration with a predominant frequency of 4–8 Hz, or to horizontal vibration of 1–2 Hz.

According to CMEA Standard No. 1932–79, the whole-body vibration is classified in the following way:

— Category 1. Transport vibration affecting drivers of mobile vehicles and other means of transport moving across rough terrain, agricultural land, and roads.

— Category 2. "In-plant" transport vibration affecting drivers of mobile vehicles moving across special surfaces in industrial premises and work sites.

— Category 3. Vibration at workplaces from stationary installations or transmitted to workplaces from other sources.

Another type of vibration is that occurring at workplaces in premises with strict vibration-proof requirements, such as plant management offices, design departments, or laboratories.

In the case of whole-body vibration, the whole of the human body is involved in the vibration process. The effects are particularly marked with low-frequency vibrations of 8–16 Hz. These produce a traumatic action, which may lead to herniation of, and damage to the intervertebral disks followed by root syndromes, including sciatica; the gastrointestinal system and other organs may also be affected.

Reflex tendency of the body to retention of the stable vertical position under prolonged exposure to low-frequency vibration (e.g., when driving self-propelled vehicles) leads to permanent muscular strain, higher energy consumption, and undue fatiguability.

It is assumed that whole-body vibration, especially low-frequency vibration, has a marked effect on the otolithic membrane of the vestibular canal. Because the latter is connected by neural pathways to the cerebral cortex, autonomic nervous system, and muscles of the trunk, exposure to whole-body low-frequency vibration may result in inability to control motor coordination and to retain body equilibrium, and may cause autonomic nervous disorders and the development of gastric disorders.

Prolonged exposure to high-frequency whole-body vibration may lead to central nervous system disorders, and to polyneuritic syndrome.

According to International Standard ISO 2631–1978 (26) the limits of exposure to whole-body vibration are as follows: "Reduced comfort boundary", "Fatigue—decreased proficiency boundary", and "Exposure limit". The last of these is equal to approximately half the level of voluntary tolerance for healthy human subjects.

Exceeding the above limits results in the development of vibration syndrome. The following stages of vibration syndrome may be caused by exposure to total vibration.

Stage 1 is characterized by moderate initial symptoms and a compensated course; the symptoms and their combinations include angiodystonia, vegetovestibular neuritis, and polyneuropathy.

Stage 2 shows a moderate symptomatology with partially compensated course; combination of the following symptoms frequently occurs: cerebro-peripheral angiodystonia, polyradiculoneuropathy, and vertebral osteochondrosis.

Stage 3 is characterized by a pronounced symptomatology and a decompensated course involving circulatory encephalopathy combined with persistent forms of polyradiculoneuropathy; the incidence is rare.

Exposure to segmental, arm-transmitted vibration may possibly result in vascular disorders, in particular capillary disorders. It is followed by impairment of different types of cutaneous sensibility (pallaesthesia, algaesthesia, discriminatory sensibility, thermaesthesia, and tactile sense). Of frequent incidence are vegetotropic changes in the skin and nails, and pathological changes in the bones, small joints, and muscles. Low-frequency segmental vibration often results in pathological changes in the muscular and osteoarticular systems, while high-frequency vibration is often characterized by changes in the peripheral neurovascular system.

There are four stages in the development of "cutaneous" disease induced by segmental vibration:

Stage 1 (initial), characterized by few symptoms and signs and full compensation of the noxious effect;

Stage 2, characterized by a moderately severe symptom complex;

Stage 3, characterized by marked vascular disorders;

Stage 4, characterized by generalization of the vascular disorders as a result of impairment of the central nervous control of the vascular system.

The principal clinical features of vibration syndrome are:

(1) angiodystonia;

(2) angiospastic disorders;

(3) neuropathy of the autonomic nervous system;

(4) muscle changes of autonomic origin;

(5) peripheral neuropathy.

Biological agents

Biological agents include substances of vegetable and animal origin, and microorganisms and products of their metabolism. Workers may be exposed to the following biological agents:

(1) vegetation and vegetable dusts;

(2) substances of animal origin;

(3) combinations of substances of vegetable and animal origin;

(4) microorganisms and products of their metabolism;

(5) insects—mites, locusts, bees, ants, mosquitos, flour weevils, etc. (27).

Biological agents may be either simple or complex in composition. The complex ones include proteins of vegetable and animal origin, disseminated by miscellaneous flora, complex vegetable dusts, (e.g., from mixed feeds) and other dusts containing numerous organic compounds, inorganic substances, silica, etc.

It is recommended that standards for exposure to biological agents should be based on their protein content. If necessary, their biological activity should be determined and, in some cases, also their silica content.

The "dose–effect" relationship for agents of this kind depends on the content of active chemical components and proteins rather than on the dust concentration. Thus, for example, the incidence of byssinosis depends mainly on the level of proteolytic enzymes per unit volume of air, but not on the volume of the inhaled dust.

The specific action of biological agents is characterized by an impairment of immune functions resulting in sensitization to allergens and decreased resistance to infection. Allergic reactions may occur in latent or manifest form, and may be of immediate or delayed type. Fungi and saprophytic bacteria are considered to be more powerful allergens than organic substances of vegetable and animal origin and than pathogenic flora (28).

Initial sensitization is manifested in the form of rhinitis, dyspnoea, bradycardia, urticaria, hyperaemia, vomiting, oedema, fever, and reflex bronchospasm caused by the retention of organic particles inside the bronchial tree followed by release of serotonin and histamine. Other frequent sequelae are dermatitis, bronchitis without initial local impairment of lung function, and asthma.

In men working with mouldy grain or hay containing fungal spores at concentrations of 300 000 to 5 million per m^3, allergic alveolitis may appear in an acute, sub-acute, or chronic form, with a symptomatic or asymptomatic picture. The reaction is of the Arthus type mediated by precipitin. It may lead to granulomatous interstitial pneumonia developing into chronic interstitial fibrosis with multiple mycetomas, cardiopulmonary failure, and emphysema (29, 30).

Sensitization is confirmed by cutaneous reaction tests and by safer and more sensitive immunological reactions in vitro. A positive response in more than 50 % of the persons examined indicates that they constitute a "risk group".

In addition to allergies, exposure to biological hazards may lead to the development of mycoses: candidiasis, aspergillosis, penicilliosis, mucormycosis, coccidiosis, histoplasmosis, chromomycosis, etc. (31, 32). The commonest route of infection is by inhalation, leading to pseudotuberculous changes in the lungs. Local and systemic diseases of the skin may be observed, as well as diseases of the mucous membranes, viscera, nervous system (cryptococcosis), and lymphatic system (sporotrichosis). Secondary septicopyaemia may also occur, with the formation of abscesses. Mycoses are more frequent in the presence of immunodepression and metabolic disorders.

Zoonoses (over 150 known) may be caused by viruses (rabies, milkers' nodules, ornithosis, Newcastle disease, viral hepatitis, etc.), bacteria (salmonellosis, brucellosis, tularaemia, tuberculosis, anthrax, etc.), rickettsiae (Q-fever, Rocky Mountain fever, etc.), fungi (trichophytosis, etc.), protozoa (toxoplasmosis, trypanosomiasis, leishmaniasis, etc.), leptospiras and spirochaetes (leptospiroses, sodoku, etc.), and mites (scabies, etc.) (33–35).

Metabolic products of fungi, microbes (toxins) and higher plants (alkaloids) may cause intoxication. About 120 kinds of poisonous fungi and 100 kinds of mycotoxins have been described.

Workers with long service often suffer from functional nervous disorders and dysfunction of the liver and kidneys. Neurotoxic effects are often due to metabolic products of saprophytic bacteria, the microflora most often found in cattle-breeding complexes and mixed-feed-producing plants. The toxins produced by these bacteria increase the permeability of the mucous membrane of the gastrointestinal tract and are the possible cause of gastritis and intestinal bacterial disease (36, 37).

The irritant action of biological agents leads to subatrophic and atrophic changes in the upper respiratory tract (in tobacco growers), keratinization of the skin (e.g., in silkworm cultivators), etc. Such effects occur rather frequently in countries with tropical and subtropical climates.

Ergonomic and psychosocial factors

The workload and the intensity of activity are the ergonomic factors that largely determine the effects on health. Overwhelming evidence is available today that heavy exhausting physical work and intense mental activity characterized by extreme nervous and emotional strain may cause early functional disorders and pathological changes in the cardiovascular and nervous systems.

Most investigators assume that the energy expenditure per unit time (minute, hour, working day) can serve as an indicator for the assessment of the workload. For this purpose, the heart rate and oxygen consumption are often used. There are a number of classifications of work according to the load imposed (38–40).

Table 6 shows one such classification, which closely resembles those suggested by other investigators.

As a result of the intensive mechanization and automation of industrial processes, the proportion of the workforce engaged in manual labour is being drastically reduced, while the number of professions characterized predominantly by brain work and by load on the central nervous system (memory, concentration, thinking, etc.) is increasing; the distinction between manual and nonmanual labour is becoming blurred. Professions of the "operator" type are proliferating. The role of the operator is limited to monitoring functions for the control of complex operations, automated processes, etc. A new problem has arisen, that of ensuring the reliability of the "man–machine" system, and such health

Table 6. Classification of work according to load imposed on the muscles involved (41).

Location of active muscles	Work intensity	Energy expenditure	
		kcal/min	kJ/min
Hand	low	0.3–0.6	1.2–2.5
	medium	0.6–0.9	2.5–3.7
	high	0.9–1.9	3.7–7.9
One upper limb	low	0.7–1.2	2.0–5.0
	medium	1.2–1.7	5.0–7.0
	high	1.7–2.2	7.0–9.2
Both upper limbs	low	1.5–2.0	6.2–8.4
	medium	2.0–2.5	8.4–10.5
	high	2.5–3.0	10.5–12.5
Whole body	low	2.5–4.0	10.5–16.7
	medium	4.0–6.0	16.7–25.0
	high	6.0–8.5	25.0–35.6

risk factors as monotony, lack of physical activity, and "mental overload" have come to the fore.

The mental workload has increased significantly as a consequence, above all, of the growing stream of information and the need to process it and introduce it into everyday practice. This may contribute to the incidence of cardiovascular diseases among nonmanual workers and their occurrence at an early age. (42–44). There is an urgent need to elaborate reliable criteria and methods for assessing nervous and emotional strain at work.

By strain is meant the functional strain on the organism during work with a high level of nervous tension. At present, there is no consensus on the need to distinguish between work intensity and workload, since whatever the working conditions the organism reacts in an integral manner.

In recent years, numerous scientific findings have been reported concerning changes in physiological processes and neuroendocrine regulation in response to various degrees of nervous and emotional strain. Analysis of these findings may be of particular interest. It should be emphasized that in jobs characterized predominantly by strain on the central nervous system, i.e., on the auditory and visual centres and on mental functions, the energy expenditure is insignificant, especially if there is little or no physical load, and is far from actually reflecting the strain on the analytical and synthetic processes in the brain and their regulatory mechanisms. Of course, determinations of metabolic rates and of the levels of transmitters in the nerve centres would give more reliable information, but the range of methods at present available to research workers does not include such determinations—at least, not among the tests that can be performed in industry at the workplace. According to a number of investigations, the degree of nervous and emotional strain can be assessed sufficiently accurately on the basis of such data as heart rate, changes in cardiac rhythm, and the levels of catecholamines, corticosteroids and their metabolic excretion products in blood and urine. Meaningful information can be obtained by such methods as electroencephalography, radioencephalography, and the determination of indices of the functional state of the brain—speed of conditioned motor reflexes, capacity of short-term memory, ability to concentrate and to process visual information, as well as other mental processes (45–47).

The suggested classification of work according to the degree of nervous and emotional strain on the basis of physiological criteria has, as yet, no practical application.

In addition to physiological classifications of the degree of nervous and emotional strain, there are other schemes that take into account specific features of the work process and include a number of criteria based on a job description and the results of a time-and-motion study (47, 48).

The most significant psychosocial factors include interest in one's work, job satisfaction, the creative character of the work, and relations with colleagues and superiors. Underestimation of the importance of these and other psychosocial factors could lead to erroneous conclusions. However, the majority of these factors cannot as yet be expressed quantitatively. In epidemiological studies, therefore, they can at least be given descriptive treatment.

Combined effects of occupational factors

The study of the combined effects on workers of various occupational factors is one of the most important aspects of epidemiology in occupational health. The harmful effects may not necessarily occur simultaneously but may be consecutive or intermittent. For example, workers in the pharmaceutical industry may be exposed initially to the action of a certain compound (for a given period of time) and may subsequently deal with the synthesis of other agents. A similar situation may be observed in agriculture when the fields are treated at different times with different pesticides.

Industrial workers are usually exposed not only to the combined effects of various chemicals, but also, at the same time, to the action of physical agents.

The majority of studies cover combined exposure to industrial chemicals and high environmental temperatures. It has been demonstrated that such combined exposure leads to faster development of toxic effects and higher sensitivity to toxic factors. At the same time, poisoning by certain chemical agents may render the organism more susceptible to overheating (49). This effect has been noted in the case of combined exposure to high temperature and the following chemicals: carbon monoxide (50); oxides of nitrogen (51); aniline (52); mercury (53); certain compounds of heavy metals (54); and other chemical compounds.

Observations made under industrial conditions have shown that high atmospheric humidity increases the sensitivity of the organism to the toxic effect of fluorine compounds (55) and some petroleum hydrocarbons (56).

Recently, the problem of combined exposure to some industrial poisons, noise, and vibration has been intensively studied. Data are available proving that industrial noise affects the development of toxic manifestations and the resulting clinical picture. Similar findings have been reported in the case of exposure to high-frequency noise and quartz dust (57).

The characteristics of combined exposure to noise and acetone—a solvent widely used in many branches of industry—have been studied at low levels of exposure. The application of dispersion analysis to the results of experiments has made it possible to demonstrate the important influence of noise on the appearance of some adverse changes in the functional state of the nervous system. Special studies have revealed that at low levels of exposure the effects of noise and acetone are largely independent (58).

The toxic effects of fluorine (59), lead (60), and other compounds have been shown to be increased in workers also exposed to whole-body vibration. Other studies have found the respiratory function in some silicosis patients to be unexpectedly prolonged, even in cases of severe pneumofibrosis. Furthermore, impairment of respiratory function has been reported recently not only in workers with silicosis, but also in silica-exposed workers without radiological evidence of such disease. Analysis of the results showed that the overwhelming majority of such patients had been simultaneously exposed to two factors, quite different in nature and site of principal action: silica-containing dust and local vibration. This example demonstrates that combined exposure can have a very complex action, which cannot be considered as a summation of the effects of two independent factors but rather as a complex integrated effect on the whole system at different functional levels (61). Tiunov & Kustov (62) came to the conclusion that the diverse physical factors to be found in the occupational environment might exert considerable influence on the toxic action of industrial poisons, which, in turn, modify the response of the organism to the altered environmental conditions. The authors consider that physical and chemical factors in the environment produce interdependent effects only when, at a given intensity of effect, one of the factors, while reducing the reactivity of the organism as a whole, increases its sensitivity to the effect of the other factors. Depending on the particular conditions, either of the factors may acquire the leading role in producing the biological effect.

At fairly low but medically significant concentrations of the toxic substance, the physical factor may increase considerably the effects of the toxic substance. In contrast, with differences in the intensity of the effect of the physical factor, the significance of the chemical agent in the production of the total biological effect generally increases with increasing concentration of the chemical agent in air.

The problem of individual sensitivity, particularly in relation to the low levels of exposure encountered today, is important in carrying out epidemiological studies of the effects of occupational risk factors on workers. With low levels of exposure, the risk of harmful effects depends much more on the sensitivity of the organism than in the case of high levels. Of interest here is not only the sex- and age-dependent individual sensitivity, but also the initial functional state of the organism. Study of total and occupational morbidity among industrial workers shows that the decrease in adaptation as a response to exposure to chemical and physical factors may take place but this seems to occur in only a few cases and in highly sensitive people.

The problems of individual sensitivity and of the aggravation of premorbid conditions are closely interrelated; aggravation of premorbid conditions is particularly important when both the agent causing the disease and the hazardous factor produce a similar effect on the organism. One of the most important ways of solving this problem is by developing criteria for the identification of individuals exhibiting hypersensitivity to a given toxicant. The importance of occupational selection (i.e., rejection of individuals who show hypersensitivity to a given toxicant) lies in the fact that it will lead to a decrease in the frequency of hypersensitivity in workers and hence to a reduction in the probability that will develop occupational diseases. Skin reactions, loading tests, and determinations of enzyme activity are among the possible methods for determining hypersensitivity.

The considerations presented above testify to the necessity of undertaking a many-sided evaluation of the effects on workers of occupational factors during the planning and carrying out of epidemiological studies.

References

1. **Henderson, Y. & Haggard, H. W.** *Noxious gases.* New York, Reinhold, 1943.
2. **Mikheyev M. I. & Lyublina E. I.** In: Tolokontsev, N. A. & Filov, V. A., ed. *Osnovy obshchej promyshlennoj toksikologii* [*Essentials of general industrial toxicology*]. Leningrad, Medicina, 1976.
3. **Sanotsky, I. V. & Ulanova, I. P.** *Kriterii vrednosti v gigiyene i toksikologii pri otsenke opasnosti khimicheskikh soedinenij* [*Criteria for assessment of hazard in hygienic and toxicological evaluation of dangerous chemical compounds*]. Moscow, Medicina, 1975.
4. **Sanotsky, I. V. et al.** *Journal of the All-Union Chemical Society D. I. Mendeleyev*, **19** (2): 146 (1974).
5. **Okhnyanskaya, L. G. et al.** In: *Rannie stadii silikoza* [*Early stages of silicosis*]. Leningrad, Medicina, 1968, pp. 50–128.
6. **Sokolov, V. V. et al.** *Metodologicheskie osnovy gigienicheskogo normirovaniya* [*Methodological basis of hygienic standardization*]. Moscow, Research Institute for Occupational Hygiene and Health, 1976, pp. 71–85.
7. **Trakhtenberg, I. M. et al.** *Gigiyena truda i profzabolevaniya*, No. 11, pp. 1–6 (1976).
8. **Alexeyeva, O. G.** *Gigiyena i sanitariya*, No. 10, pp. 82–85 (1974).
9. *Methods for studying biological effects of pollutants (A review of methods used in the USSR). Report on a working group.* Copenhagen, WHO Regional Office for Europe, 1975 (unpublished document EURO 3109(4)).
10. **Aldyreva, M. V. & Gafurov, Sh. A.** Gigiyena truda v proizvodstve iskusstvennykh kozh [Occupational hygiene of artificial leather manufacturing]. Moscow, Medicina, 1980.
11. **Gubernsky, Yu. D. & Korenevskaya, Ye. I.** *Gigiyenicheskie osnovy konditsionirovaniya mikroklimata zhilykh i obshchestvennykh zdanij* [*Hygienic essentials of air-conditioning of residential and industrial buildings*]. Moscow, Medicina, 1978.
12. **Shakhbazyan, G. H. & Schleifman, F. M.** *Gigiyena proizvodstvenogo mikroklimata* [*Hygiene of industrial microclimate*]. Kiev, Zdorov'ya Publishers, 1977, p. 136.
13. **Fanger, P. O.** *Thermal comfort. Analysis on applications in environmental engineering.* Copenhagen, 1970.
14. **National Institute for Occupational Safety and Health.** *The industrial environment–its evaluation and control.* US Department of Health, Education and Welfare, 1973.
15. **Kandror, I. S. et al.** Fiziologicheskie printsipy sanitarno-klimaticheskogo rayonirovaniya territorii SSSR [*Physiological principles of sanitary and climatic zoning in the USSR*]. Moscow, Medicina, 1974.
16. **Dell, R. A. et al.** Gigiyena odezhdy [*Clothing hygiene*]. Moscow, Legkaya Industriya, 1979.
17. **Azhayev, A. N.** Fiziologo-gigienicheskie aspekty deistviya vysokikh i nizkikh temperatur [Physiological and health aspects of the effects of high and low temperature]. In: *Problemy kosmicheskoj biologii*. Moscow, Nauka, 1979.
18. WHO Technical Report Series No. 601, 1977 (*Methods used in establishing permissible levels in occupational exposure to harmful agents. Report of a WHO Expert Committee with the Participation of ILO*).

19. **International Commission on Radiological Protection.** *Report of the Task Group.* Oxford, Pergamon Press, 1975 (ICRP Report No. 23).
20. *Acoustics–assessment of occupational noise exposure for hearing conservation purposes.* Geneva, International Organization for Standardization, 1975 (ISO Standard No. 1999–1975).
21. **Kolomijchenko, A. I. et al.** *Profilaktika professionalnykh porazhenij organa slukha u rabochikh mashinostroitelnykh predpriyatij* [*Prevention of occupational impairment of hearing in workers in the machine-building industry. Recommendations*]. Kiev, 1975.
22. **Dieroff, H. G.** *Lärmschwerhörigkeit. Leitfaden der Lärmhörschadenverhütung in der Industrie.* Munich, 1979.
23. **Peyser, A.** Audiometrischer Ermüdungstest zum Zwecke der Berufsduswahl. *Acta otolaryngologica,* **41**: 156–158 (1952).
24. **Alexeyev, S. V. & Kadyskina E. N.** Mediko-biologicheskie aspekty profilaktiki shumovoj patologii [Medicobiological aspects of the prevention of diseases caused by noise]. In: *Zvukoizolirujushchie i zvukopogloshchajushchie konstruktsii v praktike bor' by s shumom* [*Sound-proof and sound-absorbing structures in noise control*]. Leningrad, Medicina, 1977, pp. 4–7.
25. *Vibration. Permissible levels of total vibration at workplaces* Moscow, Standard Publishing House, (CMEA Standard No. 1932–79).
26. *Guide for the evaluation of human exposure to whole-body vibration.* Geneva, International Organization for Standardization, 1978 (ISO Standard No. 2631–1978).
27. **Barber, E. & Husting, E.** Biological hazards. In: *Occupational diseases.* Washington, DC, NIOSH, 1977, pp. 45–79.
28. **Pasternak, N. I. & Brysin, V. G.** *Allergennost' plesnevykh gribov* [*Allergenicity of moulds*]. Tashkent, Uzb. SSR Medicina Publishers, 1975.
29. **Stepanov, S. A.** *Etiologiya, patogenez i morfogenez pnevmokonioza vyzvannogo zernovoj pyl'yu* [*Etiology, pathogenesis and morphogenesis of pneumoconiosis induced by grain dust*]. Saratov, Saratov University, 1974.
30. **Racovianu, C. & Nicolaesci, V.** Immunologic complications of pulmonary diseases. In: *Immunobiology. Immunochemistry. Immunopathology.* Bucharest, SRR Academy Publishing House, 1977, pp. 481–501.
31. **Leshchenko, V. M.** *Aspergillez* [*Aspergillosis*]. Moscow, Medicina, 1973.
32. **Chausovskaya, M. M.** *Pnevmokoniozy* [*Pneumoconioses*]. Moscow, Medicina, 1978.
33. **Tokarevich, K. N.** *Professional'nyj faktor v epidemiologii zooantroponozov* [*Occupational factors in the epidemiology of zoonoses*]. Moscow, Medicina, 1978.
34. **Kotima, M. et al.** *Homepölyaltistus rehuntuotannossa ja krjan hoitotyössä. Osal: Karjan sisärukintakauden alussa. Tyoterveyslaitoksen tutkimutsia 141.* Helsinki, Tyotervlyslaitos, 1978.
35. **Kalina, G. P.** *Salmonellezy v okruzhayushchej srede* [*Environmental salmonelloses*]. Moscow, Medicina, 1978.
36. **Weiss, R.** Mykotoxine und ihre Bedeutung für die Gesundheit von Mensch und Tier. *Tierärztlicher Praxis,* **6**: 9–17 (1978).
37. **Dony, J.** Les endotoxines et leurs effets biologiques. *Journal de pharmacie de Belgique,* **34**: 127–133 (1979).
38. **Christensen, E. U.** Physiological valuation of work in Nykroppa Iron Works. In: *Symposium on fatigue.* London, Lewis, 1953, pp. 93–108.
39. **Soul, C. et al.** Aspects musculaires, sensoriels, psychologiques et sociaux de la fatigue. *Archives des maladies professionnelles, de médecine du travail et de sécurité sociale,* No. 22, pp. 419–446 (1961).
40. **Rozenblat, V. V.** Ob otsenke tyazhesti i napryazhennosti truda [Assessment of work load and intensity]. In: *Funktsii organizma v protsesse truda* [*Occupational functions of organism*] Moscow, Institute of Hygiene, 1975, pp. 8–30.
41. **Spitzer, H. & Hettinger, Th.** Tables donnant la dépense énergétique en calories pour le travail physique. *Cahiers du B. T. E.,* No. 302–04. Paris, vol. 1, (1966).
42. **Korkushko, O. V. et al.** Rol' serdechno-sosudistykh zabolevanij v razvitii prezhdevremennogo stareniya u lits umstvennogo truda [The role of cardiovascular diseases in premature aging of nonmanual workers]. *Vrachebnoye delo,* No. 4, pp. 7–11 (1979).

43. **Theorell, T. & Myrhed, B. F.** Workload and risk of myocardial infarction—a prospective psychosocial analysis. *International journal of epidemiology,* **6**: 17–21 (1977).
44. **Shkhratsabaya, I. K. et al., ed.** *Epidemiologiya serdechno-sosudistykh zabolevanij [Epidemiology of cardiovascular diseases].* Moscow, Medicina, 1977, p. 354.
45. **Bayevsky, R. M.** *Prognozirovanie sostoyanij na grani normy i patologii [Forecasting states bordering on the normal and the pathological].* Moscow, Medicina, 1979.
46. **Kundiev, Yu. I. et al.** Napryazhennaya umstvennaya deyatelnost i sostoyanie regulyatsi serdechno-sosudistoj sistemy [Intense mental activity and the state of regulation of the cardiovascular system]. *Fiziologia cheloveka,* **2**, (3): 433–440 (1976).
47. **Navakatikyan, A. O. & Kryzhanovskaya, V. V.** *Vozrastnaya rabotosposobnost' lits umstvennogo truda [Age-dependent work efficiency of nonmanual workers].* Kiev, Zdorov'ya Publishers, 1979.
48. **Kandror, I. S. & Dyomina, D. M.** O printsipakh i kriteriyakh fiziologicheskoj klassifikatsii vidov truda po stepeni ikh tyazhesti i napryazhyonnosti [Principles and criteria for physiological classification of work according to load and intensity]. *Fiziologia cheloveka,* **4**, (1): 136–147 (1978).
49. **Volkova, Z. A. & Matsak, V. G.** In: *Gigiyena truda v khimicheskoj promyshlennosti [Occupational hygiene in chemical industry].* Moscow, Medicina, 1967, p. 22.
50. **Tiunov, L. A. & Kustov, V. V.** *Toksikologiya okisi ugleroda [Carbon monoxide toxicology].* Moscow, Medicina, 1969.
51. **Paribok, V. P. & Ivanova, F. A.** *Gigiena truda i profzabolevaniya,* No. 7, p. 22 (1965).
52. **Solovyova, V. A.** In: *Voprosy promyshlennoj toksikologii [Problems of industrial toxicology].* Moscow, Research Institute for Occupational Hygiene and Health, 1960, p. 29.
53. **Shakhbazyan, G. H. & Savitsky, I. V.** *Vestnik Akademii Medicinskih Nauk SSSR,* **18** (2): 38, (1963).
54. **Savitsky, I. V.** In: *Farmakologiya i toksikologiya [Pharmacology and toxicology].* Kiev, Zdorov'ya Publishers, Issue 6, 1971.
55. **Krivoglaz, B. A.** *Klinika i lechenie intoksikatsij yadokhimikstsmi [Clinical picture and treatment of pesticide poisoning].* Leningrad, Medicina, 1965.
56. **Samedov, I. G.** In: *Promyshlennaya toksikologiya [Occupational toxicology].* Moscow, Research Institute for Occupational Hygiene and Health, 1960.
57. **Onopko, B. N.** In: *Gigiyena truda [Occupational hygiene].* Kiev, Zdorov'ya Publishers, 1970.
58. **Britanov, N. G.** *Gigiyenicheskaya otsenka sochetannogo deistviya shumov i atsetona na organizm [Evaluation of health effects of human exposure to noise and acetone].* Moscow, 1979 (abstract of Cand. Med. Sci. thesis).
59. **Davydova, V. I.** In: *Kombinirovannoye deistvie khimicheskikh i fizicheskikh faktorov proizvodstvennoj sredy [Combined effect of environmental chemical and physical factors].* Sverdlovsk, Research institute for occupational hygiene and health, 1972, p. 100.
60. **Tartakovskaya, L. Ya. et al.** In: *Voprosy promyshlennoj toksikologii [Problems of industrial toxicology].* Moscow, Research Institute for Occupational Hygiene and Health, 1972, p. 44.
61. **Okhnyanskaya, L. G.** In: *Metodologicheskie osnovy gigienicheskogo normirovaniya proizvodstvennykh faktorov [Methodological basis of hygienic standardization of occupational factors].* Moscow, Research Institute for Occupational Hygiene and Health of the Academy of Medical Sciences of the USSR, 1976, pp. 18–29.
62. **Tiunov L. A. & Kustov, V. V.** *Journal of the All-Union Chemical Society D. I. Mendeleyev,* **19** (2): 164–168 (1977).

3

Work, health and disease

G. Kazantzis[a] & J. C. McDonald[b]

As far as the authors are aware, no comprehensive assessment has ever been made of the contribution of occupational factors to human disease. Most accounts are "agent-orientated", that is to say they concentrate on describing and, where possible, measuring the effects in working populations of exposure to specific occupational hazards. The resulting picture is inevitably fragmented. Much has been learned, nevertheless, about the types of effect produced by a multitude of agents, sometimes with information on risk, but practically nothing is known about the total impact and relative importance of occupational factors in the major disease groups. This is illustrated by the recent controversy in the United States over the proportion of cancers related to occupation (*1*), which also served to highlight the problem of interaction between multiple agents, either in the working environment or partly outside it. Most epidemiological studies have sidestepped these very difficult questions by examining only one stimulus variable (or, at best, one at a time), treating other factors as confounders to be allowed for in design and analysis. This approach can be largely justified in terms of research priorities, but some interactions—for example between smoking and a variety of airborne dusts, between alcohol intake and solvent exposure, between occupational exposures and therapeutic agents—are of immediate practical importance and will have to be faced more urgently. Health effects of combined exposures in the work environment have been recently considered by a WHO expert committee (*2*).

In this chapter the major organ systems are examined in turn and an overview is given of the epidemiology of occupationally related diseases affecting each of them. Where possible, comments are made on the nature of the evidence but individual surveys are not described in any detail and those that are cited should be regarded as examples only. This chapter is thus concerned with the distribution of disease related to

[a] TUC Centenary Institute of Occupational Health, London School of Hygiene and Tropical Medicine, London, England.
[b] School of Occupational Health, McGill University, Montreal, Canada.

43

work and not with disease mechanisms or with clinical signs and symptoms as such. Finally, the other side of the coin, i.e., the extent to which work promotes health and wellbeing, is briefly examined. This is a very difficult area, much influenced by philosophical considerations and almost defying objective or quantitative evaluation.

Work and disease

Diseases of the respiratory system

The respiratory tract and skin are readily accessible to noxious factors in the environment; not surprisingly they account for a high proportion of all work-related disease. The vulnerability of the respiratory organs is increased by the large volume of air, readily contaminated by aerosols, gases and vapours in the workplace, that moves in and out of the lungs. Nevertheless, the possible types of response—acute reaction, chronic inflammation, progressive fibrosis, malignant disease—are limited and nonspecific.

Acute reaction

Illness of rapid onset may result from the inhalation of materials that are either active or inactive biologically, the former giving rise to irritative, inflammatory, or asthmatic symptoms and the latter to asphyxia. A number of toxic substances are absorbed through the respiratory tract (e.g., carbon monoxide, cyanides, and many solvents) but have systemic rather than local effects.

A few microbial pathogens cause serious pneumonitis in persons whose work brings them into close contact with animals or animal products. Q fever and ornithosis remain common and widespread infections in such groups as farmers, poultry keepers, slaughtermen, hide and wool workers, and employees in zoos, pet shops and animal houses. Pneumonic plague and anthrax, with similar epidemiological features, are now virtually unknown in the industrialized countries but no doubt still occur in less developed parts of the world. Soil infected with *Coccidioides immitis* and *Histoplasma capsulatum* continues to cause acute and chronic lung disease in farmers, migrant workers, and construction workers over wide geographical areas, and tuberculosis remains a risk for persons, particularly in the health services, whose work exposes them to infection. In all these diseases, the dominating factors are climatic or socioeconomic. Acute respiratory infections find conditions conducive to rapid spread where large groups of workers live in barracks; such occupational situations may require special measures from the health services (*3*, *4*).

The asphyxiants (e.g., nitrogen, methane, carbon dioxide) and respiratory irritants (e.g., chlorine, ammonia, phosgene, oxides of nitrogen and sulfur, and several heavy metal compounds) are present in a considerable variety of gases and fumes encountered in mining, diving, agriculture, firefighting, and in the chemical and metal industries. Their effects are direct and proportional to their toxicity and the amounts

44

liberated. The epidemiology is generally that of "accident occurrences", mediated primarily by human error and preventable by appropriate safety procedures.

Finally, there is the growing problem of occupational asthma and allergic alveolitis, the mechanisms of which remain obscure. Whereas most diseases reflect both the "toxicity" and the "dose" of the etiological agent(s) and the susceptibility of the individuals exposed, in the asthma-alveolitis group of illnesses susceptibility appears to be related to a state of hypersensitivity that develops after a latent period of months or years. Virtually all organic products of animal, vegetable, or microbial origin are capable of giving rise to this type of disease reaction in exposed workers; the occupations affected are therefore numerous, especially in the farming, lumber, food, textile, antibiotic and detergent industries. In addition, there are a number of chemicals not of organic origin—formaldehyde, isocyanates, platinum salts, acid anhydrides—that appear exceptionally potent causes of occupational asthma. These substances are encountered mainly in the manufacture and use of foam products, paints, plastics and resins, in soldering, and in medical laboratories. Some are so potent that virtually all persons exposed eventually develop symptoms. The occupational asthmas have received little systematic epidemiological study, partly because their full importance has only recently been recognized and partly because of difficulties in case identification and because affected workers tend to remove themselves from exposure. There is some evidence that familial atopy (often vaguely defined) may influence susceptibility; on the other hand, age, sex, and cigarette smoking, do not appear to be factors of importance. A useful review of acute reactions to inhaled agents is given by Weill & Hendrick (5).

Chronic inflammation
Workers in many heavy industries—mines, quarries, foundries, coal distillation, textile mills, building products factories, etc.—are exposed to respirable dust, some of which is fibrogenic and/or carcinogenic. These workers are often exposed also to irritant gases and fumes and to adverse conditions of temperature and humidity; most of them smoke cigarettes, some heavily. Chronic bronchitis and emphysema are highly prevalent among such workers, leading to respiratory symptoms and functional impairment, disability, and eventually to excess mortality from cardiorespiratory disease. For many years, epidemiological surveys have sought to disentangle this complex etiological maze but still without complete success.

The failure to settle these questions is largely due to the fact that (a) most of the surveys have been cross-sectional rather than longitudinal, making it difficult to interpret the sequence of preceding events, and (b) the pathological processes in question are directly observable only at autopsy (and in only a selected proportion of deaths) when the dynamic interpretation is even more difficult. Certain points are fairly clear, however. In British and American coalminers, cough, phlegm, and reduction in forced expiratory volume are related to *both* dust exposure and cigarettes smoked, with evidence that the effects are additive (6). The same appears broadly true of American and British cotton textile workers (7, 8) and up to a point, of chrysotile miners and millers (9). What is much less certain is whether dust-induced occupational bronchitis, in the absence of cigarette smoking or significant pulmonary fibrosis, leads to emphysema with its more serious effects on respiratory function and life expectancy. The reader is referred to the proceedings of an International Symposium on this subject in Warsaw in 1971 (10).

Progressive fibrosis
Fibrosis of the lung, with or without fibrosis and thickening of the pleura, results from exposure to a variety of particulate and fibrous mineral dusts. The classical example is silicosis, a serious and sometimes fatal disease of miners, quarrymen, and many other groups (foundry-men, stone-cutters, pottery workers, etc.) who are exposed to respirable crystalline silica. The risk of the disease is directly related to dust concentration, and mortality is due to the cardiorespiratory effects of massive pulmonary fibrosis resulting in part from coexistent tuberculosis. A recent report (11) illustrates the pattern of mortality experienced by hard rock miners in the past. In a cohort of 1321 South Dakota gold miners employed for 21 years or more, of whom 660 had died, there were 81 excess deaths, 37 from silicosis and 35 from silico-tuberculosis. The excess mortality was linearly related to the estimated dust concentration. At very much lower quartz levels, the prevalence of small rounded opacities in the chest radiograph was similarly related to the estimated quartz concentration in British gypsum miners (12).

Coalminers' pneumoconiosis is of comparable importance and shows epidemiological similarities. The natural history of the disease is more easily separable into (a) the prolonged and essentially benign "simple pneumoconiosis", the incidence and severity of which is determined by accumulated dust exposure (better expressed in mass than in concentration of respirable particles) (6) and (b) the rapidly progressive "complicated pneumoconiosis", which may intervene in a small proportion of men with heavy exposure. The risk of pneumoconiosis at equivalent levels of exposure differs from colliery to colliery and from country to country: "coal rank" (a mineralogical classification of coal) rather than quartz content is probably the explanation (13, 14). The attack rate does not appear to be related to smoking habit (15).

The third major group of dust diseases is associated with the fibrous silicates (asbestos, talc, and mica) and possibly with some nonfibrous silicates, such as kaolin and Fuller's earth (16). For practical purposes, asbestosis is the main problem, a diffuse interstitial fibrosis of the lung often accompanied by pleural thickening. Prevalence surveys, using chest radiography and function tests, have shown that some degree of asbestosis is present in a high proportion of workers exposed for long periods in poorly controlled conditions of asbestos production,

46

manufacturing, and application. Fibre concentration and duration of exposure seem to be the principal determinants (17); age, sex, smoking habit, and fibre type have all been shown not to be important variables. The progression of this disease is usually slow and insidious, increasing fibrosis leading to increasing breathlessness and to death from cardiorespiratory causes in a small proportion of cases. Rapid deterioration associated with advanced massive fibrosis, as seen in coalminers and persons with silicosis, is unusual, perhaps because there is no clear interaction with tuberculosis. Fibrosis has been shown to progress in a proportion of subjects after withdrawal from asbestos exposure (18); the same is probably true with the other fibrogenic dusts. There is now evidence from longitudinal surveys that, in both asbestos workers and coal miners, radiological changes at work are each in their own way good predictors of life expectancy and cause of death (19, 20).

Neoplasms

Exposures at work to a number of agents—asbestos, compounds of arsenic, nickel and chromium, coke-oven fumes, and ionizing radiation—have all been shown to increase the risk of respiratory cancer, particularly of the bronchus. Beryllium and cadmium are also under suspicion. The amphiboles, crocidolite and amosite, carry a substantial risk of malignant tumours of the pleura and peritoneum, chrysotile carries a much smaller one. In addition, some cases of cancer of the nasal sinuses and possibly of the larynx are related to occupation.

Respiratory cancer in asbestos workers has been very fully investigated (21, 22). Many well designed cohort studies have shown substantial excess mortality with all types of fibre and in a variety of industries where asbestos is produced, manufactured, and applied. Asbestos is very widely used in modern industrial societies and the fact that the exposure–response relationship appears linear and without a threshold has important implications for control (23); so too has the interaction between asbestos and cigarette smoking, which appears at least additive and, in some circumstances, multiplicative (24).

While ionizing radiation increases the risk of malignant disease at several sites, probably also in a linear manner with dose, excess lung cancer has been observed mainly in miners exposed to the inhalation of radon daughters. The risk was demonstrated first in the metal miners of Schneeberg and in the uranium miners of Joachimsthal in the German Democratic Republic; more recently, cohort and case–referent studies have provided better quantification in the fluorspar mines of Newfoundland (25), the uranium mines of Colorado (26), and the zinc-lead mines of Sweden (27). The degree of synergistic interaction with cigarette smoking seems to vary among the different populations.

The carcinogenic effects of metals and their compounds on the respiratory tract have recently been reviewed by Kazantzis (28). A clear relationship has been demonstrated in the chromate industry, in nickel refineries, and in copper-smelter workers exposed to arsenic. Surveys in the United States of coke-oven workers and in the United Kingdom of men exposed to the products of coal carbonization have also shown a definite excess of lung cancer (*6*).

Diseases of the circulatory system

Heart disease is the major cause of disability and death in industrial communities today, but the role of occupational factors in its causation has been little explored. Coronary arteriosclerotic heart disease is commoner in some occupational groups than in others. This does not necessarily imply a causal relationship to occupation, for such risk factors as cigarette smoking, physical inactivity, obesity, and, in particular, stress are often features of the lifestyle of employees in certain occupations. However, the lifestyle may also be dictated by the occupation. The hypothesis that physical activity at work affords protection against ischaemic heart disease was first investigated by Morris (*29*) and was found to hold for bus drivers and bus conductors, although it may be that these groups were inadequately matched with regard to characteristics other than physical activity. However, other studies, such as one concerned with the role of physical activity in reducing coronary mortality among American long-shoremen (*30*) show a remarkably consistent relationship between mortality and inactivity, even within the same social class. In Britain, mortality from coronary heart disease was found to be higher in professional and managerial groups than in lower grade workers in the earlier part of the century, but this social class differential now appears to have been reversed. In a longitudinal study of a large group of civil servants working in London, men in the lowest grade, mainly messengers, had more than three times the coronary heart disease mortality of men in the highest employment grade. The higher coronary heart disease mortality experienced by working class men could be only partly explained by established coronary risk factors (*31*). Employment status was found to be a stronger predictor of the risk of dying from coronary heart disease than any of the more familiar risk factors. The prevalence of angina pectoris and of electrocardiographic abnormalities was also found to be substantially higher in men in the lowest employment grade (*32*).

A randomized controlled trial was performed in industry to assess the effectiveness of a programme aimed at preventing heart disease by the control of risk factors (*33*). Advice was given on dietary reduction of plasma cholesterol concentration, stopping or reducing smoking, weight reduction, daily exercise, and treatment of hypertension. The trial showed that coronary risk factors can be changed in a working population, but the changes obtained were not large and were not sustained.

There is evidence for a possible causal association between certain toxic agents in the working environment and coronary arteriosclerotic heart disease. Workers exposed to carbon disulfide, as in the manufacture of rayon and carbon tetrachloride, have experienced an increased mortality (*34*) associated with hypertension and hyperlipidaemia (35). Carbon monoxide, which is atherogenic in rabbits and monkeys, is believed to be a causal factor in arteriosclerotic heart disease in cigarette smokers (36). In a prevalence study of angina in Finnish foundry workers (37), the highest rate was found in smokers with carbon monoxide exposure, and the lowest rate in nonsmokers without such exposure. Munitions workers exposed to glyceryl trinitrate and other organic nitrates have been shown to have an increased mortality, believed to result from coronary artery spasm, following re-exposure after a short period of absence from work (38). An increased risk of cardiovascular and cerebrovascular disease has been shown in a case–control study of Swedish explosives workers (39). Sudden death from cardiac dysrhythmia has been associated with exposure to a number of halogenated hydrocarbons, in particular trichloroethylene, trichloroethane, and fluorocarbon aerosol propellants. Many workers are exposed to these compounds, but appropriate epidemiological studies have not been performed.

Of the metals, cobalt used as a foaming agent in beer has caused epidemics of cardiomyopathy with high mortality, including some cases among brewery employees (40). An increased mortality from cerebrovascular disease was observed in lead battery workers in England (41). Although chronic nephropathy as a sequel to lead poisoning has been documented, there is no evidence of increased mortality from hypertension in lead battery and smelter workers (42). Cadmium has been implicated as a possible causal factor in hypertension, higher levels of cadmium having been found postmortem in the kidneys of hypertensive patients as compared with normotensive subjects (43). Raised blood cadmium levels have also been observed in a group of untreated hypertensive subjects compared with normotensive controls (44) but a larger study failed to show an association between blood cadmium and hypertension (45).

Extensive use of vibratory hand tools may lead to the development of Raynaud's syndrome, known in industry as vibration white finger. The condition is seen frequently in workers engaged in chain sawing, grinding, chipping with pneumatic tools, and swaging. Quarrymen, miners, and others handling pneumatic drills, hammers, and chisels are also affected. In one study, 85% of chain sawyers were affected, but prevalence rates in different groups have varied widely (46).

Diseases of the digestive system
Diseases of the digestive system are common and of great economic importance. During the present century, duodenal ulcer increased in industrialized countries until the 1940s and then declined. It has been considered to be related to stress, and an increased prevalence has been observed in occupations thought to be stressful. Doll et al. (47) found a higher than expected frequency of duodenal ulcers in foremen and executives. In a study of air traffic controllers, considered to be in a stressful occupation compared with second-class airmen taken as

controls, the air traffic controllers experienced an excess risk of developing peptic ulcer and were affected at a younger age than the controls (48). The problem of obtaining an objective measure of stress in relation to work is, however, considerable. In a longitudinal study, nervous strain at work was found to relate both to predisposition to anxiety and to the workers' own report of day-to-day activities in their job (49).

Type B hepatitis is an important health hazard in health care personnel, especially among those working in hospital laboratories, and in dialysis and oncology departments (50, 51). Cirrhosis of the liver, while not an occupational hazard in the ordinary sense, is far more common in some social groups than in others (52). In the United Kingdom, for example, standardized mortality rates are generally higher among the more affluent whereas in the Unites States, where alcohol has been cheap, unskilled workers have the highest rates (e.g., the standardized mortality rate in the United Kingdom is 96, while in the United States it is 148) (53). Among the occupations with the highest mortality from cirrhosis are barmen and publicans, seamen, company directors, and medical practiners (54).

Other risk factors for cirrhosis are certain fungal toxins such as aflatoxin, drugs that stimulate enzyme systems in the liver, and chemical exposures. Chronic arsenic poisoning has been followed by cirrhosis, which has been observed in Moselle vintners using arsenical sprays. Arsenic, thorotrast, and vinyl chloride have given rise to the rare haemangiosarcoma of the liver, of which not more than 200 cases have been reported worldwide. Haemangiosarcoma of the liver in a worker exposed to vinyl chloride monomer in the polymerization process was first reported in the United States in 1974. Since then, haemangiosarcoma and also hepatic fibrosis and cirrhosis have been reported from vinyl chloride polymerization plants in many countries. In Great Britain, a register set up in 1974 has collected 34 cases in which diagnostic criteria for haemangiosarcoma of the liver were fulfilled. Two of these occurred in vinyl chloride workers, and eight were attributable to past exposure to thorium dioxide. There was possible exposure to vinyl chloride in four other cases and the survey also suggested a possible increased risk in the electrical and plastics fabrication industries (55).

Mention has been made of primary malignant mesothelial tumours resulting from exposure to asbestos, of which the amphiboles are most hazardous. About half the occupationally related cases are peritoneal, perhaps due to ingestion of fibre at work, particularly in the insulation and heating trades (56). Asbestos work is also associated with cancer of the gastrointestinal tract; however, the evidence is much less consistent than for bronchial cancer, gastrointestinal cancers being observed in some cohorts but not in others. Possibly they occur only when there is interaction with some other as yet unidentified factor(s) (22). Finally, there are grounds for suspecting an increased risk of digestive cancers in coalminers. Occupational mortality statistics in Great Britain have shown high standardized mortality rates for many years and there is some supporting evidence from epidemiological studies in the United States (57).

Urinary tract disease

Toxic chemicals encountered in the workplace may lead to glomerular or tubular damage or, after a latent interval, to cancer of the urinary tract or bladder. Acute tubular necrosis resulting in renal failure has followed the absorption of inorganic salts of mercury and other heavy metals, ethylene glycol, tetrachlorethane, and carbon tetrachloride. In France, carbon tetrachloride was once the commonest cause of acute renal failure because of its widespread use (58). An increased prevalence of proteinuria has been observed in mercury workers compared with a control group, with a significant correlation between urinary mercury excretion and protein concentration (59). In some cases, more severe glomerular damage following exposure to mercury has led to the nephrotic syndrome, probably with an immune complex pathogenesis (60).

Following absorption, cadmium accumulates in the renal cortex giving rise to tubular proteinuria and other defects of tubular reabsorption, including hypercalciuria. Renal stone formation has been described in cadmium workers and, in a few instances, osteomalacia (61). Tubular proteinuria, with an increased excretion of β_2-microglobulin, and in some cases a glomerular type proteinuria have been observed, mainly in workers exposed to cadmium for more than 25 years, whose cadmium concentration in blood exceeded 10 μg/litre and that in urine 10 mg/kg creatinine (62). Taking urinary β_2-microglobulin excretion as the earliest indicator of an adverse effect, Kjellström et al. (63) reported a prevalence of about 20% tubular proteinuria in a group of workers in a cadmium-nickel battery plant with 6–12 years' exposure to cadmium at a level of 50 μg/m^3 air. However, atmospheric cadmium concentration may have been higher in the earlier years. Epidemiological studies have revealed widespread renal tubular dysfunction in cadmium-polluted areas of Japan. Again, taking β_2-microglobulin as an indicator of effect, Kjellström et al. (64) estimated that long-term ingestion of about 0.15 mg cadmium per day in food was associated with a higher prevalence of proteinuria than in their reference groups. Excessive lead absorption from occupational exposure has also given rise to tubular dysfunction and to interstitial nephropathy, progressing to renal failure and sometimes associated with gout. Exposure to a number of hydrocarbons has been associated with the development of chronic glomerulonephritis (65).

That certain aromatic amines may give rise to bladder cancer has been recognized for over 80 years. Those at risk include workers in the synthetic dyestuff, chemical, and rubber industries exposed to β-naphthylamine, 4-aminodiphenyl, benzidine and other chemically similar compounds. Workers in the British rubber industry showed a large excess of bladder cancer in cohort studies (66). Routine occupational mortality statistics failed to reveal the problem because of the small number of workers involved. The tumours, frequently multifocal in origin and involving the epithelium on the renal pelvis and ureters as well as the bladder, with a latent interval most frequently between 15 and 20 years, characteristically occurred at younger ages than non-occupational bladder cancer. All 15 distillers of β-naphthylamine at one plant developed bladder cancer and the risk in other workers was extremely high.

Following Case's study, the chemical industry withdrew certain rubber additives and, in the United Kingdom in 1967, the known carcinogenic chemicals referred to above were brought under strict control. A survey was subsequently conducted in which the records of over 40 000 men employed for at least one year in the rubber and cablemaking industries were observed for 8 years. In comparing the mortality pattern for 1972–1974 with that for 1968–1971, a significant excess of deaths from bladder cancer was found throughout the industry, including firms where exposure to acknowledged bladder carcinogens had not occurred (67). The excess bladder cancer deaths were found in workers in the tyre sector, and in those employed by footwear manufacturers and footwear suppliers. Occupational bladder cancer is, therefore, a continuing problem.

Kipling & Waterhouse (68) surveyed the records of 248 workers exposed to cadmium oxide dust for a minimum period of one year. They found 4 cases of prostatic cancer where the expected figure, based on regional rates, was computed at 0.58. A cohort mortality study of smelter workers exposed to cadmium fumes and cadmium oxide dust showed a significant excess mortality from prostatic cancer 20 years after initial exposure (69), and two smaller studies in Sweden showed a similar trend. To date, 14 cases of prostatic cancer have been noted in cadmium workers, compared with 5.4 expected. However, in a large cohort study of cadmium-exposed workers in England, no excess mortality from prostatic cancer has been observed (70).

Diseases of the musculoskeletal system

The extent to which occupation gives rise to disorders of the skeletal system has been little investigated in population-based studies. In addition to the varied and extensive effects of accidental trauma, low back pain and osteoarthrosis are common conditions and a major cause of lost working time. Back pain is very common in both manual and sedentary workers, but in heavy manual workers, such as miners, dockers and nurses, it is an important cause of disability. Osteoarthrosis of the spine, hips, or knees is particularly common in heavy manual workers and the interphalangeal joints are affected in workers such as tailors, where Heberden's nodes produce a characteristic deformity. Disabling osteoarthrosis of the hip, knee, or shoulder also occurs as a sequel to aseptic necrosis of bone in workers under increased atmospheric pressure, where the infarcted area is in proximity to the joint and involves the articular cartilage. In a study of septic necrosis of bone in a large group of British commercial divers, the prevalence of definite bone lesions had increased from less than 1 % in 1975 to 4.8 % in 1979. Joint damage had developed in 14.5 % of divers with potentially disabling juxta-articular lesions. While no bone damage was seen in men who had never been deeper than 30 metres, almost one quarter of the men who had dived to 300 metres had such lesions (71).

Osteoarthrosis, in particular of the elbow and wrist joints, is seen in workers who handle vibrating tools, both with and without associated Raynaud's phenomenon. These and other heavy manual workers may have multiple small areas of decalcification in the carpal bones. In one series, such changes were seen on X-ray in over 40 % of a large workforce (72).

Acro-osteolysis has occurred in workers engaged in the polymerization of vinyl chloride, in particular those whose work entailed the manual cleaning of the pressure vessels, where intermittent exposure to the monomer fume was high. Cystic lesions in the terminal phalanges of the fingers and toes and sometimes in the patella and sacroiliac bones, together with Raynaud's phenomenon and scleroderma-like changes in the skin of the hands, were found in about 3 % of exposed workers (73). An epidemiological study of over 5000 employees in polymerization plants in North America revealed only 25 definite cases, with a further 16 under suspicion (74). All employees who worked at the time of the survey, or in the past, in any capacity in plants manufacturing polyvinyl chloride were included; workers exposed to very low levels probably contributed little to the prevalence rate.

Housemaid's knee, or prepatellar bursitis, is a descriptive term for one of a large number of conditions characterized by a collection of synovial fluid in bursae subjected to repeated friction or pressure. The most important of these are the beat disorders of miners, involving the elbow or the knee, where the bursae are liable to become infected and where subcutaneous cellulitis may occur. Such a cellulitis may also involve the hand in boilermen as well as in miners. Although declining, these conditions continue to be important causes of sickness absence. In Great Britain, in 1978, there were 936 new spells of certified incapacity from the beat disorders for which injury benefit was payable under the Prescribed Diseases Regulations, compared with 1505 new spells in 1973.

Repetitive movement of the hands is required in many occupations, for example by carpenters, braiders, typists, and telegraphists (75). The tendon sheath or the musculotendinous junction of the most used muscles may become inflamed, with fluid exudation, giving rise to an incapacitating tenosynovitis. The most commonly affected are the radial extensors and the abductors of the wrist and thumb. These muscles were involved in 77 % of a group of 544 cases, the majority of which came from a motor vehicle assembly plant (75). The main etiological factors were: occupational change necessitating unaccustomed movement (27 %), resumption of work after absence (21 %), and repetitive stereotyped movement (16.5 %). In 1978, in Great Britain, there were 3428 new spells of certified incapacity for injury benefit, making tenosynovitis the second most frequent cause of certified incapacity after occupational dermatitis.

Skin disease

Industrial skin diseases are common and have major social and economic implications. In Great Britain, in 1978, there were over 7000 new spells of certified incapacity for industrial skin disease for which injury benefit was payable under the Prescribed Diseases Regulations (61 % of total), compared with nearly 11 000 (67 % of total) in 1973. Industrial skin diseases have accounted, on average, for more than twice as many working days lost as all the other prescribed occupational disorders together (77).

A wide range of agents in the working environment may be involved; these may be classified as (a) mechanical, e.g., trauma, friction, or pressure; (b) physical, e.g., temperature, radiation; (c) biological, e.g., plant or animal contact, insects, microorganisms; and (d) chemical, e.g., both organic and inorganic compounds. Whether a disorder of the skin develops is dependent not only on the pattern and intensity of exposure but also on individual susceptibility, atopy, and skin pigmentation. Ultraviolet light is a potent carcinogen when acting on the unprotected skin (78). Fair-skinned and poorly pigmented seamen, farmers, and other workers exposed to high-intensity sunlight in the tropics have a high incidence of skin cancer in comparison with pigmented races. An atopic diathesis has a complex relationship to industrial skin disease. Thus, nursing would not be a suitable occupation for a young person with a history of eczema because of the greater risk of developing not only an allergic but also an irritant contact dermatitis. Repeated friction or mechanical pressure on the skin gives rise to the beat disorders or to cellulitis (mentioned also under diseases of the musculoskeletal system). Erythema of the skin with pruritus occurs if the skin is chilled during decompression, and gangrene of the finger tips may accompany Raynaud's phenomenon associated with the handling of vibrating tools. The effects of exposure to ionizing radiation were dramatically demonstrated by the pioneer radiologists early this century, who developed chronic X-ray dermatitis, post-irradiation telangiectasis, and eventually skin cancer. Many plants, fruits and vegetables cause contact dermatitis affecting a wide range of workers. Among the most potent are those belonging to the Anacardiaceae and the Primulaceae, and certain woods such as South African boxwood. A large number of chemicals give rise not only to contact dermatitis, but also to leukoderma (hydroquinone derivatives), to acne (dioxin and other chlorinated hydrocarbons), and some to skin cancer.

Irritant contact dermatitis is more prevalent than allergic contact dermatitis, the common irritants being skin cleansers, acids, alkalis, oils, and organic solvents. Of the common industrial allergens, chromate is the most widespread affecting males, the most frequent source being cement. Nickel allergy is more common in women, sensitization often occurring from non-occupational sources, such as jewellery and fastenings. In an epidemiological study of contact dermatitis in North America, 1200 subjects were tested against 16 allergens (79). The most common sensitizers observed were nickel sulfate, followed by potassium dichromate, thiomersal, and paraphenylene diamine, the latter being a component of hair dyes. Epoxyresins and chemicals used in rubber, such as antioxidants and accelerators, are also common allergens. Contact

54

dermatitis is especially common in coalmining, metal manufacturing, and in the leather, chemical, and textile industries.

Scrotal cancer in chimney sweeps was described over 200 years ago by Percival Pott; subsequently scrotal and other skin cancers were reported in cotton mule spinners exposed to shale oil, and more recently, in automatic machine operators whose clothing and skin becomes contaminated with mineral oils (80). However, chimney sweeps may still experience an increased cancer risk. A cohort study among chimney sweeps employed for many years in Sweden has shown a significant excess mortality from cancer of the lung and oesophagus, as well as from chronic respiratory disease (81). Tar, pitch, and creosote have produced warty growths in the skin followed by neoplastic change. The common factor in these exposures is the presence of carcinogenic polycyclic hydrocarbon compounds, such as benzo[a]pyrene.

Workers exposed to arsenicals may develop contact dermatitis or, following more prolonged exposure, hyperkeratosis of the palms and soles, warts, melanosis, and patchy depigmentation commonly called "raindrop" pigmentation. Epidemiological studies have indicated an association between skin cancer, often multifocal in origin, and heavy exposure to inorganic arsenicals via medication, contaminated drinking water, or occupational exposure (82).

Diseases of the nervous system and sense organs
Occupational factors are associated with a broad spectrum of disorders affecting both the central nervous system—causing organic and also behavioural manifestations—and the peripheral nerves.

Hearing loss
The commonest adverse effects of occupational factors are mechanical and acoustic trauma, the former in construction, transport and mining, and the latter throughout heavy industry. In the United Kingdom, the construction industry accounts for about 300 cases of deafness per million workers per year (83). A survey of industrial noise exposure in Britain in 1971 revealed that between half a million and one million workers were exposed to the equivalent of more than 90 dB(A) for eight hours every working day, a noise level above which high-frequency hearing loss is known to occur.

Central nervous system effects
Work under increased atmospheric presure can have complex effects on the function of the central nervous system. Air breathed under increased pressure has adverse effects on performance and narcotic properties attributed to nitrogen. Other inert gases also give rise to narcosis, related to their lipid solubility (84). The high-pressure nervous syndrome is characterized by electroencephalographic abnormalities and tremors (85). However, the most serious effects occur in decompression sickness with nervous system involvement in 8–35% of reported cases (86). As the entire central nervous system is at risk, any neurological lesion may

occur, the most serious effect being permanent paraplegia. Not uncommon amongst divers are visual manifestations and a disturbance of vestibular function with vertigo, nausea, vomiting, and nystagmus known as "the staggers".

Carbon monoxide is the most frequent cause of death from poisoning, because of its ubiquitous nature. Those who survive may be left with permanent cortical and extrapyramidal system damage. Traffic policemen, blast furnace workers, and others are exposed to carbon monoxide at lower concentrations that may give rise to impaired performance of tasks requiring perceptual and motor skills.

Of the metals, lead, mercury, and manganese have specific central effects. Acute encephalopathy may be the presenting feature in lead poisoning, but industrial cases are now rare. Tetraethyl lead poisoning is a toxic organic psychosis that has occurred in workers handling leaded petrol, but again, because of preventive measures, such poisoning is now rare. Workers in manganese mines and mills have developed psychotic symptoms followed by extrapyramidal involvement, with akinesia, rigidity, and tremor. The relationship between the level of exposure and biological effects of manganese was studied in a group of workers employed in the production of ferro-alloys (87). Sixty-two (17%) manganese alloy workers showed some signs of neurological impairment, mainly tremor at rest, but this prevalence did not correlate well with mean manganese concentrations at the workplace.

The inhalation of mercury vapour may lead to an acute psychosis or to a Parkinsonian syndrome. More commonly, however, it causes a milder mental disturbance known as erithism, which, in the past, was so frequently seen in the felt hat industry that the expressions "mad as a hatter" and "hatter's shakes" passed into everyday speech. The short-chain aliphatic mercury compounds synthesized primarily as seed dressings produce a specific disorder characterized by selective cerebral cortical and cerebellar damage. The condition was initially described in formulators of seed dressings, but in more recent years a similar condition known as Minimata disease, caused by an environmental pollutant from an industrial source, has gained importance and a large-scale epidemic of methyl mercury poisoning has been traced to the ingestion of contaminated bread (88). Aluminium, bismuth, and organotin compounds have also given rise to encephalopathy, again in isolated cases in industry, but in epidemic proportions in other population groups.

Many organic compounds, some volatile and frequently used as solvents, have a narcotic effect when inhaled and in lower concentrations depress the level of conciousness, affecting initially the higher mental functions. Some have other toxic effects, too, such as carbon tetrachloride affecting the liver and kidneys, trichlorethylene producing cranial nerve damage, methyl bromide giving rise to encephalopathy and neuropathy, and methyl alcohol acting on the optic nerve. Subjective changes in mood and some slowing of reaction time have been observed in studies of behavioural effects in workers exposed to a variety of organic solvents, and there is some evidence that lasting brain damage may occur (89).

Effects on peripheral nerves

Toxic neuropathies account for only a small proportion of all peripheral neuropathy, most of which has an obscure etiology. Of the metals, inorganic lead gives rise to a motor neuropathy affecting principally the extensors of the wrist and fingers, the classical lead palsy. Lower exposures have caused subclinical dysfunction of peripheral nerves (*90*). Arsenic and thallium can give a mixed motor and sensory neuropathy, the former after a single large exposure and the latter delayed in onset, progressive, and accompanied by central effects. A number of organic compounds have also caused neuropathy. Workers preparing acrylamide polymer from the monomer have been affected and a mixed neuropathy has also followed exposure to carbon disulfide. *N*-hexane is a commonly used solvent that was believed to be responsible for a symmetrical, distal, largely motor neuropathy affecting a large group of Japanese workers employed on cementing sandals (*91*). A high prevalence of poly-neuropathy in shoe and leather workers in Italy was attributed to exposure to a variety of solvents which were not identified (*92*). Other workers have developed neuropathy following the practice of glue-sniffing. An outbreak of symmetrical, mainly motor, peripheral neuropathy involved 79 workers at a plant for the manufacture of plastic-coated printed fabrics where no previously known neurotoxic agent had been used (*93*). The cause of the outbreak was traced, after an epidemiological study on over 1000 workers exposed to 275 different chemicals, to the substitution in a solvent mixture of the apparently innocuous methyl *n*-butyl ketone for the previously used methyl isobutyl ketone.

Numbness and pain are frequent complaints in workers who have contracted vibration white finger. In one field study of forest workers in Finland using chain saws over a period of 10–15 years, subclinical neuropathy was found in about one half of the workers examined (*94*). Delay in nerve conduction in the forearms was considered to be due to partial demyelination of the peripheral nerves.

Blood diseases

Ionizing radiation and a large number of chemical substances encountered in the working environment may cause disorders of the formed elements of the blood and their precursors. However, host sensitivity plays an important part in determining outcome. A haemoglobinopathy or genetically determined glucose-6-phosphate dehydrogenase deficiency in certain subjects may lead to a haemolytic crisis following the absorption of doses quite harmless to normal persons. Exposure to high doses or ionizing radiation produces an immediate fall in the lymphocyte count, followed by a fall in granulocyte and platelet counts with a more gradual fall in the erythrocyte count. Aplastic anaemia or agranulocytosis may supervene, or alternatively proliferation of cellular elements may occur, in particular giving rise to acute or chronic myelogenous leukaemia. In the case of leukaemia, a linear dose–response curve has been demonstrated, although in the very low dose range quantitative data are insufficient to determine whether this relationship still holds. In the luminizing industry, the ingestion of traces of radioactive substances has resulted in aplastic anaemia and osteogenic sarcoma. Lower doses of radiation have produced chromosome abnormalities in circulating lymphocytes.

Both reduced and increased haemopoietic activity has been observed following industrial exposure to benzene, either in the pure form or as a component of solvent mixtures which, in their effects, can be described as radiomimetic. Over the past 50 years, many reports have associated benzene exposure with the subsequent development of non-lymphocytic leukaemia. Epidemiological studies suggest such an association (95, 96), but a causal relationship has yet to be convincingly demonstrated.

Many chemical exposures can give rise to haemolysis. Arsine is formed whenever arsenic, often present in traces in scrap metal recovery, comes into contact with acids or nascent hydrogen. It can cause massive haemolysis, giving rise to jaundice and secondary renal failure, following almost always an unsuspected exposure. A variety of nitro and amino organic compounds, such as nitrobenzene and analogues, trinitrotoluene and aniline also cause haemolysis. These compounds are readily absorbed through the skin. In addition to haemolysis, nitro and amino compounds of the aromatic series also produce methaemoglobinaemia, imparting a characteristic blue colour to the skin and inhibiting oxygen transport to the tissues.

Excessive exposure to inorganic lead produces a mild or moderate anaemia as a result of both haemolysis and inhibition of haem synthesis. More sensitive indicators of early and reversible lead effects are erythrocyte zinc protoporphyrin in the blood or ∂-amino-laevulinic acid in the urine. These parameters have been measured extensively in epidemiological studies on lead-exposed workers.

An important cause of a severe chronic anaemia in many parts of the world, especially in the tropics, is infestation with hookworm (ankylostomiasis), known as miners' or tunnel workers' anaemia. This parasitic infestation is especially common in South East Asia where it affects agricultural workers, in particular labourers in rice fields, in contact with damp soil. Schistosomiasis also gives rise to anaemia through chronic blood loss.

Reproductive effects

Men and women are exposed to a similar range of occupational hazards but, by virtue of their childbearing function, women experience additional occupational health risks mainly affecting the fetus. Although concern has been expressed about possible adverse effects of occupational factors on fertility in both women and men, no such effects have so far been observed with certainty. Menstrual, breast, and genital changes have been found in women who manufacture oral contraceptives (97). There are reports of delayed menarche and amenorrhoea in dancers (98) and menstrual disorders in airline stewardesses (99, 100). Like other young women who leave home to work, nurses are notoriously liable to amenorrhoea.

Although there is little direct evidence in human subjects, experimental research has demonstrated the teratogenicity of very many physical and chemical agents potentially present in the working environment (*101*). That chromosomal aberrations can occur in occupationally exposed persons is suggested by their presence in lymphocytes cultured from the blood of women who worked in laboratories and a rotoprinting factory, and from the blood of their newborn children (*102*). Hospital employees, especially those who use X-ray apparatus outside the radiology department, may be exposed to ionizing radiation, with attendant risks in pregnancy, if monitoring is not strictly applied. Another "physical" agent, hypoxia, known to be teratogenic in experimental animals, constitutes a possible hazard for female flight staff in early pregnancy.

An increase in the rate of spontaneous abortions with or without a rise in congenital malformations has been reported following a variety of chemical exposures: in anaesthetic rooms (*103, 104*), in hospital laboratories (*105*), following exposure of fathers to vinyl chloride (*106*), and in those living around a Swedish smelter (*107*). In Finland, there is now systematic surveillance of maternal occupations in relation to outcome of pregnancy and the findings to date indicate a number of statistically significant associations (*108–110*). For example, exposure to styrene and laundry work, and employment in the pharmaceutical industry came under suspicion as causes of abortion; telephone operators, teachers, gardeners, and cooks had high malformation rates; and farming and the food industry were associated with malignant disease in offspring. These associations mean little in themselves but merit further investigation. An association has also been reported between exposure during pregnancy to organic solvents and both central nervous defects (*111*) and oral clefts (*112*). Lead is known to cross the placental barrier and has long been considered toxic to the fetus; it has been suggested (*113*) that maternal exposure may affect fetal mental development. Certain infective agents—rubella virus, toxoplasma, and cytomegalovirus—cause specific developmental abnormalities; teachers, school staff, animal minders and others may thus be at increased risk in pregnancy from these infections (*114*).

Work, health and wellbeing

In a paper on the impact of unemployment, Johoda (*115*) argued that "if there is more to work than making a living, there should be more to the lack of work than the reduction of one's standard of living". Not all societies share the same work ethic, but this premise probably holds in most industrialized countries. For better or worse, there may even be some empirical correlation between the process of socioeconomic development and acquisition of this attitude to work. Johoda went on to argue that since pay and conditions are on balance negative factors, employment must have powerful compensatory advantages. She identified five possibilities: (*a*) an imposed time structure; (*b*) shared experiences and contacts; (c) transcendent goals and purposes; (d) personal status and identity; and (e) enforced activity. These same "advantages" are clearly anathema to more contemplative philosophies! However, if one accepts that "psychological supports are more or less enduring requirements for all of us", it follows that enforced unemployment is indeed a terrible evil, more serious in its direct and indirect effects than even the most soul-destroying and alienating assembly line piecework. However, unemployment also leads to poverty and this in turn to morbidity and mortality from diseases well known to be associated with it (116). In addition, there is evidence that serious emotional problems and psychosomatic disease may result directly from job loss (117–119).

The beneficial effects of the "enforced activity" mentioned above must surely be more than psychological. A considerable body of presumptive epidemiological evidence supported by common sense and the pattern of evolution all suggest that regular and demanding exercise sufficient to maintain physical fitness will retard the development of arterial disease, reduce mental stress, limit undue obesity, and promote sound sleep. It is certainly hard to believe that, within reason, hard exercise is bad. We should therefore regard with some concern the fact that, despite the obvious need to remove the back-breaking tasks of the bad old days, work in the industrialized world is becoming less and less physically demanding.

There remains the more general question of the quality of working life, a subject of growing interest in socially advanced industrial countries. Dull and repetitive work that offers no scope for personal initiative or sense of achievement is of concern on two accounts. Such jobs are associated with absenteeism and high labour turnover, strikes and low morale, poor quality work, and reduced productivity. In the present health context, there is every reason to believe that work satisfaction is a prerequisite for a full and healthy life and that its absence is associated with high accident rates, anxiety and depression, frustration and stress. However, the scientific evidence for all this is scanty, no doubt because the required epidemiological research is difficult to design and conduct. The reader is referred to a recent review of this field (120).

Contribution of occupational factors to disease

The aim of this chapter has been to put occupational health in epidemiological perspective. It has been possible to review only the salient features of present knowledge and numerous detailed and controversial questions have received little or no attention. A great many more surveys of good scientific quality are needed in both developing and industrialized countries to quantify the contribution of occupational factors to morbidity and mortality over the full range of acute, chronic, and malignant disease. Table 1 is a rather simplistic attempt to summarize the data that have been presented for industrialized countries.

Its content is quite subjective and the rough assessments shown can easily be challenged. Only in malignant disease has any serious thought been given to this question and, as already mentioned (1), initial estimates in this disease category are geographically limited, range widely, and have little foundation. For the third world, there is virtually no evidence to permit an assessment of the importance of occupation as a cause of disease and accident, although it must surely be very large.

In Table 1 the aim is to give some impression of the probable extent to which occupation, relative to all other causal factors (known and unknown), accounts for disease in the main organ systems. Ideally, the "denominator" would integrate in a rational way all the various possible measures of disease frequency and severity. The pattern displayed reflects no more than common knowledge that, to date, the main impact of the working environment has been on accessible organs and is largely traumatic or toxic. The same is generally true of the chronic inflammatory, degenerative, and fibrotic diseases, which attain comparable prevalence despite their lower incidence. So far as malignant disease is concerned, only in regard to the respiratory tract are occupational factors of more than slight importance, and this is largely because of interaction between these agents and smoking. However, the present situation may be transitional: acute occupational diseases and accidents, and indeed most chronic effects, present few etiological mysteries and their incidence should eventually be reduced by appropriate control measures. One cannot be so optimistic about occupational cancer: the rate of introduction of new materials, the growing number of possible interactions, and the long periods of latency emphasize the need for caution; the same is true of teratogenicity.

Table 1. Contribution of occupational factors in industrialized countries to the causation of various types of disease and accident (excluding genital and reproductive effects)

System affected	Importance[a]		
	Acute	Chronic	Malignant
Respiratory system	+ + +	+ + +	+ +
Skin	+ + +	+ +	+
Musculoskeletal system	+ + +	+ +	−
Digestive system	+	+	+
Blood	+	+	+
Urinary tract	+	+	+
Nervous system	+	+	−
Cardiovascular system	+	+	−

[a] + + + = major; + + = moderate; + = slight; − = insignificant.

An even more tentative analysis of current epidemiological knowledge is presented in Table 2, which examines the nature and probable order of risks associated with various types of industry. Acute occupational disease and trauma are seen to be concentrated in the heavy and primary industries. Chronic diseases follow the same pattern, whereas there is now evidence of at least some risk of malignant disease in virtually all manufacturing and production industries. Looking to the future, the challenge for epidemiology is clear. It is to produce much more precise data on the distributions of occupational disease and accidents, to detect and explain upward trends, and to evaluate whatever control measures are instituted.

Table 2. Estimated relative risk of occupationally induced disease or accident in various types of industry

Industry	Risk[a]		
	Acute disease and accident	Chronic disease	Malignant disease
Mines and quarries	+ + +	+ + +	+
Heavy metals	+ + +	+ +	+
Construction	+ + +	+	+
Agriculture and forestry	+ + +	+	+
Fisheries	+ + +	−	+
Petrochemical	+ +	+	+
Transport	+ +	−	+
Textile	+	+ +	+
Manufacture	+	+	+
Trade, commerce and service	−	−	−

[a] + + + = high; + + = moderate; + = slight; − = insignificant.

References

1. What proportion of cancers are related to occupation? *Lancet*, **2**: 1238–1240 (1978).
2. WHO Technical Report Series, No. 662, 1981. (*Health effect of combined exposures in the work environment*. Report of a WHO Expert Committee).
3. **McDonald, J. C.** Survey of acute respiratory disease in the Royal Air Force. *Journal of hygiene, epidemiology, microbiology and immunology (Prague)*, **4**: 440–446 (1960).
4. **Van der Veen, J.** The role of adenoviruses in respiratory disease. *American review of respiratory diseases*, **88** (Suppl.): 167–180 (1963).
5. **Weill, H. & Hendrick, D. J.** Acute reactions to inhaled agents. In: McDonald, J. C., ed. *Recent advances in occupational health*. Edinburgh, Churchill Livingstone, 1981.
6. **Rae, S. & Jacobsen, M.** Energy production. In: McDonald, J. C., ed. *Recent advances in occupational health*. Edinburgh, Churchill Livingstone, 1981.
7. **Merchant, J. A. et al.** An industrial study of the biological effects of cotton dust and cigarette smoke exposure. *Journal of occupational medicine*, **15**: 212–221 (1973).
8. **Berry, G. et al.** Relationship between dust level and byssinosis and bronchitis in Lancashire cotton mills. *British journal of industrial medicine*, **31**: 18–27 (1974).

9. **McDonald, J. C. et al.** The health of chrysotile, asbestos mine and mill workers of Quebec. *Archives of environmental health*, **28**: 61–68 (1974).
10. **Brzezinski, Z. et al.** *Ecology of chronic respiratory diseases.* Panstwowy Zaklad Wydawnickow Lekarskich, Warsaw, 1972.
11. **McDonald, J. C. et al.** Mortality after long exposure to cummingtonite-grunerite. *American review of respiratory disease*, **118**: 271–277 (1978).
12. **Oakes, D. et al.** Respiratory effects of prolonged exposure to gypsum dust. In: Walton, W. H., ed. *Inhaled particles V.* Oxford, Pergamon Press, 1982.
13. **Walton, W. H. et al.** The effect of quartz and other non-coal dusts on coalworkers' pneumoconiosis. Part I: Epidemiological studies. In: Walton, W. H., ed. *Inhaled particles IV.* Oxford, Pergamon Press, 1977, vol. 2, pp. 669–690.
14. **Bennett, J. G. et al.** The relationship between coal rank and the prevalence of pneumoconiosis. *British journal of industrial medicine*, **36**: 206–210 (1979).
15. **Jacobsen, M. et al.** Smoking and coalworkers' simple pneumoconiosis. In: Walton, W. H., ed. *Inhaled particles IV.* Oxford, Pergamon Press, 1977, vol. 2, pp. 759–772.
16. **Seaton, A.** Silicate pneumoconioses. In: Morgan, W. K. C. & Seaton, A., ed. *Occupational lung diseases.* London, Saunders, 1975, pp. 112–123.
17. **Berry, G. & Lewinsohn, H. C.** Dose-response relationships for asbestos-related disease: implications for hygiene standards. Part I. Morbidity. *Annals of the New York Academy of Science*, **330**: 185–194 (1979).
18. **Becklake, M. R. et al.** Radiological changes after withdrawal from asbestos exposure. *British journal of industrial medicine*, **36**: 23–28 (1979).
19. **Liddell, F. D. K. & McDonald, J. C.** Radiological findings as predictors of mortality in Quebec asbestos workers. *British journal of industrial medicine*, **37**: 257–267 (1980).
20. **Cochrane, A. L. et al.** The mortality of men in the Rhondda Fach, 1950–1970. *British journal of industrial medicine*, **36**: 15–22 (1979).
21. **McDonald, J. C.** Asbestos and lung cancer: has the case been proven? *Chest*, **78** (Suppl.): 374–376 (1980).
22. **McDonald, J. C.** Asbestos-related disease: an epidemiological review. In: Wagner, J. C., ed. *Biological effects of mineral fibres.* Lyon, International Agency for Research on Cancer, 1980, vol. 2, pp. 587–601.
23. **Acheson, E. D. & Gardner, M. J.** Asbestos: scientific basis for environmental control of fibres. In: Wagner, J. C., ed. *Biological effects of mineral fibres.* Lyon, International Agency for Research on Cancer, 1980, vol. 2, pp. 737–754.
24. **Saracci, R.** Personal-environmental interactions in occupational epidemiology. In: McDonald, J. C., ed. *Recent advances in occupational health.* Edinburgh, Churchill Livingstone, 1981.
25. **de Villiers, A. J. & Windish, J. P.** Lung cancer in a fluorspar mining community. I. Radiation; dust and mortality experience. *British journal of industrial medicine*, **21**: 94–109 (1964).
26. **Lundin, F. E. et al.** An exposure-time response model for lung cancer mortality in uranium miners: effects of radiation exposure, age and cigarette smoking. In: Breslow, N. E. & Whittemore, A. S., ed. *Energy and health.* Philadelphia, Society for Industrial and Applied Mathematics, 1979, p. 243.
27. **Axelson, O. & Sundell, L.** Mining, lung cancer and smoking. *Scandinavian journal of work, environment and health*, **4**: 46–52 (1978).
28. **Kazantzis, G.** Carcinogenic effects of metals. In: McDonald, J. C., ed. *Recent advances in occupational health.* Edinburgh, Churchill Livingstone, 1981.
29. **Morris, J. M. et al.** Incidence and prediction of ischaemic heart disease in London busmen. *Lancet*, **2**: 553–559 (1966).
30. **Paffenbarger, R. S. & Hale, W. E.** Work activity and coronary heart mortality. *New England journal of medicine*, **292**: 545–550 (1975).
31. **Marmot, M. G. et al.** Employment grade and coronary heart disease in British civil servants. *Journal of epidemiology and community health*, **32**: 244–249 (1978).
32. **Rose, G. & Marmot, M. G.** Social class and coronary heart disease. *British heart journal*, **45**: 13–19 (1981).
33. **Rose, G. A. et al.** Heart disease prevention project: a randomized controlled trial in industry. *British medical journal*, **1**: 747–751 (1980).

34. **Tiller, J. R. et al.** Occupational toxic factors in mortality from coronary heart disease. *British medical journal*, **4**: 407–411 (1968).
35. **Tolonen, M. et al.** A follow-up study of coronary heart disease in viscose rayon workers exposed to carbon disulphide. *British journal of industrial medicine*, **32**: 1–10 (1975).
36. **Astrup, P. & Kjeldsen, K.** CO, smoking and atherosclerosis. *Medical clinics of North America*, **58**: 323–350 (1973).
37. **Hernberg, S. et al.** Angina pectoris, ECG findings and blood pressure of foundry workers in relation to CO exposure. *Scandinavian journal of work, environment and health*, **2**, (Suppl. 1): 54–63 (1976).
38. **Hamilton, E. & Hardy, H. L.** Esters. In: *Industrial toxicology*, 3rd ed. Acton, MA, Publishing Sciences Group, Inc., 1974, pp. 317–319.
39. **Hogstedt, C. & Axelson, O.** Nitroglycerine-nitroglycol exposure and the mortality in cardio-cerebro-vascular diseases among dynamite workers. *Journal of occupational medicine*, **19**: 675–678 (1977).
40. **Morin, Y. & Daniel, P.** Quebec beer drinkers' cardiomyopathy: etiological considerations. *Canadian Medical Association journal*, **97**: 925–928 (1967).
41. **Lane, R. E.** Health control in inorganic lead industries. *Archives of environmental health*, **8**: 243–250 (1964).
42. **Cooper, W. C. & Gaffey, W. R.** Mortality of leadworkers. *Journal of occupational medicine*, **17**: 100–107 (1975).
43. **Lener, J. & Bibr, B.** Cadmium and hypertension. *Lancet*, **1**: 970 (1971).
44. **Glauser, S. C. et al.** Blood cadmium levels in normotensive and untreated hypertensive humans. *Lancet*, **1**: 717–718 (1976).
45. **Beevers, D. et al.** Blood cadmium in hypertensives and normotensives. *Lancet*, **2**: 1222–1224 (1976).
46. **Taylor, W. & Pelmear, P. L.,** ed. *Vibration white finger in industry*. London, Academic Press, 1975.
47. **Doll, R. et al.** *Occupational factors in the aetiology of gastric and duodenal ulcers*. London, H. M. Stationery Office, 1951 (Medical Research Council special report Series No. 76).
48. **Cobb, S. & Rose, R. M.** Hypertension, peptic ulcer and diabetes in air traffic controllers. *Journal of the American Medical Association*, **224**: 489–492 (1973).
49. **Cherry, N.** Stress, anxiety and work: a longitudinal study. *Journal of occupational psychology*, **51**: 259–270 (1978).
50. **Williams, S. V. et al.** Epidemic viral hepatitis, type B, in hospital personnel. *American journal of medicine*, **57**: 904–911 (1974).
51. **Denes, A. E. et al.** Hepatitis B infection in physicians—results of a nationwide seroepidemiologic study. *Journal of the American Medical Association*, **239**: 210–212 (1978).
52. **Chalmers, T. C.** Potential contributions of multiple risk factors to the etiology of cirrhosis. In: Lee, D. H. K. & Kotin, P., ed. *Multiple factors in the causation of environmentally induced disease*. London, Academic Press, 1972.
53. **Terris, M.** Epidemiology of cirrhosis of the liver: National mortality data. *American journal of public health*, **57**: 2076–2088 (1967).
54. **Fox, J.** *Occupational mortality 1970–1972. Population trends*. London, H. M. Stationery Office, 1977.
55. **Baxter, P. J. et al.** Angiosarcoma of the liver: Annual occurrence and aetiology in Great Britain. *British journal of industrial medicine*, **37**: 213–221 (1980).
56. **McDonald, J. C. & McDonald, A. D.** Epidemiology of mesothelioma from estimated incidence. *Preventive medicine*, **6**: 426–446 (1977).
57. **Ames, R. G.** Gastric cancer in coal miners: Some hypotheses for investigation. *Journal of the Society of Occupational Medicine*, **32**: 73–81 (1982).
58. **Hamburger, J.** Les anuries par inhalation de tetrachlorure de carbone. *Acquisitions médicales récentes*, **15**: 568 (1958).
59. **Joselow, M. M. & Goldwater, L. J.** Absorption and excretion of mercury in man. XII: Relationship between urinary mercury and proteinuria. *Archives of environmental health*, **15**: 155–159 (1967).

64

60. **Kazantzis, G.** Carcinogenic effects of metals. In: McDonald, J. C., ed. *Recent advances in occupational health.* Edinburgh, Churchill Livingstone, 1981.

61. **Kazantzis, G.** Renal tubular dysfunction and abnormalities of calcium metabolism in cadmium workers. *Environmental health perspectives*, **28**: 155–159 (1979).

62. **Bernard, A. et al.** Renal excretion of proteins and enzymes in workers exposed to cadmium. *European journal of clinical investigation*, **9**: 11–22 (1979).

63. **Kjellström, T. et al.** Dose-response analysis of cadmium-induced proteinuria. A study of urinary β_2-microglobulin excretion among workers in a battery factory. *Environmental research*, **13**: 303–317 (1971).

64. **Kjellström, T. et al.** Urinary β_2-microglobulin excretion among people exposed to cadmium in the general population. *Environmental research*, **13**: 318–344 (1977).

65. **Zimmerman, S. W. et al.** Hydrocarbon exposure and chronic glomerulonephritis. *Lancet*, **2**: 199–201 (1975).

66. **Case, R. A. M. et al.** Tumours of the urinary bladder in workmen engaged in the manufacture and use of certain dyestuff intermediates in the British chemical industry. *British journal of industrial medicine*, **11**: 75–104 (1954).

67. **Fox, A. J. & Collier, P. F.** A survey of occupational cancer in the rubber and cablemaking industries: analysis of deaths occurring in 1972–74. *British journal of industrial medicine*, **33**: 249–264 (1976).

68. **Kipling, M. D. & Waterhouse, J. A. H.** Cadmium and prostatic carcinoma. *Lancet*, **1**: 730–731 (1967).

69. **Lemen, R. A. et al.** Cancer mortality among cadmium production workers. *Annals of the New York Academy of Sciences*, **271**: 273–279 (1976).

70. **Armstrong, B. G. & Kazantzis, G.** *Mortality of cadmium workers. Lancet*, **1**: 1425–1427 (1983).

71. Report from the Decompression Sickness Central Registry and Radiological Panel. Aseptic bone necrosis in commercial divers. *Lancet*, **2**: 384–388 (1981).

72. **James, J. B. et al.** An investigation of the prevalence of bone cysts in hands exposed to vibration. In: Taylor, W. & Pelmear, P. L., ed. *Vibration white finger in industry.* London, Academic Press, 1975.

73. **Wilson, R. H. et al.** Occupational acro-osteolysis. Report of 31 cases. *Journal of the American Medical Association*, **201**: 577–581 (1967).

74. **Dinman, B. D. et al.** Occupational acro-osteolysis. 1. An epidemiological study. *Archives of environmental health*, **22**: 61–73 (1971).

75. **Ferguson, D.** An Australian study of telegraphists cramp. *British journal of industrial medicine*, **28**: 280–285 (1971).

76. **Thompson, A. R. et al.** Peritendonitis crepitans and simple tenosynovitis: a clinical study of 544 cases in industry. *British journal of industrial medicine*, **8**: 150–160 (1951).

77. **Newhouse, M. L.** Trends in morbidity due to industrial dermatitis. *Proceedings of the Royal Society of Medicine*, **65**: 257–258 (1973).

78. **Scott, E. L. & Straf, M. L.** Ultraviolet radiation as a cause of cancer. In: Hiatt, H. H. et al., ed. *Origins of human cancer.* Cold Spring Harbor, NY, Cold Spring Harbor Laboratory, 1977.

79. Report. Epidemiology of contact dermatitis in North America: 1972. *Archives of dermatology*, **108**: 537–540 (1973).

80. **Cooke, M. A. & Kipling, M. D.** Occupation and cancer of the skin in Great Britain. *Archives des maladies professionnelles*, **34**: 244–246 (1973).

81. **Hogstedt, C. et al.** A cohort study on mortality among long-time employed Swedish chimney sweeps. *Scandinavian journal of work, environment and health*, **8** (Suppl.): 72–78 (1982).

82. **Tseng, W. P.** Effects and dose-response relationships of skin cancer and blackfoot disease with arsenic. *Environmental health perspectives*, **19**: 109–119 (1977).

83. **Pochin, E. E.** The acceptance of risk. *British medical bulletin*, **31**: 184–190 (1975).

84. **Bennett, P. B.** Inert gas narcosis. In: Bennett, P. B. & Elliott, D. H., ed. *The physiology and medicine of diving and compressed air work*, 2nd ed. London, Balliere-Tindall, 1975, pp. 207–230.

85. **Bennett, P. B.** The high pressure nervous syndrome man. In: Bennett, P. B. & Elliott, D. H., ed. *The physiology and medicine of diving and compressed air work*, 2nd ed. London, Balliere-Tindall, 1975, pp. 248–263.

86. **Elliott, D. H. et al.** Acute decompression sickness. *Lancet*, **2**: 1193–1199 (1974).
87. **Saric, M. et al.** Occupational exposure to manganese. *British journal of industrial medicine*, **34**: 114–118 (1977).
88. **Kazantzis, G.** Mercury. In: Waldron, H. A., ed. *Metals in the environment*. London, Academic Press, 1980.
89. **Cherry, N. & Waldron, H. A.** Behavioural tests in human toxicology. In: McDonald, J. C., ed. *Recent advances in occupational health*. Edinburgh, Churchill Livingstone, 1981.
90. **Catton, M. J. et al.** Subclinical neuropathy in lead workers. *British medical journal*, **2**: 80–82 (1970).
91. **Inoue, T. et al.** A health survey on vinyl sandal manufacturers with a high incidence of N-hexane intoxication. *Japanese journal of industrial health*, **12**: 73–84 (1970).
92. **Buiatti, E. et al.** Relationship between clinical and electromyographic findings and exposure to solvents, in shoe and leather workers. *British journal of industrial medicine*, **35**: 168–173 (1978).
93. **Billmaier, D. et al.** Peripheral neuropathy in a coated fabrics plant. *Journal of occupational medicine*, **16**: 665–671 (1974).
94. **Seppäläinen, A. M.** Peripheral neuropathy in forest workers. A field study. *Scandinavian journal of work, environment and health*, **9**: 106–111 (1972).
95. **McMichael, A. J. et al.** Solvent exposure and leukemia among rubber workers: an epidemiologic study. *Journal of occupational medicine*, **17**: 234–239 (1975).
96. **Infante, P. F. et al.** Leukaemia in benzene workers. *Lancet*, **2**: 76–78 (1977).
97. **Harrington, J. M. et al.** Occupational exposure to synthetic estrogens: a survey of employees. *Archives of environmental health*, **33**: 12–15 (1978).
98. **Frisch, R. E. et al.** Delayed menarche and amenorrhoea in ballet dancers. *New England journal of medicine*, **303**: 17–19 (1980).
99. **Preston, P. S. et al.** Effects of flying and of time changes on menstrual cycle length and on performance in airline stewardesses. *Aerospace medicine*, **44**: 438–443 (1973).
100. **Iglasias, R.** Disorders of menstrual cycle in airline stewardesses. *Aviation space environmental medicine*, **51**: 518–520 (1980).
101. **Messite, J. & Bond, M. B.** Reproductive toxicology and occupational exposure. In: Zenz, C., ed. *Developments in occupational medicine*. Chicago, Year Book Medical Publishers, 1980, pp. 59–129.
102. **Funes-Cravioto, F. et al.** Chromosome aberrations and sister-chromatid exchange in workers in chemical laboratories and a rotoprinting factory and in children of women laboratory workers. *Lancet*, **2**: 322–325 (1977).
103. Report by American Society of Anesthesiologists. Occupational disease among operating room personnel. *Anesthesiology*, **41**: 321–340 (1974).
104. **Tomlin, P. J.** Health problems of anaesthetists and their families in the West Midlands. *British medical journal*, **1**: 779–784 (1979).
105. **Strandberg, M. et al.** Spontaneous abortions among women in a hospital laboratory. *Lancet*, **1**: 384–385 (1978).
106. **Infante, P. F. et al.** Carcinogenic, mutagenic and teratogenic risks associated with vinyl chloride. *Mutation research*, **41**: 131–141 (1976).
107. **Nordström, S. et al.** Occupational and environmental risks in and around a smelter in N. Sweden. The frequency of spontaneous abortion. *Hereditas*, **88**: 51–54 (1978).
108. **Hemminki, K. et al.** Spontaneous abortions among female chemical workers in Finland. *International archives of occupational and environmental health*, **33**: 12–15 (1978).
109. **Hemminki, K. et al.** Congenital malformations by parental occupation in Finland. *International archives of occupational and environmental health*, **46**: 93–98 (1980).
110. **Hemminki, K. et al.** Transplacental carcinogens and mutagens: Childhood cancer, malformations and abortions as risk indicators. *Journal of toxicology and environmental health*, **61**: 1115–1116 (1980).
111. **Holmberg, P. C.** Central-nervous-system defects in children born to mothers exposed to organic solvents during pregnancy. *Lancet*, **2**: 177–179 (1979).
112. **Holmberg, P. C. et al.** Oral clefts and organic solvent exposure during pregnancy. *International archives of occupational and environmental health*, **50**: 371–376 (1982).

113. **Bridbord, K.** Occupational lead exposure and women. *Preventive medicine, 7:* 311–321 (1978).
114. **Browne, K. S. & Rosenberg, G. J.** Risk of cytomegalovirus exposure to staff in a preschool for the retarded. *Journal of Indiana State Medical Association, 72:* 418–420 (1979).
115. **Johoda, M.** The impact of unemployment in the 1930s and 1970s. *Bulletin of the British Psychological Society, 32:* 309–314 (1979).
116. Does unemployment kill? *Lancet, 1:* 708–709 (1979).
117. **Kasl, S. V. & Cobb, S.** Some physical and mental health effects of job loss. *Pakistan medical forum (Karachi). 6:* 95–106 (1971).
118. **Kasl, S. V. et al.** Reports of illness and illness behaviour among men undergoing job loss. *Psychosomatic medicine, 34:* 475 (1972).
119. **Kasl, S. V. et al.** The experience of losing a job: reported changes in health, symptoms and illness behaviour. *Psychosomatic medicine, 37:* 106–122 (1975).
120. **Edström, R.** Quality of working life: a Scandinavian view. In: McDonald, J. C., ed. *Recent advances in occupational health.* Edinburgh, Churchill Livingstone, 1981.

Evaluation of the long-term effects of harmful occupational factors

M. I. Mikheev[a]

Evaluation of the adverse health effects of long-term exposure to low levels of harmful factors at work is currently the most pressing problem in occupational health. It is also highly complex because the effects appear either after a long period of exposure or when the person is no longer exposed. The neurotoxic effects of lead or mercury may not appear until 5–10 years after the onset of exposure, depending on the level of the exposure. Carcinogenic effects, such as those of some metals (Ni, Cr) and polycyclic aromatic hydrocarbons, may be delayed until 15–25 years after initial contact with these substances.

Animal studies provide information that allows, to a limited extent, the prediction of "safe" levels of human exposure to harmful factors. The difficulties are known, the main ones being interspecies differences and differences in dose and duration of exposure. Because of these difficulties, there is still a need for epidemiological studies to validate the predicted effects as such and to reappraise occupational exposure limits. Epidemiological studies provide data for the determination of so-called health-based occupational exposure limits. As far as toxic chemicals are concerned, these levels were defined by a WHO study group as "levels of harmful substances in workroom air at which there is no significant adverse health effects; this does not take into account technological and economic considerations . . . " (*1*).

This definition of health-based exposure limits can be easily extended to other occupational factors as long as the meaning of the adverse effect is understood. Adverse effects were defined by the above-mentioned WHO study group as effects that:

— indicate early stages of clinical disease;

---

[a] WHO Regional Office for Europe, Copenhagen, Denmark.

— are not readily reversible and indicate a decrement in the body's ability to maintain homoeostasis;

— enhance the susceptibility of the individual to the deleterious effects of other environmental influences;

— cause relevant measurements to be outside the "normal" range, if they are considered as an early indication of decreased functional capacity; and

— indicate important metabolic biochemical changes.

When planning epidemiological studies, it should be remembered that adverse health effects can be caused by both occupational and non-occupational factors and that the latter, if not controlled, lead to bias and erroneous results. Therefore, it has to be borne in mind that the health status of the working population is influenced not only by health hazards at work but also by lifestyle factors (smoking, alcohol, nutrition, etc.), genetics, etc. (see Fig. 1).

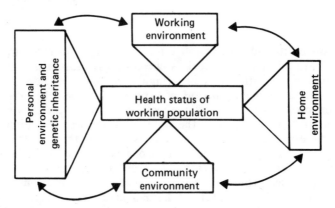

Fig. 1. Health status of working population in relation to factors present in different environments and to genetic inheritance

Undetected biological effects may occur when epidemiological studies are carried out during the latency period, i.e., when the effects of the harmful factor have not yet been revealed. Therefore, occupational exposures, particularly those relevant to new chemicals, should be considered potentially dangerous until reliable negative results have been obtained.

Epidemiological approaches and methods

Health problems posed by hazardous factors in the work environment involve two major issues, which are closely linked. The first is the nature of the health problem itself; the second is the nature of the hazardous factors and the level of exposure to them. Association between those issues is explored by descriptive epidemiology (see Chapter 7), which studies the patterns of disease in human populations and its relationship

70

to basic characteristics such as age, sex, race, occupation, social class and geographic location. The major parameters of descriptive epidemiology that are applicable to occupational health include occupation, morbidity and mortality, workplace and work conditions, harmful factors, and level of exposure to them. The data needed for descriptive epidemiology can be obtained from health surveillance of workers and monitoring of workers' health and the work environment (2). Hypotheses can also be developed on the basis of data from screening programmes, which are usually concerned with chronic effects of harmful factors and aim to detect the early stages of health impairment or disease before it has come under medical care.

The formulated hypotheses are then tested by means of analytical or etiological studies, which attempt to establish associations between disease occurrence and possible causative factors, the individuals in the group under study being classified according to the presence or absence of specific disease (or future development) and exposure (see Chapters 8–10).

The cross-sectional (or prevalence) study examines the relationship between diseases and other characteristics or variables, such as occupational factors and the level and period of exposure to them, behavioural factors, and other factors in the occupational environment to which a defined working population is exposed at one particular time (see Chapter 8). The relationship between exposure to a certain occupational factor or other variable and disease can be examined either:

(1) in terms of the prevalence of disease in different occupational subgroups defined according to the presence or absence (or level) of exposure to the occupational factor or other variable and

(2) in terms of the presence or absence (or level) of exposure to the occupational factor or other variable in diseased versus non-diseased persons.

In a cross-sectional study, disease prevalence rather than incidence is normally recorded.

The cohort study (also called prospective, longitudinal, follow-up, or incidence study) uses a representative group of the working population that is, has been, or may be in the future exposed to defined levels of risk for a factor or factors hypothesized to be associated with disease or death (see Chapter 9). Incidence rates or deaths from conditions under investigation are measured and compared in occupational groups with known different exposures to presumed causal factors, in order to test hypotheses about such factors and/or to estimate the probability of getting the disease or dying from the disease in relation to the amount of exposure to risk. Follow-up of the group over time (usually some years) is required to accumulate a sufficient number of cases.

The historical prospective cohort study identifies a cohort of people who have been working under similar conditions at particular times in the past and uses analysis of existing data to establish their subsequent morbidity or mortality experience. Unlike the purely prospective study, the historical study uses longitudinal information covering a time interval extending from past to present rather than present to future. The historical prospective study can also be extended by adding contemporaneous data to those assembled retrospectively. It then becomes a retrospective follow-up study.

The case–control study (also called retrospective, case–history, case–referent, or case–comparison study) is a hypothesis-testing study that starts with identification of persons with the disease of interest and a suitable control group of workers without the disease (see Chapter 10). The relationship between exposure to suggested factors and the disease is examined by comparing diseased and non-diseased persons with regard to frequency of exposure or level of exposure in the two groups. The case–control study could be set up retrospectively, starting after the onset of disease and looking back to prior events and experiences. It could also be set up prospectively, accumulating cases and controls as each new case is diagnosed. Nevertheless, such a study may still be called "retrospective" because it is concerned with prior events and experiences.

Fig. 2 provides a schematic illustration of two of the main epidemiological study designs for exploring a relationship between exposure and health effects (3).

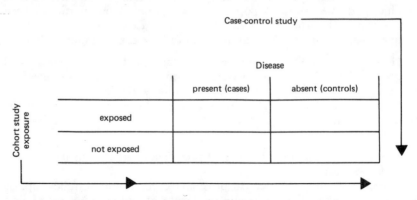

Fig. 2. Epidemiological study designs to explore the relationship between exposure and disease

The cohort study envisages the observation of groups of persons who differ in exposure and then determines if they differ in the health impairment (disease) being investigated. In Fig. 2 this approach can be read horizontally.

The case–control study envisages the selection of groups that differ in incidence of the health impairment of interest (cases and controls) and then determines whether or not they differ in exposure. In Fig. 2 this approach can be read vertically.

Modern epidemiological methods have the advantage of applying to "real life" complex situations, being able to take account of numerous confounding variables, such as age, sex, socioeconomic factors, and lifestyle variables.

It should be emphasized that the results of cohort studies provide direct measures of the risk of disease in exposed and non-exposed groups, whereas case–control studies primarily explore estimates of relative risk. Cohort studies, however, require a larger number of subjects and a longer period of follow-up. Case–control studies have great utility in the study of rare diseases, such as cancer. The need for a comparatively small number of cases and the short period of time make this approach the method of choice in many circumstances.

Cross-sectional studies have an important drawback: they include only living cases; those who died before the studies were undertaken are not included (4). This restriction may seriously bias the results. For example a sudden cardiac death caused by a particular chemical would not be included in such a study, which would fail to reveal a cardiotoxic effect. Another difficulty is to determine the temporal relation between first exposure to the factor and the onset of the disease. Thus, a latency effect and the length of the latency period cannot usually be determined in cross-sectional studies.

It is evident that each study design has its merits and its limitations. The choice depends upon an assessment of the available data and careful weighing of the advantages and disadvantages of each type of study.

Simple mathematical treatment of the data obtained from the epidemiological studies mentioned above provides some indices that are of great importance for the evaluation of the long-term effects of occupational hazards.

Epidemiological indices of risk estimation

As previously mentioned, a cohort study gives the data needed to calculate incidence rates. These in themselves are an estimation of the risk that an individual will develop the adverse health effect under study. The ratio of the cumulative incidence rates of those exposed to harmful factors to the incidence rate of those not exposed is the relative risk (RR) (also called relative ratio, risk ratio, or cumulative incidence ratio).

Incidence rates are not determined in a case–control study, and the measure of association is the ratio of odds (cross-product ratio, relative odds, or incidence–density ratio). It is the ratio of the odds of exposure among the cases to that among the controls (5, 6).

This is illustrated in the following example:

Risk factors	Disease	
	present	absent
exposed	a	b
unexposed	c	d

Here, the odds ratio is $\dfrac{ad}{bc}$

73

When the need is to estimate the extent to which a particular health impairment is due to a particular occupational factor, the population *attributable risk* (AR) can be used (7). It is computed as:

$$ARe = \frac{Ie - Iu}{Ie} = \frac{RR - 1}{RR}$$

where:

    ARe = attributable risk due to exposure
    Ie   = incidence rate among those exposed
    Iu   = incidence rate among those unexposed
    RR = risk ratio, or Ie/Iu

Attributable risk is a useful criterion in the use of epidemiological studies for planning, particularly in situations of multifactorial exposure.

Excess rate ratio (ERR) is the proportion of an observed rate ratio that is due to the effect of the exposure under study. It is computed as:

$$ERR = \frac{Ie - Iu}{Iu} = \frac{Ie}{Iu} - 1 = RR - 1$$

where:

    Ie = incidence rate in the exposed persons
    Iu = incidence rate in the unexposed persons

Standardized mortality ratio (SMR) and proportionate mortality ratio (PMR) are indices commonly used to measure differential mortality risks in epidemiological studies (8). The SMR is defined as the ratio of the number of deaths observed in the study group to the number of deaths expected if it had the same demographic structure as the standard population. Also used are standardized morbidity and incidence ratios. The PMR is the number of deaths from a given cause in a specified time period, per 100 or 1000 total deaths in the same period.

The SMR is the index of choice when the age, sex, and ethnic composition of the population at risk are known. It provides information on a study group's overall risk of dying as well as risks for specific causes of death. The PMR is a tool for estimating cause-specific risks when the available data consist only of deaths without knowledge of the population characteristics mentioned above.

It should be emphasized that the indices described above are not intended to represent a complete set of statistical indices to be used for estimating health risks. They are, however, both simple to use and informative and are adequate for all the purposes for which this manual is intended.

Evaluation of carcinogenic effects

Evidence of carcinogenicity in humans usually comes from: (*a*) case reports of individual cancer patients who were exposed to a particular chemical or industrial process and (*b*) epidemiological studies in which the incidence of cancer in a human population exposed to a particular factor was higher than in the control group not exposed to the suspected carcinogenic agent. Of the epidemiological studies described above, the case–control study is the most appropriate for assessing the carcinogenicity of occupational factors (9).

In the absence of sufficient evidence from epidemiological studies alone, evaluation of the carcinogenic risk to humans could be based on the combined consideration of epidemiological and experimental evidence. The structure of chemical compounds in relation to the structure of known carcinogens should be taken into account.

Exposure to many potential carcinogens is obviously greater in the workplace than elsewhere, making it relatively easier to identify the adverse effects of such agents in an epidemiological study of occupational risks, provided that an exposure-oriented recording system is employed (10).

For a causal association to be inferred between exposure and human cancer in an epidemiological study, the three criteria elaborated by the International Agency for Research on Cancer must be met (11):

(1) no identified bias that could explain the association;
(2) the possibility of confounding factors explaining the association has been ruled out;
(3) the association is unlikely to be due to chance.

In general, although a single study may be indicative of a cause–effect relationship, confidence in inferring a causal association is increased when several independent studies are concordant in showing the association, when the association is strong, when a dose–response relationship occurs, or when a reduction in exposure is followed by a reduction in the incidence of cancer.

Evaluation of reproduction effects

Effects on reproduction have recently become an important aspect of the evaluation of long-term exposure to occupational factors, particularly chemical agents. The main effects to be determined and evaluated are embryotoxicity, mutagenicity and teratogenicity. In an epidemiological study it is possible to monitor the following outcomes of such effects: reduction of fertility; genetic effects; various kinds of reproductive wastage, including spontaneous abortions, congenital malformations, stillbirths, and neonatal deaths; intrauterine growth retardation; and conditions that develop later in life, such as developmental disabilities, behavioural disorders, and malignancies (12). This list makes no pretence to be complete, but it is sufficient to show that the possible adverse outcomes are different in type and involve many processes and causes.

An embryotoxic effect early in development could result in unrecognized wastage but later in pregnancy could result in spontaneous abortion. Monitoring of the latter is being considered as one of the most sensitive methods for detecting the adverse effects of environmental factors on reproduction (13).

Abortion is defined as a loss of nonviable products of conception from the uterus. The period when abortions are observed spans the time from conception to about week 20 of gestation; thereafter, the miscarriage is classified as a stillbirth (14). Spontaneous abortion may result not only from embryotoxic effects of occupational factors but also from abnormal morphogenesis, which may be due to genetic factors or to adverse effects of environmental factors on embryonic or fetal development resulting in deviation from normal morphology or function. In evaluating the teratogenic effects of occupational exposure, it is also important to consider subtle effects, such as growth retardation, developmental abnormalities, behavioural disorders, and structural defects. Teratogenic effects of occupational exposures on reproductive outcome also include carcinogenic effects of intrauterine exposure (transplacental carcinogenesis).

It is well known that teratogens, mutagens, and carcinogens may act either directly or after metabolic activation to form covalent DNA adducts (15). These adducts, as well as chromosomal aberrations and mutagenic activity detected by bacterial mutagenicity in urine assays, are promising criteria for biological monitoring of exposure to chemicals possessing carcinogenic or mutagenic effects (16) and could also be used for screening purposes. However, much basic research is needed before these methods can be applicable for routine monitoring of exposure. At present, cytogenic approaches appear to be the most promising for routine surveillance to detect early biological effects of mutagenic agents (17).

It must be borne in mind that reproductive impairments are rare events from the statistical point of view. To detect moderate increases in the frequency of such events (for instance, doubling), data are needed from large samples of both exposed and unexposed individuals (18).

Cohort study and case–control study designs can be applied to evaluate the association between occupational factors and adverse reproductive outcomes. Birth certificates and employee records are useful sources of data. However, it must always be remembered that working people often have combined exposures that are difficult to determine and quantify. Nonoccupational factors, such as smoking and alcohol consumption, should be taken into account because they have clearly been shown to influence pregnancy outcome (18). In every epidemiological study, the selection of a control group, as well as the temporal relationship between an exposure and an outcome, is of fundamental importance. Validation of cases using medical records is also an element that should be included in the plan of the study.

To study a reproductive outcome, a case–control study begins with the selection of a group of cases and a group of controls. The cases may consist of infants with congenital abnormalities, or fetal or infant deaths where at least one of the parents was exposed to the factor being investigated. The aim of the study is to determine whether groups that differ in terms of outcomes also differ in terms of exposure to the factor of interest.

The cohort study explores two occupational groups that differ in terms of their exposure and determines whether or not they have differences in reproductive outcomes. The easiest type of cohort study is one in which groups of interest are identified and followed through time. The specific aims of such a study (12) would be to determine whether or not:

(1) rates of spontaneous abortion differ according to the levels of a given factor to which women or their husbands have been exposed;

(2) there are reported differences in "fertility" according to the levels of a given factor to which women or their husbands have been exposed;

(3) an exposure-associated risk of cancer or birth defects is detectable at any time in offspring of exposed workers; and

(4) there are effects of exposure on other pregnancy outcomes, such as perinatal mortality or prematurity.

The questionnaire approach can identify spontaneous abortions and disturbances in fertility that could not be ascertained using other study designs.

A second type of cohort study follows the population of children born to the defined groups during a specified period of time. Births could be identified from vital and hospital records and grouped according to the degree of parental exposure. Comparisons could then be made between exposure group for variables such as sex ratio, birth weight, gestational age, and congenital malformation rates. A particular group of these infants could be identified and followed into childhood using a questionnaire given to mothers and concentrating on child development, thus attempting to identify birth defects, behavioural abnormalities, etc. As previously mentioned, the data obtained should be verified using medical records.

Exposure–response relationship

An important task of the epidemiological study is to attempt to establish the exposure–response relationship, i.e., the quantitative association between environmental factors and their adverse health effects. The exposure–response relationship is determined as the relationship between a quantified exposure (level and duration) and the proportion of individuals exhibiting an effect. It should be distinguished from the exposure–effect relationship, which is the relationship between quantified exposure and quantitative severity of an adverse health effect in an individual or group (1).

The exposure–response curve reflects the interindividual variation in susceptibility, metabolism, and all other factors involved in the organism's response. The relationship between dose and concentration in critical organs receives special consideration because it determines many aspects of the response (19). There are considerable differences in the types of biological response to various actions of occupational factors. The shape of the curve in the low-exposure region is difficult to estimate, and it depends very much on the assumptions made and the mathematical model used for extrapolation (20). In the case of toxic effects, a threshold model is used, while in the case of carcinogenic effects, a non-threshold model is often advocated, as, for example, with asbestos (21). According to another point of view, however, the threshold model is also applicable to carcinogenic effects (22). With carcinogenesis, the dose of the carcinogenic agent is expressed as a product of time and intensity of exposure. Animal studies have shown that many carcinogenic agents display a linear exposure–response relationship. This type of relationship has also been demonstrated in an epidemiological study of asbestos (21).

The simplest way to express the exposure–response relationship is by means of the so-called relative risk for multiple categories. Statistical data for exposed and control groups are treated as a series of 2×2 tables. The relative risk for the control group is considered as 1. Relative risks for groups with a different degree of exposure or different time of exposure are compared with that of the control group. An example of how such a relative risk calculation is made is given in Table 1.

Table 1. Relative risk for multiple categories

Time of exposure (years) (could be level of exposure)	Groups of workers		Relative risk
	With specific health effects or diseases (cases)	Without specific health effects or diseases (controls)	
0 (control group)	10	100	1.0
1–4	30	150	2.0
5–9	75	250	3.0

The relative risk for workers exposed for 1–4 years is:

$$RR\,(1\text{--}4 \text{ years}) = \frac{30 \times 100}{10 \times 150} = \frac{3000}{1500} = 2.0$$

The relative risk for workers exposed for 5–9 years is:

$$RR\,(5\text{--}9 \text{ years}) = \frac{75 \times 100}{10 \times 250} = \frac{7500}{2500} = 3.0$$

A significance test for this relative risk has been developed by Cochran (23).

Health risk assessment

The risk in epidemiological terms is the probability that individuals will become ill or die within a stated period of time or age. In terms of occupational health, the level and length of exposure to harmful occupational factors become the main elements. Health risk assessment of occupational factors is based on all available and useful information resulting from physics, chemistry, toxicology, the practice of occupational medicine, epidemiological studies, etc., which are helpful in evaluating the harmful occupational factors and their health effects.

Health risk assessment is a complex methodology comprising three main stages: (1) risk estimation, (2) risk evaluation and (3) risk management. Epidemiological studies and the epidemiological indices resulting from them are the most important part of the risk estimation stage, which predicts the identification and quantification of the risk from a given technological process or machinery. The risk estimation for chemicals generally involves three sources of data: short-term toxicity and mutagenicity testing, long-term animal studies, and epidemiological data.

Risk evaluation assesses the importance of adverse effects liable to be suffered by those exposed to particular hazards. It considers the social and economic aspects and the advantages and disadvantages of a given technology for society, and investigates the attitudes of society with respect to perception of given risks.

Risk management includes actions to be taken to regulate the defined level of risk. Based on the information obtained from evaluation of short-term and mainly long-term effects of occupational hazards, occupational exposure limits should be established and adequate monitoring of workers' health and the work environment defined and carried out.

References

1. WHO Technical Report Series, No. 647, 1980 (*Recommended health-based limits in occupational exposure to heavy metals.* Report of a WHO Study Group).
2. WHO Technical Report Series, No. 535, 1973 (*Environmental and health monitoring in occupational health.* Report of a WHO Expert Committee).
3. **Lilienfeld, A. M. & Lilienfeld, D. E.** *Foundations of epidemiology*, 2nd ed. London, Oxford University Press, 1980.
4. **Pell, S.** The epidemiologic approach. *Environmental health perspectives*, **26**: 269–273 (1978).
5. **Gart, J. J.** Statistical analysis of the relative risk. *Environmental health perspectives*, **32**: 157–167 (1979).
6. **Fleiss, J. L.** Confidence intervals for the odds ratio in case–control studies: the state of the art. *Journal of chronic diseases*, **32**: 69–77 (1979).
7. **Walter, S. D.** Calculation of attributable risk from epidemiological data. *International journal of epidemiology*, **7**: 157–182 (1978).
8. **Decouflé, P. et al.** Comparison of the proportionate mortality ratio and standardized mortality ratio risk measures. *American journal of epidemiology*, **111**: 263–269 (1980).
9. **Breslow, N. E. & Day, N. E.** Statistical methods in cancer research. Vol. 1. The analysis of case–control studies. Lyon, International Agency for Research on Cancer, 1980 (IARC Scientific Publications No. 32).
10. **Moor, Sh. K. et al.** An occupation and exposure linkage system for the study of occupational carcinogenesis. *Journal of occupational medicine*, **22**: 722–726 (1980).

11. *Some monomers, plastics and synthetic elastomers, and acrolein.* Lyon, International Agency for Research on Cancer, 1979 (IARC Monographs on Evaluation of the Carcinogenic Risk of Chemicals to Humans, Vol. 19), pp. 13–33.
12. **Sever, L. E.** Reproductive hazards of the workplace. *Journal of occupational medicine,* **23**: 685–689 (1981).
13. **Haas, J. F. & Schottenfeld, D.** Risk to the offspring from parental occupational exposure. *Journal of occupational medicine,* **21**: 607–613 (1979).
14. **Carr, D. H.** Detection and evaluation of pregnancy wastage. In: Wilson, J. G. & Fraser, E. C., ed. *Handbook of teratology,* Vol. 3. New York, Plenum Press, 1977, pp. 189–225.
15. **Hemminki, K. et al.** Genetic risks caused by occupational chemicals. *Scandinavian journal of work, environment and health,* **5**: 307–327 (1979).
16. **Draper, M. H. et al.** Occupational cancer control in association with theory of carcinogenesis. In: *Prevention of Occupational Cancer—International Symposium.* Geneva, International Labour Office, 1982, pp. 576–581.
17. **Sorsa, M. et al.** Biologic monitoring of exposure to chemical mutagens in the occupational environment. *Teratogenesis, carcinogenesis and mutagenesis,* **2**: 137–150 (1982).
18. **Kline, J. et al.** Epidemiologic detection of low dose effects on the developing fetus. *Environmental health perspectives,* **42**: 119–126 (1981).
19. **Nordberg, G. F. & Strangert, P.** Fundamental aspects of dose–response relationships and their extrapolation for noncarcinogenic effects of metals. *Environmental health perspectives,* **22**: 97–102 (1978).
20. **Guess, H. A. & Crump, K. S.** Best-estimate low-dose extrapolation of carcinogenicity data. *Environmental health perspectives,* **22**: 149–152 (1978).
21. **Nicholson, W. J.** The dose and the time dependence of occupational cancer. In: *Prevention of Occupational Cancer—International Symposium.* Geneva, International Labour Office, 1982, pp. 44–67.
22. **Sanotski, I. V.** [Basic questions of the problem of delayed effects of occupational poisons.] In: Sanotski, I. V., ed. [*Problems of hygienic standardization in the investigation of delayed effects on industrial substances.*] Moscow, Medicina, 1972 (in Russian).
23. **Cochran, W. G.** Some methods of strengthening the common χ^2 test. *Biometrics,* **10**: 417–451 (1954).

5

Sources of data

R. S. F. Schilling[a]

Sources of data for possible use in epidemiological studies fall into two groups. First, there are data collected routinely for other purposes, which can be of immense value in descriptive epidemiology and may be useful in case–control or prospective cohort studies. Second, there are specially recorded data on work people and their environmental exposures, which are made in cross-sectional surveys and prospective cohort or experimental epidemiological studies. Only data of the former type will be described in this chapter.

The three main sources of data routinely collected for other purposes are: (1) official statistics and records kept by corporate bodies; (2) records of employment and other data in workplaces; and (3) records kept by occupational health services.

As these three sources of data are not collected primarily for epidemiological purposes, it is essential to know how the information was obtained, what it means, and whether it is accurate enough to be used.

Official statistics and records kept by corporate bodies

A wide variety of mortality and morbidity data is obtainable from government statistics and from records kept by corporate bodies.

National mortality data

Several countries provide mortality rates for the trades and professions in which large numbers are employed. For example, in England and Wales these rates are calculated from the number of deaths occurring in a particular occupation in a three-year period around the decennial census, which provides the number at risk in that population. As crude rates make no allowance for differences in age distribution, age-standardized indices of mortality are necessary. There are direct and indirect methods of standardization.

---

[a] 46 Northchurch Road, London, England.

In the *direct* method, age-specific death rates from the study population are applied to corresponding age groups in the standard population to give the number of deaths expected to occur if the standard population had experienced the same death rates as the study population. The expected deaths in each age group are totalled and compared with the observed deaths in the standard population. The comparative mortality figure (CMF) and the direct standardized death rate (SDR) are calculated as follows:

$$CMF = \frac{\text{total expected deaths in standard population}}{\text{total observed deaths in standard population}} \times 100$$

$$\text{Direct SDR} = \frac{CMF}{100} \times \text{crude death rate in standard population}$$

In the *indirect* method of standardization, age-specific death rates from the standard population (the total male or female population of England or Wales) are applied to corresponding age groups in the study population, which is an occupational group such as agricultural workers or cotton textile workers. The expected deaths in each age group are totalled to show how many deaths would have occurred in the study population if it had experienced the death rates of the standard population. The total is then compared with the observed deaths in the study population and the standardized mortality rate (SMR) is calculated as follows:-

$$SMR = \frac{\text{Total observed deaths in study population}}{\text{Total expected deaths in study population}} \times 100$$

These calculations can be made for deaths from all or particular causes.

The most recent data published for England and Wales are for 1970–1972, the years around the census in 1971 (*1*). Examples of SMRs for certain occupations reveal wide ranges of mortality risk, which may be directly related to occupational or socioeconomic factors. For example, butchers have high death rates from cancer of the lung and maxillary sinus. Their exposure to sawdust may account for this as similar excesses are found in woodworkers. In social class I, self-employed professional workers have consistently higher death rates than employees in this class.

Limitations of national mortality data

In the nineteenth century, the Registrar General's occupational mortality data revealed gross mortality excesses in hazardous occupations and were instrumental in improving working conditions (*2*). In present times, they may still be useful indications of undiscovered occupational hazards that are fatal. They do, however, have serious limitations as they may be based on incorrect or incomplete information about the cause of death and the occupation. The information given by the individual about his or her occupation at the National Census is likely to be correct. That given by next of kin about a deceased person's occupation was found in one study to be not infrequently wrong for British coalminers, the deceased being afforded a higher status than that actually attained (*3*). A more recent investigation (*4*) has shown that next of kin are able to give accurate information about deceased persons.

People now live longer and change their jobs more frequently. Thus, their occupation at death may be different from their main lifetime occupation. Some workers dying of pneumoconiosis have been employed, as a result of their disability, in sedentary "end occupations" (Table 1). Another misleading feature may be that national mortality rates are often given for a whole industry and not for its separate trades. In the United States, the mortality and morbidity rates for chronic respiratory disease in textile workers in 1950 were similar to those for all occupied males (*5*). This was taken as evidence against there being a risk of byssinosis. Surveys of cotton workers have since shown a high prevalence of byssinosis in those working in card room or engaged in other dusty processes, and higher levels of respiratory disability in textile workers than in the community. The national rates were misleading because the number of workers employed in the dusty departments of cotton spinning mills represented a small proportion of all textile workers, who probably had a lower risk than all males of contracting respiratory disease. This is because many of them work in warm clean work rooms where smoking is prohibited because of the risk of fire. Also, there may be unsuspected anomalies in methods of recording deaths. In England and Wales, the SMRs for occupational accidents among deep-sea fishermen have always been high, yet they have grossly underestimated the risk because deaths at sea are recorded by a different government department and no account of them is taken in calculating SMRs. If deaths at sea are included, the SMR for accidents in 1959–1963 is increased from 466 to 1726 (*6*).

Another serious and growing problem in handling mortality data is that, in many countries, records of individual deaths are regarded as confidential and are not available to research workers.

In summary, a significantly raised SMR in a particular occupation will indicate a need for further investigation. However, the absence of a significant mortality excess cannot be regarded as reliable evidence that no occupational risk exists.

Table 1. Pneumoconiosis deaths in "end occupations" in England and Wales (1970/1972) at age 15–74[a]

Occupation	Number of deaths
Warehousemen and storekeepers	28
Stationary engine drivers	23
Clerks	11
Guards and related workers	9
Caretakers	6
Salesmen	4

[a] Reproduced from Schilling, R. S. F., ed. *Occupational health practice*. London, Butterworths, 1981.

Other mortality data

Records of deaths available from local or regional sources, such as registers, hospitals, company and trade union pension schemes, are used most frequently in analytical epidemiological studies to test hypotheses. Unlike national data, they are seldom analysed routinely as a means of identifying possible hazards.

Local or regional death registers

The works manager of a small factory making arsenical sheep dip thought that an unduly high proportion of his workers were dying of cancer. Over the previous 40 years the factory had kept a record of their employees who died, but the cause of death was frequently missing. There were no figures for the population at risk, so ordinary death rates could not be calculated. A search of the local death registers of the town in which the factory was situated made it possible to identify the causes of death of the sheep dip workers. By using proportional mortality rates, it was found that 29.3 % of male sheep dip workers had died of cancer compared with 12.9 % of men in other occupational groups living in the same town. A further analysis (Table 2) of the sheep dip workers' deaths revealed a significant excess of cancer deaths among chemical process workers (39 %), who would be most heavily exposed, compared with non-process workers (12.5 %) such as printers and watchmen (7). This is a good example of the combined use of local death registers and factory records as sources of data for identifying an occupational hazard.

Table 2. Proportion of cancer deaths of males by occupation in sheep-dip factory[a]

	Total deaths	Cancer deaths	Percentage due to cancer
Chemical workers	41	16	39[b]
Engineers and packers	10	3	30
Others[c]	24	3	12.5[b]

[a] Reproduced from Schilling R. S. F., ed. *Occupational health practice*. London, Butterworths, 1981.
[b] $\chi^2 = 3.95$; $P = 0.047$.
[c] Includes non-process workers, such as printers, watchmen, boxmakers and carters.

In Sweden, local death registers in a semi-rural area, where copper smelting was the principal industry, provided information for a case–control study of the association between lung cancer and arsenic exposure. Other local sources of information, i.e., the medical files of patients, confirmed the diagnosis clinically and histologically. Those who had been exposed in the smelter could be identified from employment records. Out of 29 men who died of lung cancer, 18 had worked in the smelter. Out of 74 men dying of other diseases, only 18 had worked in the smelter, giving a fivefold excess (or crude risk ratio) of lung cancer in the smelter workers (8).

Hospital records

One of the first epidemiological studies of the relationship between mesothelial tumours and asbestos exposure was based on autopsy records kept at the London Hospital and filed under diagnosis. This hospital is situated in an area where asbestos materials are made and used and asbestos is handled in nearby docks. From the hospital records, 83 cases of mesothelioma were confirmed after the histological specimens had been reviewed. In all but 7 of the cases it was possible to get occupational and domestic histories. The study revealed that 53 % of the mesothelioma patients had a history of occupational or domestic exposure to asbestos, compared with 12 % of the patients in the control group (9). If hospital records contain details of patients' occupational histories and work exposures, they can be used for case–control studies without having to interview relatives or the patient, if alive.

Pension and funeral benefit records
Pension records of firms have provided valuable information for investigating certain types of occupational risks, such as cancer and cerebrovascular disease. They have the disadvantage of dealing only with the pensionable age groups and, thus, give no information about the risk among younger people. Doll (*10*) used gas company pension records to identify the risk of lung cancer in workers employed in horizontal retort houses. A study of the pension records of a group of companies making lead batteries revealed a significant mortality excess (Table 3) from cerebrovascular disease among workers who had been heavily exposed to lead before the enforcement of regulations (*11*).

Table 3. Observed and expected deaths from cerebrovascular disease in retired lead battery workers

Lead exposure	No. of men	No. of deaths	
		Expected	Observed
none	158	7.9	6[a]
negligible	80	3.4	6
high	187	9.3	24[a]

[a] $\chi^2 = 21.7$; $P < 0.001$

Trades unions who run their own funeral benefit schemes have records of death certificates of those for whom benefits are paid. Such records have been exploited by Selikoff in his investigations of the risk of malignant disease in asbestos workers in the United States (*12*).

Records of morbidity

National morbidity data
National morbidity data are available in most countries, but generally are less valuable to the epidemiologist than national mortality data. This is because the denominators (populations at risk) are inaccurate or not available. The diagnoses on sickness absence certificates are often vague

or inaccurate and some of the absences are due to causes that are not related to health impairment. National data on work injuries and occupational diseases are known to be incomplete. National data, however, can provide a crude overall picture of morbidity among the gainfully employed and may stimulate epidemiological enquiry into particular occupational groups.

Sickness absence

In the United Kingdom, sickness absence rates for the insured population vary substantially between regions. They are high in Wales, Northern Ireland, and the North East, and low in London and the South East. These differences could be due to occupational, economic, climatic, cultural, or even genetic factors. The Chief Medical Officer of the Post Office, whose workers are not included in the national figures, examined the Post Office figures for regional differences (13). It might have been concluded that high national rates for Wales and the North East could have been due to a preponderance of heavy industry, such as mining, and the low rates for London and the South East could have been caused by the greater numbers employed in light industry and offices. Nevertheless, variations between regions could not be explained by differences in types of industry, since the same trends occurred in the Post Office absence figures, which are effectively standardized for occupation.

Such studies of national trends and differences in sickness absence rates are more relevant to socioeconomics than to the investigation of occupational health.

In summary, national morbidity figures give a broad picture of the health of the gainfully employed, but without enough detail to be of much value to the epidemiologist, except as a means of initiating more detailed investigations.

Accidental and occupational disease

Accident data are usually kept separately from sickness data by the government department responsible for health and safety at work. They are based on compulsory reporting of lost-time accidents and they provide information on the frequency and severity of risks in various types of industry and occupation.

National figures may highlight dangerous practices. For example, a yearly occurrence of 10–15 serious or fatal injuries to farm workers, caused by tractors overturning, led to new regulations. These required tractors to have reinforced roofs.

Figures for compensation for work injuries are generally more reliable than those based on the reporting of accidents (14). They are reasonably accurate since it is on such figures that levels of insurance premiums, paid by employers, are based. However, they do not give complete figures for all lost-time accidents. In Great Britain, they are incomplete for work injuries to women and self-employed persons (15).

The figures for the number of workers receiving compensation for occupational diseases may be used to indicate the degree of risk in a particular industry. Secular trends in these numbers show successes or failures in preventive measures. Such figures give only a very crude index of secular trends in risk because they are influenced by changes in the numbers eligible for compensation, increasing awareness of compensatable diseases, the action of trades union officials, and changes in diagnostic criteria.

Comparison between the prevalence of coal workers' pneumoconiosis and the numbers of miners compensated reveals the considerable disparity between these two sets of figures (16) (Table 4).

Table 4. Comparison of certification rates for pneumoconiosis and radiological prevalence of disease

Pit and year of survey	Certification rate per 100 workers per year[a] (I)	Radiological prevalence rate per 100 workers (II)	Ratio II/I
B 1949	9.69	59.3	6.1
C 1949	0.21	26.4	125.7
D 1950	1.45	29.5	20.3

[a] Average for 5 years before survey.

In most countries, strict criteria must be fulfilled before an occupational disease is included in the compensation scheme. In the United Kingdom, the disease must be shown to be a risk of a certain occupation and the attribution of particular cases to the nature of employment must be established with reasonable certainty. Thus, there are bound to be delays in including an occupational disease in the compensation scheme and, in some countries, the less obvious occupational diseases are never made compensatable.

In summary, compensation figures are likely to be incomplete and cannot be used by the epidemiologist, with any degree of confidence, as an index of risk.

Notification of occupational disease
Government departments responsible for occupational health have instituted statutory notification of occupational diseases in order to ensure that they are investigated and controlled. In the past, such notification has not been a reliable measure of their extent and severity as they may not be correctly diagnosed and many physicians do not fulfil their obligation to notify. In the Manchester area, out of 130 patients attending hospital with squamous epithelioma, 81 were likely to have had work exposures to known carcinogens and, thus, their cases were statutorily notifiable. Only 3 of these cases were notified (17).

The British Health and Safety Commission (*18*) has proposed new regulations putting the onus for notification of occupational ill health entirely on the employers. They have to notify (1) uncontrolled escape of harmful substances liable to damage health, (2) acute ill health caused by inhalation or absorption of any harmful substance and requiring medical attention or causing disability lasting more than 4 years, (3) acute ill health from occupational exposure to pathogens or infected material. These regulations may provide a useful source of information for a limited number of new or uncontrolled health risks.

A statutory instrument of this kind must be limited in its scope by its definitions, by failure of workplaces to recognize such events and, possibly, by a reluctance to notify in order to avoid liability at common law. Thus, it is unlikely to give anything like an accurate picture of the risk of occupational ill health, but the notification of a minority of such incidents should be helpful in providing clues for further epidemiological investigation.

Other morbidity records

Special registers

Finland (*19*) has a National Occupational Disease Register at its Institute of Occupational Health. It obtains information from three different sources: (1) insurance companies who give data on every case reported to them as an occupational disease, (2) cases of occupational disease diagnosed at the Institute, (3) physicians in Finland are required by law to report cases of occupational disease or work-related illness to the National Board of Labour Protection, which sends this information to the Register. Annual incidence estimates for various types of industry, commerce and services, and occupations are based on population statistics of the census years. This information has shortcomings, mainly because it is incomplete. Some physicians neglect their responsibility to report or fail to connect diseases with working conditions. Nevertheless, the register has a value in increasing the effectiveness of health care and improving standards of health and safety. It is a crude epidemiological tool that can highlight health problems for further investigation.

Special registers are also used to record information on occupational cancer and diseases related to beryllium and asbestos. For example, in Finland workers exposed to carcinogens are registered.

In the United States, a Beryllium Case Registry was established in 1952 (*20*). It has been an invaluable source of information regarding the assessment of risk and the natural history of disease. Neighbourhood cases were found to be more common in women than in men, possibly because women were more exposed to handling and cleaning contaminated clothing. As with other registries of occupational disease, the data are incomplete.

Sickness benefit schemes

Bradford Hill has made use of the records of trades unions sickness benefit schemes. For respiratory disease, sickness rates of card room workers were more than three times higher than similar rates for workers less exposed to cotton dust (21). In another enquiry, London bus drivers and conductors had an excess of stomach complaints compared with tramway drivers and conductors (22).

Records of employment and other data held at workplaces

Records of employment
Dates of employment are essential in the historically prospective type of cohort study in which the population under study entered employment many years ago. A typical example is the one by Newhouse & Berry (23), who from the employment records of an asbestos factory, were able to define a cohort of 2887 men first employed between 1 April 1933 and 31 December 1964 for a period of more than 30 days. They were separated into those with up to two years employment and those with more. There were no records of levels of dust exposure but from the employment records, which gave details of occupations, it was possible further to divide the cohort into those with "low to moderate" and those with "severe" exposures.

Employment records are unlikely to include data and cause of death, particularly concerning workers who have changed jobs or have retired. If there is a record of the worker's date of birth and an identification number (social security or national health) information with regard to mortality can be obtained from other sources. This is now possible in many countries. In the Newhouse & Berry study, the date and cause of death were obtained from the Office of Population Censuses and Surveys, which contains a national depository of all recorded deaths and their causes. Results revealed the serious nature of the risk among workers who had had severe exposures, both for short and long periods (Table 5).

Other work records
These records include reasons for leaving employment, which are generally unreliable and of little value in epidemiological enquiries into health risks, although the labour records of lateness and absence from work may be used as indices of working conditions, of morale, and of work satisfaction. An enterprising study by Bjerner et al. (24) used mistakes made by workers to measure diurnal variation in mental performance. The subjects were men in a Swedish gas works who did simple calculations from meter readings and recorded them in a ledger at hourly intervals. These were checked and, where necessary, corrections were made. They worked in 8-hour shifts which began at 6 h.00, 14 h.00, and 22 h.00. Over a period of 20 years there were over 175 000 entries, of which about 75 000 had errors. The distribution of errors, which had marked peaks at 15 h.00 and 3 h.00, indicated that mental performance was lowest at these times and generally lower during the night.

Table 5. Mortality experience of male asbestos factory workers from 1933 to end of 1975[a]

Cause of death	Exposure category[b]							
	Low to moderate				Severe			
	< 2 years (884)		> 2 years (554)		< 2 years (937)		> 2 years (512)	
	Observed	Expected	Observed	Expected	Observed	Expected	Observed	Expected
All causes	118 (4)	118.0	89 (7)	95.3	162[c] (16)	122.2	176[c] (19)	102.5
Cancers of lung and pleura (ICD 162, 163)	17 (3)	11.01	16[c] (1)	9.0	31[c] (6)	12.8	56[c] (7)	10.4
Gastrointestinal cancer (ICD 150–158)	10	9.0	9 (4)	7.3	20[d] (6)	9.5	19[d] (8)	8.2
Other cancers	6	7.4	8 (1)	5.8	16[d] (3)	7.9	16[c] (4)	6.3
Chronic respiratory disease	19	17.5	16	14.7	20 (1)	17.6	28[d]	15.9

[a] Reproduced, with slight modifications, from Schilling, R. S. F., ed. Occupational health practice. London, Butterworths, 1981.
[b] Figures in brackets indicate the number of mesothelial tumours in each category.
[c] P < 0.001.
[d] P < 0.01.
[e] P < 0.05.

Records kept by occupational health services

Occupational health services usually keep details of medical examinations, consultations, environmental monitoring, treatments at work, and sickness absences. Such records are essential for the management and treatment of patients and for environmental investigations, as well as for the investigation of compensation claims. They provide the basis for annual reports, which give an account of the activities of the service to its consumers. They have a further value as a source of epidemiological data.

Medical examinations and environmental monitoring

The preplacement examination is primarily used to place new employees in suitable occupations. It makes it possible to identify persons who are likely to be vulnerable to certain exposures. It provides baseline data that, together with the results of periodic examinations, make it possible to measure early adverse effects of exposure and to detect susceptible subjects. Such data, which are designed to protect the individual, can also be used epidemiologically to evaluate environmental control measures and personal protective equipment. If they are used in conjunction with levels of exposure, it becomes possible to determine exposure–response or uptake–response relationships from which exposure limits may be derived. Epidemiological studies of work people and their exposure to contaminants provide the most valid data on which to base occupational exposure limits. If occupational health services, in their routine work, use reliable methods of the same high standard as required in cross-sectional epidemiological studies, they can make a valuable contribution to the derivation and reappraisal of occupational exposure limits.

Treatment records

Attendances for treatment for work injuries are required to be kept by workplaces in many countries. They enable employers to fulfil their duty with respect to notification of work accidents. If the facts recorded are accurate and reasonably detailed (see Fig. 1) treatment records may be used to indicate dangerous practices and risks of occupational disease, or to identify persons prone to accidents. The use of simple epidemiological methods may reveal hazards that would be difficult to recognize from a perusal of individual occurrences. Stott (25) examined the records of treatment at the dispensaries of an African sisal factory and found that the attendance rate for chest complaints was twice as high for workers in the card room, where dust exposures were excessive, as for other departments. Further analysis of hospital records revealed that 25 % of all workers admitted for acute chest conditions were card room workers, who represented 12 % of the factory population. This preliminary enquiry was followed by a full investigation of card room workers, in whom a risk of occupational pulmonary disease was identified.

Name

Personal No.

Department

Address

Supervisor/Foreman

Occupation

Date of birth

Patient's own doctor

Date	Treatment	Occup.	Non-occup.	Disposal

Fig. 1. Example of an individual treatment record card

Reproduced from: **Schilling, R.S.F., ed.** *Occupational health practice.* **London, Butterworths, 1981.**

Sickness absence at the workplace

Sickness absences are recorded by occupational health services because they are helpful in the management of individual health problems. They can also be used epidemiologically to assess sickness and behavioural characteristics of working groups and to study effects of working conditions, such as toxic exposures and different shift systems. They are of more value in identifying and appraising health hazards than national records. The basic data required are shown on an individual record card (Fig. 2).

In sickness absence studies, two rates are commonly used. They are as follows:

$$\frac{\text{Mean days per person}}{\text{(a severity rate)}} = \frac{\text{Total days of absence in period}}{\text{Average population at risk in period}}$$

$$\frac{\text{Mean spells per person}}{\text{(a frequency rate)}} = \frac{\text{Total no. of new spells of absence in period}}{\text{Average population at risk in period}}$$

For small groups of workers calculations can be made from the individual record cards, which are kept for everyone in the group irrespective of whether or not they incur an absence. For large groups the data can be computerized.

Sickness absence data, recorded in the workplaces, have been used to identify a hazard of chronic bronchitis in gas workers (26) and to study the effects of different shift systems on sickness absence and lateness (27, 28).

Records collected by an occupational health service can provide valuable information for evaluating the effectiveness of control measures and for determining how services are used and how they can be made more relevant to health needs. Treatment records in particular can indicate needs and deficiencies in health care. In an engineering factory where eye injuries were common, treatment records revealed hazards that could be prevented by improved eye protection and the employment of a highly skilled nurse with ophthalmic training was recommended. After her appointment, the number of eye injuries treated more than doubled, but the proportion sent to the eye hospital was reduced from 16 % to 4 % (29).

Need for a higher standard in the maintenance of employment and health records

In prospective cohort and experimental epidemiological studies, the types of data to be recorded are decided during the planning of each study.

Descriptive, case–control, or historically prospective studies depend on routinely collected records of employment, medical examinations, treatment, and exposures to environmental contaminants.

Employee records should include such information as full names, date of birth, social security number, and maiden name of married women. This assists in identifying individuals from past records.

Name	SMITH	Christian names	JOHN HENRY		D of B	30-7-33		Company number	682731		NHS number	OCEK 31741
Sex	M	Marital status	M	Address	3 Acacia Road		General practitioner	DR JONES		Registered disabled	NO	

Occupations/Working hours

Sickness absences

Date	New job etc.		Date off	Date return	Days lost	Certif. Uncertif.	Diagnosis	ICD	Occupational Non-occup.
15.3.76	Storeman	Days	3.1.77	10.1.77	5	C	Influenza	487.1	NO
			15.6.77	6.7.77	15	C	Sprained lumbar spine	847.2	O
24.10.77	Machinist Alt. Day/Night		13.2.78	14.2.78	1	UC	Cold	460	NO
		Shift	18.9.78	19.9.78	1	UC	Diarrhoea	009.3	NO
			15.5.79	18.5.79	3	C	Pharyngitis	462	NO
			21.8.79	25.9.79	25	C	Lumbar disc lesion	722.1	O
15.10.79	Clerk safety	Days	7.1.80	21.1.80	10	C	Influenza	487.1	NO

Fig. 2. Example of a personal sickness record card

Reproduced from: Schilling, R.S.F., ed. Occupational health practice. London, Butterworths, 1981.

Generally, employers are not obliged to keep employment or health records of their workplace. However, more complete and more accurate records would enhance the value of such records for epidemiological studies. There is a need for compulsory record keeping to be considered at two levels. For persons exposed to recognized hazards, detailed information should be kept on occupational histories, health examinations, illnesses, and environmental exposures. For others, particularly those in types of employment that involve exposures to possible, but unrecognized, physical, chemical, and biological hazards, less detailed information may be kept; but it should at least include employment records, dates of birth, and any obvious adverse effects on health due to work.

Where routinely collected social security data can be made available for epidemiological research, the names of employers on each individual's record would be useful in case–control and historically prospective studies.

Overemphasis on the confidentiality of records will hamper research and will not be in the best interests of work people.

References

1. **Office of Population Censuses and Surveys.** *Occupational mortality. Decennial Supplement, England and Wales, 1970–72.* London, H. M. Stationery Office, 1978.
2. **Health of Munition Workers Committee.** *Final report on industrial health and efficiency.* London, H. M. Stationery Office, 1918, p. 13.
3. **Heasman, M. A. et al.** The accuracy of occupational vital statistics. *British journal of industrial medicine,* **15**: 141–146 (1958).
4. **Pershagen, G. & Axelson, O.** A validation of questionnaire information on occupational exposure and smoking. *Scandinavian journal of work, environment and health,* **8**: 24–28 (1982).
5. **Schilling, R. S. F.** Epidemiological studies of chronic respiratory disease among cotton operatives. *Yale journal of biological medicine,* **37**: 55–73 (1964).
6. **Schilling, R. S. F.** Hazards of deep sea fishing. *British journal of industrial medicine,* **28**: 27–35 (1971).
7. **Hill, A. B. & Faning, E. L.** Studies in the incidence of cancer in a factory handling inorganic compounds of arsenic. *British journal of industrial medicine,* **5**: 1–6 (1948).
8. **Axelson, O. et al.** Arsenic exposure and mortality—a case reference study from a Swedish copper smelter. *British journal of industrial medicine,* **35**: 8–15 (1978).
9. **Newhouse, M. L. & Thompson, H.** Mesothelioma of pleura and peritoneum following exposure to asbestos in the London area. *British journal of industrial medicine,* **22**: 261–269 (1965).
10. **Doll, R.** The causes of death among gas workers with special reference to cancer of the lung. *British journal of industrial medicine,* **9**: 180–185 (1952).
11. **Dingwall-Fordyce, I. & Lane, R. E.** A follow-up study of lead workers. *British journal of industrial medicine,* **20**: 313–315 (1963).
12. **Selikoff, I. J. et al.** Mortality experience of asbestos insulation workers. In: *Proceedings, International Conference on Pneumoconiosis, Johannesburg, 1969.* London, Oxford University Press, 1970, pp. 180–186.
13. **Taylor, P. J.** Occupational and regional associations of death, disablement and sickness absence among Post Office staff, 1972–75. *British journal of industrial medicine,* **33**: 230–235 (1976).
14. **Case, R. A. M. & Davies, J. M.** On the use of official statistics in medical research. *The statistician,* **14**: 89–119 (1965).

15. **Royal Commission on Civil Liability and Compensation for Personal Injury.** *Statistics and costings*, vol. 2. London, H. M. Stationery Office, 1978.
16. **Cochrane, A. L.** *Methods of investigating the connexion between dust and disease in the application of scientific methods to industrial and service medicine.* London, H. M. Stationery Office, 1951.
17. **Murray, R.** Occupational cancer of the skin. In: Raven, R. W., ed. *Cancer*, vol 3. London, Butterworths, 1958, pp. 334–342.
18. **Health and Safety Commission.** *Consultative document on proposals for the notification of occupational ill health.* London, H. M. Stationery Office, 1978.
19. **Vaaranen, V. & Vasama, M.** *Occupational diseases in Finland in 1978– 1979. Occupational diseases reported to the Finnish Occupational Disease Register.* Helsinki, Institute of Occupational Health, 1980.
20. **Tepper, L. B. et al.** *Toxicity of beryllium compounds.* Amsterdam, Elsevier, 1961.
21. **Hill, A. B.** *Sickness amongst operatives in Lancashire cotton mills.* London, H. M. Stationery Office, 1930 (Industrial Health Research Board, Report No. 59).
22. **Hill, A. B.** *An investigation into the sickness experience of London Transport workers, with special reference to digestive disturbances.* London, H. M. Stationery Office, 1937 (Industrial Health Research Board, Report No. 79).
23. **Newhouse, M. L. & Berry, G.** Patterns of mortality in asbestos factory workers. *Annals of the New York Academy of Science,* **330**: 53–60 (1979).
24. **Bjerner, B. et al.** Diurnal variation in mental performance. *British journal of industrial medicine,* **12**: 103–110 (1955).
25. **Stott, H.** Pulmonary disease among sisal workers. *British journal of industrial medicine,* **15**: 23–27 (1958).
26. **Gregory, J.** Occupational factors in the incidence of bronchitis. *Transactions of the Association of Industrial Medical Officers,* **5**: 2–9 (1955).
27. **Taylor, P. J.** Shift and day work—a comparison of sickness absence, lateness and other absence behaviour at an oil refinery from 1962–1965. *British journal of industrial medicine,* **24**: 93–102 (1967).
28. **Taylor, P. J. et al.** Absenteeism of shift and day workers—a study of 6 types of shift system in 29 organizations. *British journal of industrial medicine,* **29**: 208–213 (1972).

6

Screening in the assessment of health risks

T. Popov[a]

Mass screening is aimed at identifying and placing under medical supervision any case that requires diagnostic (clinical) examination. The practical significance of mass screening lies in the opportunity it affords to detect the initial stage of an illness. Early case finding may be expected in most cases to improve prognosis, since treatment is more likely to be effective if it is instituted when the disease is at an initial stage. In addition, the results obtained from mass screening provide evidence of the prevalence of various diseases among the population. Such information provides a basis for further investigations on the dependence of disease on environmental factors.

The health surveillance of specific occupational groups may be undertaken at three levels: (1) screening procedures; (2) periodic medical examination; (3) diagnostic (clinical) examination. The first and third levels are largely used in public health. Level two is mainly used in occupational health.

Typical examples of mass screening examinations are chest fluorography for early diagnosis of tuberculosis and cancer of the lung, and tuberculin tests for detection of tuberculosis. These examples clearly show that mass screening substantially differs from a classical diagnostic examination and some distinctions between these two approaches are listed in Table 1.

Screening in occupational health

Public health practice is concerned principally with the health problems common to population groups rather than to individuals. There is thus a need for regular access to such groups for large-scale epidemiological

[a] Institute of Hygiene and Occupational Health, Medical Academy, Boulevard D. Nestorov 15, Sofia, Bulgaria.

Table 1. Distinctions between mass screening and diagnostic (clinical) examinations[a]

Mass screening examinations	Diagnostic (clinical) examinations
1. Used in population studies	1. Performed on individuals presenting for medical consultation
2. Performed in the absence of medical indications	2. Performed in the presence of medical indications
3. Low-cost and simple	3. Costly and at times highly sophisticated
4. Diagnosis not very reliable	4. Permit a well-grounded diagnosis
5. Results cannot be used to prescribe treatment	5. Results permit therapy to be initiated
6. Subjects investigated can be divided into two groups: (a) likely to be ill, and (b) likely to be free from the disease in question	6. Precisely classify health state

[a] From **Jedrychowski, W.** [*Methods for epidemiological surveys in industrial medicine*]. Moscow, Medicina, 1980.

studies, for early case finding, and for attack on the problems by a mass approach. In occupational medicine, screening and other epidemiological methods are applied to a specific fraction of the general population, namely, subjects exposed to occupational hazards.

For the successful prevention of occupational health hazards, it is necessary to know the quantitative—or at least the semiquantitative— relationship between the magnitude (intensity and duration) of the work factors (chemical, noise, physical load, etc.) and the resulting health effects. This relationship can be successfully estimated by application of screening procedures.

Occupational screening has at least two functions:

(1) to detect the early manifestations of disease in the individual, and

(2) to provide epidemiological data on the effects of exposure to work factors in the specific fraction of the general population, to check the efficiency of standards, or to improve information on the suspected effects of exposure.

Screening in occupational health is aimed at revealing early deviations from health, as related to occupational exposure, and at assessing health risk to the population exposed. Early detection of health impairment is meant to imply the detection of disturbances of homoeostatic and compensatory mechanisms while biochemical, morpho- logical, and functional changes are still reversible.

The prevalence of deviations from health observed in mass screening provides one of the basic arguments in assessing the risk incurred by subjects exposed to occupational factors. The positive impacts of early detection of such deviations include:

— referral of specified subjects for medical supervision, with application of the full range of relevant diagnostic procedures and, where needed, timely initiation of treatment;
— identification of subjects particularly vulnerable to a certain health hazard;
— evaluation of the effectiveness of preventive measures;
— revelation of trends in the health status of groups of workers;
— provision of data for establishing, or revising, occupational exposure limits.

Each of these impacts will contribute to reducing the risk of occupational disease among the exposed population. Because periodic medical check-ups have a similar purpose, these two types of examination have sometimes been confused. What they have in common are the subjects and the purpose of the investigation. They may be regarded as constituent parts of occupational health monitoring aimed at assessing the state of health and preventing occupational disease. The difference lies in the investigation methods employed and their diagnostic potential.

As a rule, the specialists participating in periodic check-ups make use of routine medical examination methodology. This permits the diagnosis of sufficiently manifest forms of illness (whether general or occupation-related). Screening, on the other hand, requires use of one or more diagnostic tests that may be performed by technicians. The interpretation of the results will, of course, need to be performed by a specialist competent in occupational health.

Mass screening tests can be done while workers are at their job or with brief absence from work (usually less than one hour). Periodic medical check-up examinations generally take much more time to carry out. Moreover, they are performed at fixed times (once every one to three years), whereas, because of their low cost and simplicity, screening procedures can be performed with optimum frequency.

Screening is aimed at detecting deviations from health in asymptomatic subjects, i.e., at diagnosing premorbid states. In periodic check-up examinations, on the other hand, a target-oriented search is undertaken for clinical signs characteristic of the occupational disease caused by a specific noxious agent; they are thus aimed at disclosing disease at an early clinical stage. In this context, it may be pointed out that "current practices in occupational health, even in the most highly industrialized countries, seldom meet the goals set forth above. In many instances the major emphasis is on the finding of cases of illness and providing the necessary medical care, and true preventive medicine plays only a minor role, if any, in day to day operations. Indeed, it is the practice of general medicine in industry and not the practice of industrial medicine *per se* that characterizes today's efforts"(*1*).

Inasmuch as periodic check-ups are done using routine medical examination methodology, findings will be considered pathological if any deviation is observed, whether structural (e.g., an enlarged liver) or functional (e.g., cardiac arrhythmia). In epidemiological screening with the aid of a diagnostic test, the risk to health will be assessed by comparing investigation results between the exposed subjects and a control group (subjects with no experience of exposure to the agent concerned). In addition, the exposed group may be broken into two or more subgroups according to exposure levels.

In order to fulfil its intended purpose, a screening test is required—apart from being simple to execute—to be reproducible as well as valid. The reproducibility of the test results will determine to what extent a single screening examination may be relied upon to justify a preliminary diagnosis. The variability of an end-point may be biologically determined, as in the case of urinary hippuric acid level, or may result from shortcomings of, or differences in, the methods employed for its measurement, as in the case of serum cholinesterase activity. There is thus a need for standardizing the techniques used and for establishing a critical (threshold) figure of the selected parameter as measured by the specific procedure.

The validity of a screening test is its most essential characteristic. The choice of a valid test calls for in-depth knowledge of the mechanism of action of each particular noxious agent in the occupational environment. An additional difficulty arises from the fact that, in the overwhelming majority of situations, biological effects develop under low-level and prolonged exposure conditions. To be able to appraise the health risk sustained in such circumstances, clear-cut differentiation has to be made between the normal and the pathological, between health and premorbid state, between physiological adaptation and overloading of the body's compensatory mechanisms.

Selection of an adequate test for screening

Problems relating to adaptation
One major constraint that limits the possibilities of detecting early deviations from health in an occupationally exposed population is the difficulty of distinguishing the physiologically normal from a compensated pathology.

The normal is a natural state of the human body and its vital activities. This state is relatively steady and at the same time flexible, for it reflects the dynamics of environmental change.

In industry, the occupational environment typically contains multiple factors whose qualitative and quantitative characteristics differ appreciably from those encountered outside that environment. Depending upon the intensity and duration of exposure, as well as upon the body's adaptive potential, occupational environmental agents may produce varying degrees of change in functional parameters, which may or may not remain within the limits of physiological fluctuations.

With occupational exposure occurring at low intensity over a long period of time, a gradual transition will take place, spanning the range from health to illness. Under these conditions, detection of early deviations from health is complicated, above all, by the fact that the recorded figure for any functional parameter reflects a combination of two processes: the action of the noxious agent and the counteraction of the organism (adaptation). Hence, delimitation of the latter is only theoretically possible. Such an approach has been used to explore a chemical toxic action in "pure form", i.e., making no allowance for adaptation (2). Relating the theoretical cumulative effect (TCE), as calculated for a given parameter by way of mathematical modelling, to the experimentally observed value (OV) enabled the level of adaptation to be expressed in relative units.

$$\text{Adaptation (relative units)} = \frac{\text{TCE} - \text{OV}}{\text{TCE}}$$

Using the formula above, curves describing the process of adaptation were shown to assume the form of a series of fluctuations that gradually faded away (Fig. 1). Moreover, the general pattern was found to be maintained with various toxic agents over the whole range of doses compatible with life (2, 3). These findings lead to the conclusion that bodily adaptation follows general rules.

The oscillatory nature of adaptation dynamics is conceivably related to the predominance of different defence and adjustment mechanisms and their synchronization. Depending on the type of exposure, various adaptation mechanisms appear to come into play. Thus, adaptation to toxic exposures is accomplished by both general mechanisms (increased rate of regenerative processes, synchronization of physiological and biochemical processes with environmental challenge, etc.) and more or less specific mechanisms (changes in the biotransformation rate of toxic substances).

In the adaptation curve, the portion showing fluctuations that gradually subside characterizes transitional adaptive processes leading to attainment of a steady level (Fig. 1). The portion where this steady level

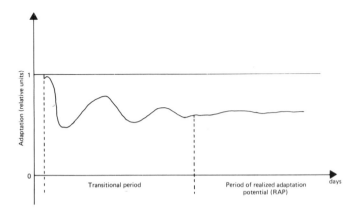

Fig. 1. Typical adaptation curve in repeated exposure
to chemical substances

has been attained corresponds to the successful adaptation of the body to a particular type of exposure, i.e., realization of the body's adaptive potential. A plot of exposure rate versus realized adaptive potential (RAP) assumes the shape of a decreasing exponent. Such a curve is valid for exposures that are compatible with life, i.e., that do not exceed the body's adaptive potential. It follows that adaptation is operative in both maintaining the balance with environmental factors and in the transitional process leading to the establishment of such a balance.

The limits of a body's adaptive potential and the dynamics of the transitional processes determine which of the following types of adaptation to a particular exposure will result:

(1) *steady state adaptation*—manifest in maintenance of steady equilibrium at the expense of structural and functional changes that remain within the range of physiological fluctuations;

(2) *compensation* or *compensatory adaptation*—manifest in recovery from imbalance to attain a qualitatively new and relatively steady state at the expense of structural and functional changes that indicate straining of compensatory mechanisms.

In the above two types of adaptation, RAP is within the limits of physiological fluctuations. Transitional processes, however, remain within the limits of physiological fluctuations for steady state adaptation only, whereas for compensatory adaptation these limits are exceeded.

(3) *Adaptation to high-intensity exposures* is achieved and maintained at the expense of structural and functional changes going beyond the limits of physiological fluctuations. The extent of RAP deviation from the adaptation indicator in relative units is dependent upon the rate, multiplicity, and pattern of exposures to the noxious factor. Any substantial change in the latter characteristics will provoke renewal of transitional processes. Prolonged straining of compensatory mechanisms may result in their impairment (adaptation failure) or "exhaustion", according to the classification of H. Selye. Other possible causes of failure include intercurrent illness, malnutrition, and mental stress of extreme intensity or duration.

Establishment of equilibrium between the body and exposure to hazardous agents in the occupational environment may produce the impression of "habituation" in the subjects concerned. However, it will be seen from the above discussion that apparent wellbeing in "habituation" is attained at the expense of straining the defence and adjustment mechanisms of the body. Actually, all cases of "habituation" fall either into the category of compensatory adaptation or into that of adaptation to high intensity exposures. The former is a case of pseudohomoeostasis; the latter, of heterostasis, i.e., an abnormally high level of equilibrium between the body and the high-intensity noxious exposure.

"Habituation" is known to be frequently associated with increased nonspecific tolerance (4). This condition takes the form of an extension of the quantitative characteristics of balance between the body and the environment; for instance, adaptation to xylenes is associated with increased tolerance to a number of other organic solvents. The condition may develop in compensatory adaptation as well as in adaptation to high-intensity exposures. Increased nonspecific tolerance is a consequence of the fact that the body utilizes but a minute part of its immense reserves of theoretically possible adjustment reactions (5). As a result, an existing compensatory strain on adjustment mechanisms due to exposure to a certain agent will also "work" for other noxious agents that are usually its congeners; a case in point is that of alcoholics who are well known to exhibit increased tolerance to narcotic drugs used in anaesthesiology. Also, increased nonspecific tolerance may develop for factors of a different nature, the so-called "cross-adaptation". Thus, subjects exposed to carbon monoxide typically display increased tolerance to hypoxia, so that they are able, for instance, to work at high altitudes. Conversely, highland dwellers show increased tolerance to carbon monoxide exposure.

Adaptation to high-intensity exposures and compensatory adaptation are conditions of the body that deviate more or less from complete health. Nevertheless, for a time-interval of varying length, these conditions produce no typical features of specific occupational disease. This is the period known as the premorbid state. To diagnose it is the objective of epidemiological screening in occupational medicine. It is most readily demonstrable within the first days of exposure, when fluctuations in functional parameters are relatively marked, as this is the period of transitional adaptive processes. More difficult to diagnose are premorbid conditions once equilibrium between the causative agent and the response to it has become established. However, in both instances the mechanisms involved will need to be well understood in order to be able to select an appropriate diagnostic test.

Problems relating to features of pathogenesis
Because of the great variety of occupational factors influencing health in exposed individuals, chemical agents will be taken as examples to illustrate the possibilities and limitations of procedures for selecting a pathogenetically adequate diagnostic test for screening purposes.

The specific nature of intoxication as a pathological process depends on its chemical etiology and related features of its pathogenesis. In contrast to other types of pathogenesis, intoxication mechanisms permit the differentiation of two basic processes: toxicokinetics and toxico-dynamics. These two processes characterize the interactions occurring between poisons and the organism. Toxicokinetics is concerned with patterns of poison transfer through the body, in other words, with the impact of the body upon the poison. As opposed to this, toxicodynamics deals with the impact of the toxic substance upon the body. Toxico-kinetics is one of the main determinants of toxicodynamics, since the effects of the poison are dependent on the amount and form in which it is available to the target structural elements of the body, the so-called receptors.

Toxic substances damage living organisms by producing disorganiz-ation of structures and impairment of functions at the various hierarchical levels (from biomolecules to the organism as a whole). Very often, when mechanisms of action are discussed in the literature, what is actually examined are toxic effects at various levels of biological organization (mostly the higher levels: organ, system, organism). It should be emphasized that, if primary effects are ignored, both the cause–effect relationship and the sequential development of intoxication are lost sight of. Lately, an increasing number of authors have conceded that, to be able properly to account for interrelations between toxic substances and the body, it is necessary to acquire a knowledge of primary effects, that is, of action at the molecular level. It is at this level that an insight may be gained into the chemical nature of the primary toxic action, which is the "triggering mechanism" for a whole chain of structural/functional changes producing various manifestations. The aggregate of these reactions is the province of toxicodynamics, which in turn, conditions the clinical picture of intoxication. Fig. 2 illustrates schematically the development of a primary toxic effect in a critical organ (6). It may be seen that structural/functional changes at a lower hierarchical level are responsible for changes in structure and function at the next higher hierarchical level.

The choice of an adequate screening test in occupational medicine will depend upon how much is known about the mechanisms underlying development of an adverse effect and about the adjustment response elicited by it. For exposure to chemical hazards, the selection of an adequate diagnostic test may be based on a functional parameter manifesting the primary toxic effect or the direct result of the toxic effect on the critical organ. It is also possible to select a parameter reflecting the functional straining of adjustment mechanisms induced by the primary toxic effect. On the other hand, recording functional changes that occur as a result of the involvement of higher regulatory centres is not appropriate for diagnostic screening because such changes lack specificity; moreover, because of the manner in which they develop (triggered by the exhaustion of the regulatory capacity of the mechanisms concerned) they are unsuitable for early detection of deviations from health in exposed workers.

From what has been said about selection of an adequate test, it is evident that selection is limited, in the first place, by lack of the necessary knowledge of primary effects. Where such information has been gained, the limiting factors may be biological (biopsy necessary for collecting the biological substrate of interest) or technical (complexity of determination). It may be expected that with advances in scientific knowledge, these difficulties will be overcome in the near future, at least in so far as important hazards are concerned. Such optimism is not unfounded; success has lately been achieved in developing diagnostic screening tests for exposure to organophosphorus compounds, lead, and some other chemicals.

At present, the application of valid biological screening procedures is limited, but it is likely that they will increasingly become available for

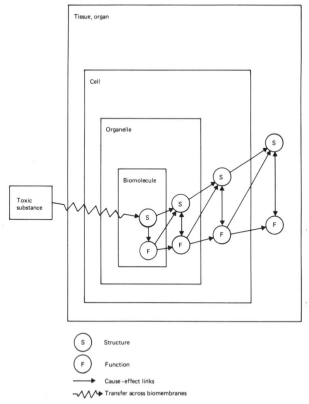

Fig. 2. Development of a primary toxic effect in a critical organ

routine practical use. Screening tests have a useful role to play within the health surveillance system, providing they are valid procedures for detecting biological changes at an early stage, so that risks to workers are minimized.

Toxicokinetic indicators for screening purposes

Unfortunately, knowledge of primary toxic effects and of critical cells and critical organs will not necessarily ensure the selection of a functional parameter suitable for detecting early changes in the health of exposed subjects. In the case of mercury, for instance, the primary toxic effect is known to consist of blocking SH-groups in the biomolecules of cell membranes. The critical organ for mercury vapour is the brain; and for inorganic mercury compounds, the kidneys. The nephrotoxic action of mercury is due to inhibition of succinyl dehydrogenase (7) and damage to epithelial cells in proximal renal tubules. In spite of this, no diagnostic criterion has so far been devised for the early demonstration of toxic effects (8). In the absence of functional parameters suitable for detecting early deviations from health—and even when such functional parameters are available—toxicokinetic indicators in the form of exposure tests may serve as additional parameters for screening purposes.

Exposure tests consist in determining chemical agents or their metabolites in biological media (blood and urine) obtained from exposed subjects. The theoretical basis for exposure tests and their practical application have been described in detail in the literature (9–13), and they will be considered here only as diagnostic screening tests.

It is well known that measurement of an airborne chemical at the site of work is indicative of its concentration in air only at the specific time and place. More informative are the data obtained by continuous monitoring of air concentrations or through the use of personal dosimetry. However, none of these techniques will ensure a good estimate of the amount actually absorbed by the body. This is particularly true where routes of entry other than inhalation are available (dermal, oral). This drawback may be circumvented by using exposure tests that are indicative of overall exposure, irrespective of the routes of entry.

Quantitative exposure tests have so far been developed and introduced into practice for about 50 chemicals. Some of them are listed in Table 2.

Table 2. Quantitative exposure tests for selected hazardous chemicals[a]

Chemical to which exposed	Substance determined	Limit value	Medium
Aniline	p-aminophenol	10 mg/litre	urine
Benzene	phenol	80 mg/litre	urine
Nitrobenzene	nitrophenol	5 mg/litre	urine
Xylene	methylhippuric acid	300 mg/litre	urine
Styrene	mandelic acid	104 mg/litre	urine
	phenylglyoxalic acid	17.2 mg/litre	urine
Trichlorethylene	trichloroethanol	100 mg/litre	urine
	trichloroacetic acid	50 mg/litre	urine
Toluene	hippuric acid	1000 mg/litre	urine
Perchlorethylene	trichloroethanol	30 mg/litre	urine
	trichloroacetic acid	15 mg/litre	urine
Carbon disulfide	dithiocarbamates	500 μg/litre	urine
Lead	lead	35 μg/litre	blood
		70 μg/litre	urine
Mercury	mercury	10 μg/litre	urine
Cadmium	cadmium	50 μg/litre	urine

[a] From Sedivec, V. Unpublished data, 1978.

In assessing the health risk, the results obtained from quantitative exposure tests are compared with the biological limits (14). Theoretically, biological limits correspond to an exposure level that will produce no adverse effect on the health of exposed subjects. In practice, they are concentrations of chemicals or their metabolites that correlate with the occupational exposure limits in air at workplaces. Higher levels indicated by exposure tests are referred to by some authors as "therapeutic levels". They are associated with the appearance of some signs and symptoms of intoxication and are indications for resorting to therapeutic measures.

From the definition of biological limits, it is evident that they will be dependent on occupational exposure limits in air as established by regulatory authorities in individual countries. For instance, the biological limit for trichloroethanol in urine as a test for exposure to trichloroethylene varies from 50 mg/litre (14) to 300 mg/litre (9), with a number of authors citing intermediate figures. The disparity is even greater for biological limits for phenol, toluene, and furfural. The interpretation of tests is particularly difficult at low-level exposures, where most of the amount measured represents the physiological level. The normal levels of these metabolites in individual subjects are dependent on their diet and are subject to diurnal and seasonal variations. It is thus necessary to investigate a sufficient number of control subjects in order to determine the normal mean and its range of variation.

Toxic substances or their metabolites that do not normally appear in the blood and urine likewise show considerable fluctuations in level. Their concentrations vary not only in the course of exposure but also after its discontinuation. Hence, it is essential in interpreting results to know the time of sampling (beginning, midpoint, or end of the working shift or working week).

The accuracy of exposure tests is also dependent on a number of additional factors, such as:

— exposure level;
— rate of metabolism and elimination;
— mode of sample storage and transportation;
— analytical methods used;
— terms in which the results are expressed.

In order to reduce deviation in the variations in quantitative exposure tests, corrections are introduced according to the relative density of the urine, its creatinine content, etc.

Use of exposure tests as diagnostic tools for screening exposed subjects is limited by the fact that in many cases laborious chemical analyses are involved. Moreover, with a number of substances it is impossible, for various reasons, to develop quantitative exposure tests. This applies, for example, to chemicals that are highly reactive (chlorine, sulfur dioxide, hydrogen chloride, hydrogen fluoride) or have a low solubility (silicon dioxide, ferrous oxide, titanium dioxide).

In interpreting results obtained from exposure tests, it should be remembered that the biological limits proposed by a particular author are based on his personal experience and reflect the regulations in force in his country. Moreover, even when it is realistic, a biological limit is not to be regarded as a sharp transition from harmless to harmful levels of exposure. In all cases, it will represent an average of different individual values. Consequently, when interpreting the results of screening tests it will be necessary to know not only the average value of the biological limit but also some of its statistical characteristics (90% confidence limits, standard deviation, etc.). Much more reliable and diagnostically significant are the therapeutic limits of exposure tests, as recorded at high and intermediate levels of exposure.

Whenever a biological indicator of exposure is used as a diagnostic test for screening purposes, comprehensive assessment of the health risk should be made. It is particularly useful to relate the results to those of a current diagnostic screening test based on the toxicodynamics of the chemical in question.

Examples of epidemiological screening

Assessment of the health risk of exposure to lead provides a good example of the use of results of epidemiological screening, since its toxicokinetics and toxicodynamics have been thoroughly studied and a wealth of data is available from examinations of subjects exposed to it.

The critical organ in exposure to inorganic lead compounds is blood-forming bone marrow. The critical cells are those of the erythropoietic series, and the primary toxic effect is inhibition of delta-aminolevulinate dehydratase (ALAD) (E.C. 4.2.I.24) and of the complex processes involved in binding iron to protoporphyrin IX, with resulting impairment of haeme synthesis. Manifestations include decreased ALAD activity, increased "free erythrocyte porphyrins" (FEP), and increased urinary delta-aminolevulinic acid (ALA-U) and coproporphyrin III (CP-U) (15).

Adjustment of the organism to the primary toxic effect and to changes directly related to it is expressed in activation of erythropoiesis. This is evidenced by increased peripheral reticulocyte counts.

At higher levels of exposure, lead produces blocking of sulfhydryl groups, inhibition of a number of enzymes, and impairment of the endogenous synthesis of pyridine nucleotides from chinolinic acid (16). These effects likewise develop at the molecular level, but, in contrast to the primary toxic effect, they first appear at higher exposures. Most sensitive to them are the cells of the central and peripheral nervous systems. This explains the relatively early appearance of peripheral nerve conduction impairment and psychoneurological symptoms.

There is an abundance of data from examinations of exposed subjects indicating that various manifestations of adverse action correlate with lead levels in the blood (Pb-B). Despite some existing disagreements between a number of authors, it may be assumed that Pb-B levels have already been specified above which various potential effects on the body can be observed. As may be seen from the data presented in Table 3,[a] the indicators most sensitive to the adverse effects of lead are those reflecting: (1) the primary toxic effect and the response it elicited within the critical cells (changes in ALAD activity and in FEP, ALA-U, and CP-U content) and (2) impairment of nerve conduction.

By the use of these indicators, it was also possible to determine the highest percentage of individuals who responded with an effect exceeding the cut-off point for a given exposure (dose)–response relationship (Table 4).[a]

Examination of the data shown in Tables 3 and 4 indicates that for epidemiological screening of subjects exposed to lead, determination of the following indicators may constitute an adequate diagnostic test: ALAD activity, FEP, ALA-U, and CP-U levels, nerve conduction velocity, and blood lead content. Screening will be more effective when Pb-B measurement is combined with determination of one of the toxicodynamic indicators. The choice should be made according to ease of interpreting the results in making a health risk assessment. For instance, the CP-U level being inferior to the ALA-U level in specificity and their sensitivities being about equal, preference should be given to determination of the ALA-U level. Absence of a definite dose–effect relationship between exposure level and the degree of impairment of nerve conduction makes this determination difficult to interpret.

In assessing the health risk to exposed subjects, use may be made of

Table 3. Blood lead levels above which various potential effects on the body may be recorded

Potential effect	Blood lead level (μg/litre)
ALAD inhibition	
above 40%	150–200
above 70%	250–300
FEP elevation	200–250 (females)
	250–300 (males)
ALA-U elevation	
above 5 mg/litre	300–400
above 10 mg/litre	400–500
CP-U elevation	400
Reticulocytosis	600–700
Haemoglobin decrease	600–700
Anaemia	700–800
Other biochemical parameters	400–600
(aminoaciduria, inhibition of enzymes,	
dysproteinaemia, etc.)	
Nerve conduction impairment	400
First subjective complaints	500
Psychoneurological symptoms	500
Gastrointestinal symptoms	800
Encephalopathy	1500
Lead colic	1500

[a] Tables 3 and 4 have been compiled using data from the literature (8, 16, 17).

Table 5. This table has been prepared by drawing on data from the literature, on adverse effects of lead. Allowance has been made for adjustment responses of the body and the possibility of spurious homoeostasis (compensatory adaptation). Moreover, the clearly defined dose–effect and dose–response relationships for all indicators reflecting the primary toxic effect and the responses elicited within critical cells (impairment of haem synthesis) are sufficient reason to assume that deviations exceeding the normal values by 25–30% are an indirect sign of overloading of compensatory mechanisms.

In Table 5, the intra-group subdivision by sex is based on the higher sensitivity of females, as confirmed in numerous investigations. In addition, the adoption of a limit of 300 μg/litre for Pb-B level is intended to prevent any adverse effect on the fetus (more susceptible to the action of lead) in females of reproductive age.

The figures shown in Table 5 for the various indicators (FEP in particular) vary considerably with the measurement methods employed.

Table 4. Dose–response relationship between blood lead levels and percentage of subjects with effect exceeding the cut-off point

Effect	Blood lead level (μg/litre)				
	< 300	300–399	400–499	500–599	600–699
ALAD inhibition					
above 40%	60%	90%	100%	100%	100%
above 70%	10%	20%	60%	80%	100%
FEP elevation					
above 826 μg/litre (females)	50%	90%	100%	100%	100%
above 708 μg/litre (males)	15%	40%	60%	100%	100%
ALA-U elevation					
above 5.44 mg/g creatinine (females)	5%	40%	50%	100%	100%
above 4.76 mg/g creatinine (males)	5%	15%	40%	50%	88%
CP-U elevation (above 80 μg/litre)					
females	—	—	50%	100%	100%
males	—	—	20%	50%	—
Nerve conduction velocity below 50 m/s	0%	0%	27%	32%	42%
Visual intelligence and visual-motor functions	0%	0%	0%	20–30%	20–30%

— = No accurate data available

Table 5. Criteria for assessing health risk in epidemiological screening of subjects exposed to lead

Diagnostic indicator[a]	Non-exposed subjects (mean value of indicator)	Risk groups in exposed subjects					
		Group I		Group II		Group III	
		females	males	females	males	females	males
Pb-B level (μg/litre)	150	<300	<400	300–500	400–600	>500	>600
ALA-U level (mg/litre)	4	<6	<6	6–10	6–10	>10	>10
ALAD activity (% inhibition)	100	<40	<50	40–70	50–80	>70	>80
FEP level (μg/litre)	600	<900	<900	900–1500	900–1500	>1500	>1500

[a] Absence of a distinct exposure–effect relationship for impairment of nerve conduction prohibits grading for this indicator.

This makes it necessary to run control tests in nonexposed subjects concurrently with the screening.

According to the degree of risk incurred, exposed subjects may be divided into three groups.

Group I ("the no risk level"). The exposed subjects are in a state of steady equilibrium (homoeostasis). The biological effects of lead are compensated by structural/functional changes within the range of physiological fluctuations (physiological adaptation).

Group II ("potential risk"). The exposed subjects are in a relatively steady state of equilibrium with the environment (compensatory adaptation), which is maintained at the expense of overloading of compensatory mechanisms (for instance, enhanced excretion of lead). For each indicator of lead exposure, an over 40% increase from the upper reference limit value is recorded. A premorbid state is present (no clinical manifestations). The existence of potential risk is confirmed by the recording, in 10–20% of the exposed subjects, impairment of nerve conduction as well as of visual intelligence and visual-motor functions. These are manifestations of disturbed equilibrium owing to reduction in adaptive capacity (illness, mental stress, etc.)

Group III ("actual risk"). The exposed subjects are in a state of unsteady equilibrium with the environment, which is maintained at the expense of structural and functional changes exceeding the range of physiological fluctuations (heterostasis or adaptation in pathology). Biological indicators of lead exposure deviate from normal values by more than 80%. Clinical manifestations are observed (subjective

111

symptoms, psychoneurological signs, anaemia, etc.). Impairment of nerve conduction velocity, visual intelligence, and visual-motor functions is demonstrable in more than 20–30 % of the exposed subjects. This labile balance maintained by considerable straining of compensatory mechanisms may be disturbed by intercurrent illness, malnutrition, mental stress, etc. (adaptation failure). Straining of compensatory mechanisms, in turn, causes reduction in tolerance of the organism to pathogenic factors, thus increasing the likelihood of development of "para-occupational morbidity".

The probability of occurrence of adaptation failure (overt lead intoxication) is abruptly increased by exposures resulting in Pb-B levels exceeding 700 μg/litre.

It should be pointed out that the Pb-B values listed in Table 5 to distinguish risk groups are not to be regarded as constituting distinct boundary lines. They are averages of individual values that vary within a more or less wide range. Because of this it is essential in interpreting the results of epidemiological screening to relate the levels of environmental (lead in air) and biological (Pb-B) indications of exposure to the figures obtained for the examined indicator. Such comparisons help to identify subjects with increased sensitivity (susceptibility) as conditioned by genetic factors or as a result of deviations from the state of health (preceding or accompanying illness), etc. In such cases, it is mandatory to discontinue exposures and to undertake a comprehensive diagnostic examination of the subject. Such examinations are indicated for all subjects falling into risk group III. Subjects in risk group II are placed under medical surveillance.

Screening and interpretation of results are much easier to perform if the deviation from health and the degree of risk can be studied using a single indicator. Such a pathognomic indicator of adverse effects is cholinesterase (ChE) for exposure to organophosphorus compounds (OPCs).

The primary toxic effect of OPCs is inhibition of ChE. OPCs compete with acetylcholine (ACh) for the active centre in ChE. Their interaction with the enzyme consists in phosphorylation of the hydroxyl group of serine, in analogy to acetylation of the esterase moiety by acetylcholine. However, in contrast to deacetylation, which occurs in microseconds, restoring the initial structure of the enzyme, dephosphorylation proceeds at an extremely slow rate, the speed being that of spontaneous hydrolysis (18). The critical organ for OPC exposure is the nervous system; critical cells are the neurons or, more exactly, nerve synapses. The critical effect is ChE inhibition.

With inhibition of ChE, ACh accumulates and exerts a pathologically enhanced action, resembling that of muscarine and nicotine, upon effector organs. Spontaneous recovery of enzyme activity occurs very slowly, at a rate of about 1.5 % per day. Affected in parallel with erythrocyte ChE (true ChE), is serum ChE (pseudocholinesterase).

The extent of ChE inhibition by OPCs is directly related to the intensity and duration of exposure. By virtue of this direct relationship, determination of ChE activity is a conclusive diagnostic test in the epidemiological screening of subjects exposed to OPCs.

Inactivation by up to 25–30 % of the normal value is considered to be within the range of biological variability (19, 20). A 25–50 % drop in ChE activity is indicative of the premorbid state. In this case, exposed subjects have no clinical manifestations; they are in relatively steady equilibrium maintained by straining of compensatory mechanisms (compensatory adaptation). As opposed to the situation with lead, the mechanisms instrumental in compensating adverse effects of OPCs have been elucidated to a considerable extent.They have been found to comprise both toxicokinetic (changes in absorption, biotransformation, and excretion) and toxicodynamic adjustments (enhanced ChE synthesis, increased tolerance of newly formed cells) (21).

With more than 50 % inactivation of ChE, clinical manifestations of OPC toxic action will be recorded. For inhibition ranging from 50 % to 80 %, symptoms of poisoning may be mild (general weakness, anorexia, nausea, vomiting, abdominal pain, diarrhoea, respiratory discomfort, slight ataxia). In a few subjects, labile balance may be temporarily established at the expense of considerable straining of compensatory mechanisms (heterostasis, adaptation in pathology).

If inhibition attains the range of 80–90 %, intoxication of intermediate severity will be observed (marked signs and symptoms of affection of the central and autonomic nervous systems, cardiovascular system, and gastrointestinal tract). It is typical of this degree of intoxication that despite muscular weakness the ability to move is preserved.

Inactivation of ChE above 90 % is associated with severe intoxication. Consciousness is blurred and the patients are unable to move. Convulsions with clonic and tonic contractions appear.

In contrast to epidemiological screening of subjects exposed to lead, in the case of OPC intoxication both evaluation of results and indication of measures for persons with recorded deviations are clearly defined. The rapid course of OPC intoxication dictates immediate action.

Subjects showing ChE inactivation within the range 25–50 % should be temporarily removed from work. When this is done, ChE activity is relatively promptly restored.

For subjects with 50–80 % ChE inactivation, in addition to cessation of exposure, antidote therapy is indicated, with hospitalization where needed.

Mixed function oxidases in epidemiological screening

Upon uptake by the body, substances that are foreign to it are largely metabolized to less toxic products. This biotransformation of xenobiotics is primarily accomplished within the endoplasmic reticulum of the hepatocytes, the site of the hydroxylating enzyme system (HES).

Schematically, oxidation involving the HES may be represented as follows:

$$S + \text{cytochrome } P - 450 + O_2 \rightarrow S - OH + \text{cytochrome } P - 450 \text{ (oxid.)} + H_2O$$

Subsequently, cytochrome P–450 (oxid.) reduces to cytochrome P–450 with the aid of the remainder of the HES components.

In this scheme, the compound being metabolized (substrate "S") may be any xenobiotic: pesticides, organic solvents, drugs, ethanol, and many other substances foreign to the body; it may also be one of certain endogenous substrates: saturated or unsaturated fatty acids, cholesterol, steroid hormones, bile acids, prostaglandins, etc.

Absence of substrate specificity, on the one hand, and the diversity of reactions involving the HES, on the other, have led to this system being termed mixed function oxidases (MFOs).

One typical feature of MFOs is the possibility of their induction. It has been demonstrated that over 200 chemical compounds increase MFO activity. They are characteristically lipophilic and have a relatively slow rate of biotransformation. Notable inducers are xenobiotics differing in structure: barbiturates, organochlorine pesticides, steroids, halogenated benzenes, alcohols, etc.

The number of known MFO inhibitors is much smaller. Among them are: diethylaminoethyldiphenylpropyl acetate (or SKF 525 A), ethionine cyclohexamide, tetrachloromethane, tetramethylthiuram disulfide (or TMTD), carbon disulfide, actinomycin D, some steroids, and allyl alcohol. For most of the compounds listed, the mechanism underlying inhibitory action involves interference with protein synthesis.

The fact that MFOs are implicated in mammalian adjustment to environmental chemical agents has been emphasized by many authors (21, 22). Our experience with repeated exposures to organophosphorus and organochlorine pesticides, carbon tetrachloride, carbon disulfide, TMTD, etc., has indicated that changes in MFO activity associated, in some cases, with rearrangement of biotransformation pathways represent one of the principal mechanisms of adaptation to xenobiotics (23). This is true, in particular, for organophosphorus pesticides with P=S bonds (thiophosphates).

Evidence obtained by a number of authors in animal studies has shown that species, age, sex, and individual differences in sensitivity to xenobiotic exposure are related to differences in MFO activity (24, 25). MFOs appear to be implicated, to some extent, in the teratogenic, gonadotropic, and carcinogenic action of chemicals (26, 27). Measurements of MFO activity in single and in combined exposure to chemical agents has been useful in disclosing mechanisms of potentiation and antagonism.

It is evident from the above that MFO activity determinations may contribute to solving a wide range of current problems in both the theory and the practice of toxicology and occupational medicine. Of special interest are opportunities for diagnosing premorbid states and such opportunities are beyond doubt in the case of marked MFO inductors or inhibitors, for instance, organochlorine pesticides, carbon disulfide, dithiocarbamates, polychlorobiphenyls, etc.

Advances in the biochemistry of endoplasmic reticulum and their applications to occupational health practice are dependent on the ability to determine MFO activity. Such a method was first proposed by Brodi & Axelrod in 1950; the authors administered aminophenazone and subsequently measured its metabolites in urine. This method does not satisfy contemporary requirements because of poor reproducibility, entailing multiple repeat metabolite extractions. Techniques proposed more recently are based on oral administration of aminophenazone, phenylbutazone, or phenazone, followed by observation of the time course of blood level changes. Common drawbacks of these techniques are poor accuracy and precision owing to repeated extractions from blood as well as the necessity of venipunctures (at least two), which makes them inappropriate for epidemiological screening purposes.

Prompted by these considerations, we have developed a method for estimating MFO activity in humans and animals, in which oral administration of aminophenazone is followed by direct colorimetric determination of its major metabolites in urine (28).

Examples of MFO activity estimation in epidemiological screening

We have used the MFO activity estimation test in epidemiological screening of subjects exposed to chemical hazards (carbon disulfide, polycyclic aromatic hydrocarbons, vinyl chloride, etc.). Results will be presented here from an investigation covering 70 workers employed in the synthetic fibre industry. Based on annual mean concentrations of carbon disulfide in air at workplaces, these subjects were grouped as follows: Group I, $1-7\,mg/m^3$; Group II, $9-18\,mg/m^3$; Group III, $30-60\,mg/m^3$; Group IV, above $60\,mg/m^3$. The majority of the subjects were in the age range 25–45 years; the durations of occupational exposure did not exceed 5 years in most cases. Concurrently investigated were 30 age-matched subjects with no exposure to carbon disulfide.

In the treatment of the statistical data, analysis of variance was used. The significance of differences from controls was tested by the t-test.

The extent of MFO inactivation showed a distinct relationship to atmospheric concentrations of carbon disulfide at the workplace (Table 6). In subjects exposed to $6\,mg/m^3$, urinary 4–aminoantipyrine (4–AAP) excretion was no different from that in controls. The second group ($9-18\,mg/m^3$) showed a 17% decrease in 4–AAP excretion, the difference being, however, insignificant ($P > 0.05$). With exposures in the range $30-60\,mg/m^3$ (Group III) and above $60\,mg/m^3$ (Group IV), the figures dropped by 38% and 85%, respectively ($P < 0.001$).

The data were analysed to determine the relation between exposure level and the prevalence of deviations from normal values ($\overline{X} \pm 2S$) as ascertained in the control group (dose–response relationship). It may be seen from the figures given in Table 6 that the number of subjects with low urinary 4–AAP values increased with rise in exposure to attain 80% in Group IV. On the other hand, the proportion of subjects with

Table 6. Dost–effect and dose–response relationships between atmospheric concentrations of carbon disulfide at the workplace and MFO activity

Exposure level (mg/m³)	MFO activity, judged by urinary 4-AAP as percentage of administered aminophenazone (mg)	Prevalence of urinary 4-AAP levels outside the normal range (% of subjects)	
		< 0.15%	> 0.80%
1–7 (n = 19)	0.487 ± 0.071	5	21
9–18 (n = 17)	0.386 ± 0.077	18	6
30–60 (n = 22)	0.289 ± 0.043	27	0
> 60 (n = 12)[a]	0.092 ± 0.024	80	0
Nonexposed	0.446 ± 0.020	0	0

[a] Only 12 workers were screened because of the small number of subjects in this category of exposure.

elevation of urinary 4–AAP above 0.80% decreased from 21% for Group I to 6% for Group II, and no such subjects were found at the higher exposure levels. These isolated cases of elevation above 0.80% were apparently a manifestation of the phenomenon of "overadaptation" during the period characterized by transitional processes (as recorded within the first months of exposure experience), which precedes attainment of balance, the RAP level (see Fig. 1).

The significant extent of MFO inactivation with exposures to carbon disulfide concentrations in the range 30–60 mg/m³ and the frequency of urinary 4–AAP values below 0.15% occurring even at > 10 mg/m³, strongly suggest that this is a case of compensatory adaptation.

For exposures to levels above 60 mg/m³, marked MFO inactivation and the large proportion of subjects with urinary 4–AAP levels below 0.15% indicate the occurrence of adaptation in pathology (heterostasis). Comprehensive diagnostic examination of these subjects showed changes in parameters of lipid metabolism (elevation of serum cholesterol, total lipids, triglycerides, and non-esterified fatty acids). Most of the subjects had complaints such as headache, anorexia, and intolerance to alcohol.

These screening results indicate that atmospheric carbon disulfide concentrations exceeding 10 mg/m³ are not harmless to the organism, thus pointing to a need for updating current exposure limits.

Screening undertaken in subjects exposed to polycyclic aromatic hydrocarbons showed that MFO changes in the opposite direction (induction) likewise have a distinct diagnostic value.

For exposure to vinyl chloride, epidemiological screening revealed that this chemical, in contrast to other carcinogens, is an inhibitor of MFOs. The results of the MFO test were consistent with the results of comprehensive clinical investigations and showed MFO activity estimation to be a promising diagnostic test for screening subjects exposed to vinyl chloride.

References

1. WHO Technical Report Series, No. 535, 1973 (*Environmental and health monitoring in occupational health*. Report of a WHO Expert Committee), p. 9.
2. **Popov, T.** [*Biological and mathematical modelling of intoxication*]. D.Sc. Thesis, Institute of Toxicology, Kiev, 1977 (in Russian).
3. **Glushkov, V. M., et al.** [*Mathematical modelling of body response to repeated chemical exposures*]. Kiev, Institute of Cybernetics, 1978 (in Russian).
4. **Lyublina, E. I. et al.** [*Adaptation to industrial poisons as a phase of intoxication*]. Leningrad, Medicina, 1971 (in Russian).
5. **Hochachka, P. W. & Somero, G. H.** *Strategies of biochemical adaptation*. New York, Holt, Rinehart & Winston, 1973.
6. **Popov, T. & Zapryanov, Z.** [*Principles of toxicodynamics*]. In: Kaloyanova, F., ed. [*Hygienic toxicology*]. Sofia, Medicina i Fizkultura, 1981, General Part, pp. 77–105 (in Bulgarian).
7. **De Bruin, A.** *Biochemical toxicology of environmental agents*. Amsterdam, Elsevier, 1976.
8. WHO Technical Report Series, No. 647, 1980 (*Recommended health-based limits in occupational exposure to heavy metals*. Report of a WHO Study Group).
9. **Elkins, H. B.** Excretory and biologic threshold limits. *American Industrial Hygiene Association journal*, **28**: 305–314 (1967).
10. **Piotrowski, J.** [*The application of metabolic and excretion kinetics to problems of industrial toxicology*]. Moscow, Medicina, 1971 (in Russian, translation from English).
11. **Immamura, T. & Ikeda, M.** Lower fiducial limit of urinary metabolite level as an index of excessive exposure to industrial chemicals. *British journal of industrial medicine*, **30**: 28 (1973).
12. **Gehring, P. J. et al.** Pharmacokinetic studies in evaluation of the toxicological and environmental hazard of chemicals. In: *Advances of modern toxicology*. Washington, Hemisphere Publishing Corporation, 1976.
13. **Spasovski, M. & Benchev, I.** [*Biologic maximum permissible concentrations*]. In: Kaloyanova, F., ed. [*Hygienic toxicology*]. Sofia, Medicina i Fizkultura, 1981, General Part, pp. 243–252 (in Bulgarian).
14. **Vigliani, E. C.** The so-called "maximum allowable biological concentrations". *Pure and applied chemistry*, **3**: 285 (1961).
15. **Zapryanov, Z. & Popov, T.** [*Metals and metalloids.*] In: Kaloyanova, F. ed. [*Hygienic toxicology*]. Sofia, Medicina i Fizkultura, 1982, Special Part, pp. 7–46 (in Bulgarian).
16. **Zapryanov, Z.** [*Changes in tryptophan metabolism under lead exposure and their importance for devising new exposure tests*]. Thesis, Sofia, 1978 (in Bulgarian).
17. **Zielhuis, R. L.** Dose–response relationships for inorganic lead. *International archives of occupational health*, **35**: 1–35 (1975).
18. **Kagan, Y. S.** [*Toxicology of organophosphorus pesticides*]. Moscow, Medicina, 1977 (in Russian).
19. **Kaloyanova-Simeonova, F.** [*Pesticides, toxic action and prevention*]. Sofia, Bulgarian Academy of Science, 1977 (in Bulgarian).
20. **Namba, T.** Cholinesterase inhibition by organophosphorus compounds and its clinical effects. *Bulletin of the World Health Organization*, **44**: 289–307 (1971).
21. **Netter, K. & Siedel, G.** An adaptively stimulated O-demethylation system in rat liver microsomes and its kinetic properties. *Journal of pharmacology and experimental therapeutics*, **146**: 61–65 (1964).
22. **Kappas, A. & Alvarez, A.** [Transformation of xenobiotics in the liver]. In: [*Molecules and cells*]. Moscow, Mir, 1977, No. 6, pp. 287–303 (in Russian, translation from English).

117

23. **Popov, T. & Zadorozhnaya, T. D.** [Cadence of structural/functional changes in hepatocytes in exposure to pesticides]. *Byulleten' eksperimental'noi biologii i mediciny,* **83** (3): 273–275 (1977) (in Russian).

24. **Sell, J. & Danson, K.** Changes in the activities of hepatic microsomal enzymes caused by DDT and dieldrin. *Federation proceedings,* **32**: 2003–2009 (1973).

25. **Chadwick, R. et al.** The effect of age and long-term low-level DDT exposure on the response to enzyme induction in the rat. *Toxicology and applied pharmacology,* **31**: 469–480 (1975).

26. **Fahim, F. et al.** Induced alteration in biologic activity of estrogen by DDT. *American journal of obstetrics and gynecology,* **108**: 1063–1067 (1970).

27. **Gelboin, H.** Studies on the mechanism of microsomal hydroxylase induction and its role in carcinogen action. *Revue canadienne de biologie,* **3**: 39–60 (1972).

28. **Popov, T. & Leonenko,** O. [Method for assessing the activity of liver oxidases]. *Gigiena i sanitariya,* No. 9, 56–59 (1977) (in Russian).

7

Descriptive epidemiology

J. Indulski[a]

The major health problems in the economically developed countries are morbidity, invalidity, and death from chronic noninfectious diseases, which include diseases of the cardiovascular system, malignant tumours, mental disorders, and nonspecific diseases of the respiratory tract. Also important are accidental injuries. For the rational prevention and control of these diseases and injuries the first requirement is to specify priorities. For this purpose, it is necessary to study the epidemiology of these diseases in particular populations using descriptive, analytical, and experimental methods (*1, 2*). The knowledge so gained also provides a basis for the development of appropriate strategies.

The major tasks of epidemiology are evaluation of the health status of the population, elucidation of the etiology of diseases, identification of hazards, and the formulation of principles for the effective prevention and control of these diseases and hazards (*3*). To achieve these aims, two approaches may be used:

(*a*) description of the frequency of occurrence of health phenomena, e.g., morbidity, sickness, absenteeism, invalidity, and mortality, in the general population and its subgroups (descriptive epidemiology);

(*b*) comparisons of the data collected among different populations and in different periods leading up to analytical epidemiology, i.e., the formulation of epidemiological hypotheses and the testing of these hypotheses in specifically designed studies (experimental and analytical epidemiology). Fig. 1 presents the relationships between the methods of descriptive and analytical epidemiology.

Descriptive epidemiology is concerned with the collection of information on the health status of a population and its changes with time, in order to provide a basis for organizational action, the planning and management of health care and the formulation of hypotheses on cause–effect relationships (*4–7*). The occurrence (incidence and prevalence) of diseases in a population is described in relation to certain

---

[a] Institute of Occupational Medicine, Lodz, Poland.

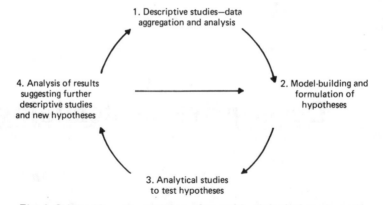

Fig. 1. Schematic representation of an epidemiological study cycle

Reproduced from: Mausner, J.S. & Bahn, A.K. *Epidemiology: an introductory text.* Philadelphia, Saunders, 1974.

personal characteristics, such as sex, age, education, occupation, socioeconomic status, and characteristics of the workplace, as well as seasonal, cyclic, and long-term changes.

Descriptive epidemiology is of great importance in occupational health. Without its aid, it would be impossible to set up effective programmes of health protection for the working population (8). Naturally, descriptive studies may pose some difficulties, such as those due to changes in work processes and changes in the population resulting from high labour turnover. Nevertheless, these difficulties do not diminish the value of such studies, if the findings obtained are subjected to due criticism and appropriate caution is observed in drawing conclusions from them.

Measurement of health phenomena

Epidemiological description of various health phenomena in a population being investigated most frequently covers negative phenomena: morbidity, deaths, invalidity, etc. Attempts to include positive measures of a population's health status are less well developed but should be regarded as supplementing the negative measures. Examples of positive measures could include life expectancy, good nutrition (as expressed, for example, by the ratio between body weight and height), immunity to infectious diseases, good results of ergometric tests, and appropriate level of protein in the blood.

Expressing single phenomena in absolute numbers is not always sufficient, although commonly applied in making simple comparisons between different populations or in the same population at different periods of time, since such calculations are essential for evaluating the actual intensity of health phenomena or in planning the preventive and treatment procedures within the health service. The occurrence of five pneumoconiosis cases in a mine employing 500 miners exposed to coal dust has a different significance from the occurrence of five cases in another similar mine with 1500 miners. Clearly, if the aim of a study is to establish whether the miners in a given mine are more or less likely to be afflicted with pneumoconiosis than those in another mine, the use of simple numbers of cases is inadequate. Instead, the frequency of occurrence of the disease must be expressed in terms of relative numbers or rates.

For the purposes of descriptive epidemiology, it is also necessary to specify the time at which the cases of disease or death were observed. The general formula for calculating rates describing the frequency of health phenomena in a population at an exactly specified time (year, month, day) is:

$$\frac{\text{number of observed phenomena}}{\text{size of population likely to be affected by those phenomena}} \times K$$

where K is a constant indicating the size of the population to which the rate applies (usually 100, 1000 or 10 000, but more generally 10^n).

It is essential that there should be a close correspondence between the numerator and the denominator. This becomes clear when attempts are made to construct so-called specific rates, i.e., rates for the frequency of occurrence of the phenomena concerned in strictly defined groups, classified by age, occupation, residence, etc. Thus, the number of cases in a particular age group should be related to the population fraction in the same age group. The primary aim of descriptive epidemiology is to compare the frequency of occurrence of health phenomena in different populations at a given point in time or to investigate changes in frequency in the same population over a given time period. This is traditionally done by calculating the ratios of the rates, but differences between rates are also of interest. In calculating the ratios, one of the rates is taken as a reference and assigned the value unity or 100, the other rates being then related to it. A distinction must be made between new events of disease over a period of time, i.e., incidence rates, and measures of prevailing morbidity at a particular point in time, i.e., prevalence rates (see below).

Negative health measures

As has been mentioned, descriptive epidemiological studies deal mostly with negative health phenomena, i.e., diseases and their effects, such as incapacity for work, invalidity, or death. The most useful measures describing the frequency of diseases are incidence and prevalence (whether period or point) rates.

$$\text{incidence rate} = \frac{\text{total of newly recorded cases of a disease during a given time}}{\text{average size of the population likely to be affected by that disease}} \times K$$

(i.e., cases per population per year, month, or other period)

$$\text{prevalence rate} = \frac{\text{total of cases of a disease at a given time}}{\text{average size of the population likely to be affected by that disease}} \times K$$

A prevalence rate expresses the frequency of disease either at a given point in time (point prevalence) or during a defined period (period prevalence).

If the incidence and duration of a disease remain constant at consecutive periods of time it means that the disease is stable in a population. Thus, the prevalence level is affected not only by the incidence but also by the effectiveness of treatment procedures expressed by shorter or longer duration of a disease or increased or decreased mortality rate.

In comparative studies of the incidence of particular diseases in different populations or in long-term studies, it is advisable to calculate rates standardized for age, sex, occupational exposure, or other characteristics, in addition to the rates mentioned above.

In evaluating the health status of a working population, information on absences due to sickness may be of great value. This measure permits assessment of differences between groups of workers standardized for age, sex, and kind of work, and between different industrial plants, with respect to both the frequency of particular diseases (causes of absence) and disease duration (length of absence from work). There are many factors influencing the level of absenteeism due to sickness and its structure. Apart from such factors as age, sex, and family status, attention should be paid to characteristics of the work environment and non-occupational factors. A careful analysis of sickness absence figures can yield many conclusions regarding the health status of the population and provide working hypotheses. Moreover, it may be an important source of information on the worker's health needs, e.g., specialist care requirements. However, the first question to be answered by an occupational physician is whether or not diseases that occur in an industrial plant might be associated with the working environment and how far the data on sickness absence permit such diseases to be identified. Most occupational diseases (pneumoconiosis, hearing impairment, intoxications) are detectable only by periodic medical examinations, but many other work-related but nonspecific diseases might be detectable by analysing sickness incidence, prevalence, and absence rates. If any specific disease or group of diseases occurs much more frequently in a particular group of workers than in those engaged in other work, the possibility that the working environment has an influence on the disease should be considered.

In the analysis of sickness absenteeism over longer periods of time, three major indices are of primary importance:

(1) number of absence spells resulting from temporary incapacity for work;

(2) number of days of incapacity for work due to disease; and

(3) average duration of incapacity for work.

Index 1, calculated per 10, 100 or 1000 workers, is, as a rule, an incidence rate. It does not reflect total incidence, however, but only the incidence resulting in absence from work. It is calculated as a ratio of the number of absence spells to the number of observed persons.

Index 2 may be calculated as follows:

(*a*) ratio of number of days (or hours) of work lost due to a disease to the total number of calendar days (or hours) comprising the period of observation;

(*b*) ratio of the number of days of incapacity for work to the number of persons under observation at a given time.

Because it is difficult to establish which days were in fact working days, the following simplified index has been proposed:

$$x = \frac{a}{b \times c} \times 100 \,(\text{or } 1000)$$

where x = sickness absence index

a = number of days of incapacity for work (including weekends)

b = average number of workers during a given period

c = number of calender days during the same period.

Index 3 is important for absence analysis and denotes the average duration of sickness caused by a particular disease, calculated from the formula:

$$x = \frac{a}{z}$$

where a is the number of days of absence due to disease and z is the number of spells of absence.

Permanent disability associated with partial or total invalidity is an important health problem. On the other hand, epidemiological analysis of permanent disability and invalidity, as a method of health measurement, is not as commonly applied as studies of disease incidence and death rates. The information and statistical data on the number of invalids and diseases or accidents resulting in invalidity are usually collected by social security institutions, which have access to information kept mainly for administrative, not medical purposes, and thus not adequate for epidemiological studies. An important contribution to improving the analysis of invalidity data is the publication by WHO of an international classification of impairments, disabilities, and handicaps (9).

The extent of the individual and social effects of diseases and invalidity may also be estimated by sociomedical measurements, i.e., by studying how effectively individuals function in the community. In this context, health is the ability of a person to carry out his/her appropriate vital and social functions. Health is a positive attribute conditioning efficient activity, whereas a disease, as a negative phenomenon, diminishes this efficiency. Consequently, the question "what are the diseases with which individuals are afflicted?" is less important than the question "what are the consequences of the diseases?" (10).

The study of mortality statistics and causes of death is the oldest method of health status evaluation. Its aim is to define the probability of death in a given population at a given time. Death rates must be analysed in relation to particular population groups because the risk of death depends on many factors, such as sex, age, occupation, marital status, and environmental factors (11, 12).

The annual crude mortality rate for the study population is calculated in the following way:

$$\frac{\text{annual total deaths due to all causes}}{\text{average size of exposed population}} \times K$$

Specific mortality rates are either calculated for population subgroups standardized for such characteristics as age, sex, residence, and occupation, or they may be related to deaths due to some specific cause or group of causes, e.g., myocardial ischaemia or malignant tumours. Specific rates calculated for diseases of high probability of death (high fatality) are a reliable measure of their occurrence in a population. An example of a specific mortality rate is:

$$\frac{\text{annual total deaths in women aged 40–49 due to cancer of the cervix}}{\text{average number of women aged 40–49}} \times K$$

However, the use of data on death rates in evaluating the health status of industrial workers demands caution because each occupational group is subject to some selection, whether natural or artificial.

When comparing mortality rates due to different causes at different points in time or between different countries, it should be remembered that errors result not from the methods of calculation used but from differences in nomenclature, diagnostic criteria, fashion, medical terminology, and the frequency of correct identification of cause of death. Those factors should always be taken into account by an epidemiologist undertaking descriptive studies and using secondary information sources. Similar considerations apply to data on the incidence and prevalence of diseases and on sickness absenteeism.

Standardization of rates

The need for standardization arises when the rates for two or more populations that differ with regard to the distribution of a disease determinant (e.g., age) are compared and where the frequency of health phenomena is defined only by crude rates that may depend strongly on selected characteristics, such as sex, age, occupation, or environmental factors. Comparisons of crude rates may be very misleading. Crude death rates, for example, will not be comparable if in one population there is a higher proportion of individuals in the older age groups, since mortality depends strongly on age (13). Likewise, a comparison of sickness absence rates in two industrial plants might be invalid because of differences in the sex and age distribution of employees. In practice, it is not infrequent that the populations being compared differ with respect to the distribution of characteristics influencing the level of the phenomena concerned. Obviously, such comparisons should not be made.

Since differences in composition of the population will influence the total rates it is preferable to use specific rates. However, it is sometimes convenient to have a statistical summary of such comparisons that takes into account the differences in distribution of the population. This is accomplished by a computational process known as "standardization".

Standardization makes it possible to compare rates in populations of different distributions. The basic idea in performing standardization is to introduce a standard population with a fixed distribution. The specific rates for observed populations are then adjusted to allow for discrepancies in the distribution between the standard and the observed population. There are two common standardization methods: direct and indirect. They have already been described in Chapter 5. Here, some illustrative examples will be given.

Direct standardization

Table 1 shows a comparison of crude accident rates in two industrial plants. Comparison of crude rates shows a higher rate in plant A. Almost all specific rates (according to length of employment) in plant A are lower than corresponding rates in plant B.

For the purpose of direct standardization, the combined population of the two plants was used as a standard, as shown in Table 2.

The expected number of accidents means the number of accidents that would have been expected if, for a given length of employment, the specific accidents rates of the studied population had obtained in the corresponding groups of the standard population.

It is now possible to calculate the comparative mortality figure (CMF), which is the ratio of the standardized rates for plants A and B.

Table 1. Crude accident rates in two industrial plants

Length of employment in occupation (years)	Plant A			Plant B		
	number of workers	number of accidents	rate per 100 workers	number of workers	number of accidents	rate per 100 workers
<1	300	24	8	50	5	10
1–2	200	12	6	50	5	10
3–5	100	6	6	350	18	5.1
>5	50	2	4	450	18	4
Total	650	44	6.8	900	46	5.1

Table 2. Direct standardization of rates in Table 1.

Length of employment	Standard population	Expected number of accidents in standard population	
		at plant A rates	at plant B rates
<1	350	$\frac{350 \times 8}{100} = 28$	$\frac{350 \times 10}{100} = 35$
1–2	250	$\frac{250 \times 6}{100} = 15$	$\frac{250 \times 10}{100} = 25$
3–5	450	$\frac{450 \times 6}{100} = 27$	$\frac{450 \times 5.1}{100} = 23$
>5	500	$\frac{500 \times 4}{100} = 20$	$\frac{500 \times 4}{100} = 20$
Total	1550	90	103
Standardized rates		$\frac{90}{1550} \times 100 = 5.8$	$\frac{103}{1550} \times 100 = 6.6$

$$\text{CMF} = \frac{\text{rate for plant B}}{\text{rate for plant A}} = \frac{6.6}{5.8} = 1.14 \text{ (or, if multiplied by } 100 = 114\%)$$

The above figure means that if each of the plants had had the same work force structure according to length of employment as the combined population there would have been 14 % more accidents in B than in A. It can be concluded that accident rates in compared plants are affected by the composition of the workforce according to length of employment.

The direct method of standardization is feasible only if the actual specific rates in subgroups of the observed population are available, along with the number of individuals in each subgroup.

Indirect standardization
In practice, not all the data indispensable for direct standardization are always available. Often only crude total rates for the compared populations are available, and specific rates in the subgroups remain unknown. Such a situation permits only indirect standardization to be carried out. In this case, the standard adopted is a population in which not only the distribution of the studied characteristic but also the specific rates in the subgroups are known.

The procedure consists in using the specific rates in the standard population for the calculation of the expected numbers of observed events and, hence, the standardized rates. The ratio of observed to expected numbers of events (usually expressed as a percentage) is known as the standardized morbidity/mortality ratio. The ratio can be defined as the number of events occurring in a given population expressed as a percentage of the number of events that might have been expected to occur if the given population had experienced within each subgroup the same rate as that in the corresponding subgroup of a standard population.

The calculation of indirect standardization will be exemplified by comparison of sickness absenteeism rates in two industrial departments. In department A, 6300 days of absence were observed, which gives a rate of 21 days per employee as compared with 15 days per employee in department B (total 6000 days).

The question is whether the differences in absenteeism rates are real or are only due to differences in population structure, e.g., according to sex (women have higher absenteeism than men).

Table 3 shows the relevant data for the standard population, which is the total workforce of the factory.

Table 3. Absenteeism rates (standard rates) in a factory of 5000 employees

	Number of employees	Total number of days absent	Number of days absent per employee
Men	3500	52 500	15
Women	1500	30 000	20
Total	5000	82 500	16.5

The standard rates for men and women are then applied to each department and the numbers of days absent that would have resulted if the department had had these rates is calculated (expected numbers) (Table 4). The ratio of observed to expected days absent is the standardized absence ratio (Table 5).

Table 4. Calculation of expected number of days absent in two factory departments, using data in Table 3.

	Department A		Department B	
	Number of employees	Expected number of days absent	Number of employees	Expected number of days absent
Men	40	$40 \times 15 = 600$	300	$300 \times 15 = 4500$
Women	260	$260 \times 20 = 5200$	100	$100 \times 20 = 2000$
Total	300	5800	400	6500

The results indicate that the absence experience in department A was only 92 % of that of the standard population, while in department B it was 108 %. This means that if the sex structure of the workforce in the two departments compared was the same as in the standard population, the absence rate in department A would be lower than in department B and that the observed rates are strongly affected by sex distribution.

Factors affecting health indices
Several characteristics must be considered, such as sex, age, marital status, occupation, time and place.

127

Table 5. Calculation of the standardized absence ratios for the two departments considered in Table 4.

	Absence rates		
	Observed	Expected	Observed/expected
Department A	21	5800:300 = 19.3	92
Department B	15	6500:400 = 16.2	108

Sex

Differences are often encountered in the frequency of disease occurrence according to sex. From the point of view of epidemiology, the most important are those that do not result from anatomical–physiological differences. When investigating such sex-related differences in disease frequency, the possibility of unequal exposure should be considered, e.g., differences in the occurrence of lung cancer in men and women may be due to differences in the proportion of smokers in these groups, or differences in the occurrence of occupational intoxications may be attributable to the fact that women are less likely to be employed on particularly hazardous jobs. Many demographic phenomena show a significantly higher rate in one or other of the sexes, e.g., higher mortality in men in most age groups. Although significant sex-related differences may be noticed in mortality distribution depending on the specified cause, as a rule, the rates for men are higher than those for women.

Distinct differences are observed in morbidity between men and women, but these have not yet been fully elucidated. Prevalence rates, unlike mortality, are, for the majority of chronic diseases, higher among women than among men. Only a few health problems are more frequent in men — peptic ulcer, accidental injuries, ischaemic heart disease, and lung cancer. It should be remembered, however, that indirect influences, e.g., different psychosocial situations and sometimes differences in the economic or occupational situation of women, may play an important role in certain countries.

Age

The relationship between disease occurrence and age is highly important in epidemiology.

A number of demographic and health phenomena are clearly related to age. The dependence of general mortality on the age structure of the population is well known, as are the great differences in causes of death in various age groups. The incidence of some diseases (e.g., malignant neoplasms) is higher in particular age groups and lower in the others.

In interpreting the dependence of disease occurrence upon age with regard to etiology and pathogenetic mechanisms, acting alone or in combination, the following considerations should be taken into account:

(*a*) age is an index of the developmental stage of the organism and some diseases may develop in a specific period of life, e.g. congenital defects;

(*b*) dependence on age may also be an index of first exposure of the organism to a noxious factor, e.g., infectious diseases of childhood may produce immunity for later life;

(*c*) if the risk of falling ill depends on the cumulation of exposure to pathogenic factors or long-lasting noxious substances in the working environment, it will increase with time and will thus also be related to age (Fig. 2).

The frequency of occurrence of specific diseases does not always increase proportionally with age. This may be illustrated by the incidence and prevalence of, and mortality from, multiple sclerosis according to age (Fig. 3).

The shape of the graph showing the relationship of the disease to age will depend on whether it is based on incidence, prevalence, or mortality. If for a fatal disease the lapse of time between falling ill and death increases with age, the mortality curve according to age will be flatter than the analogous one for incidence. In chronic, long-term diseases, the peak of prevalence in comparison with incidence will be shifted towards older age groups.

It should be remembered that the frequency of the health phenomena investigated may be distorted by the lack of precise data on populations in the oldest age groups. This is especially important, for example, in

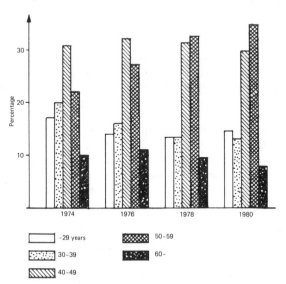

Fig. 2. Structure of a population afflicted with occupational diseases in Poland in the years 1974–1980 according to age

Reproduced from: Indulski, J. et al. [Occupational diseases in Poland in the years 1971–1977]. *Medicina pracy,* **29**: 507 (1978).

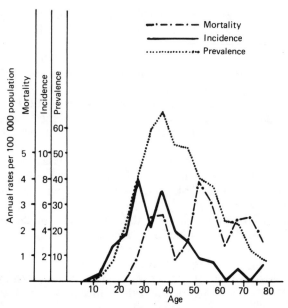

Fig. 3. Incidence, prevalence and mortality of multiple sclerosis in the male
population: Boston, 1939-1948

Reproduced from: **McMahon, B. & Pugh, T.F.** *Epidemiology: principles and methods.*
Boston, MA, Little Brown & Co., 1970.

studies on neoplasms of occupational origin occurring in a significant
percentage of older persons, especially in those over reproductive age.

The occurrence of two or more peaks in the curve of disease
frequency distribution according to age group suggests the influence of
various etiological factors.

The studies on the relationship between frequency of disease in a
population and age may be carried out in two ways, i.e., based on
information from cross-sectional examinations of a population, including
persons of different ages at a specific point in time, or from prospective
observations of groups of individuals (cohorts) composed of persons
born during a given period.

Marital status

The relationship between marital status and different health phenomena,
although it has been demonstrated in numerous studies, often creates
difficulties in interpretation. For example, specific mortality rates have
been found, in some cases, to be lower in married persons but higher in
the widowed or divorced. This relationship has been noted for both
sexes. With a view to elucidating this phenomenon it may be pointed out
that:

130

(*a*) ill and disabled persons are more likely to remain single and if they contract marriage the probability of a divorce is presumably greater than for healthy persons;

(*b*) the lifestyle of single persons differs significantly from that of married ones, which may be a contributory cause of some diseases.

Marital status may also be interpreted as an indication of the way of life of an individual. It finds expression in different habits, different eating patterns, different sharing of familial duties, etc.

Studies in which allowance is made for marital status facilitate the search for causal factors of different diseases, especially if they are common to both husband and wife and are related to their socioeconomic conditions and way of life. For instance, the occurrence of mesothelioma of the pleura has been reported, in wives whose husbands worked in contact with asbestos (*14, 15*).

Occupation

Like the variables described above, occupation may serve several different purposes in descriptive epidemiology. For instance it may be:

(*a*) an indication of the risk of occurrence of a particular occupational disease related to harmful exposure;

(*b*) a criterion for selecting individuals with regard to their health status; and

(*c*) one of the principal factors in the measurement of socioeconomic conditions.

The classification of occupations for epidemiological purposes creates great difficulties. The division of employees into blue-collar and white-collar workers is not satisfactory, as it does not distinguish the activities performed by the two groups. Usually, the division is based on the occupation in which an individual is actually engaged, rather than on the one for which he was trained. Nevertheless, the differences in production and working conditions that now exist within a particular occupation create the possibility that a classification of this kind may also be seriously at fault. Hence, in epidemiological studies on groups of employees, the criterion for the classification is, more and more frequently, not the occupation itself but rather the magnitude of the exposure to noxious or strenuous factors related to the work being performed. Nevertheless, in descriptive studies, occupation is used as a criterion when classifying the population whose health characteristics are being studied. Therefore, the standardized mortality rates for the selected occupations provide valuable clues to the existence of specific exposures, and this can then be verified in subsequent studies.

When describing the frequency of diseases in particular occupations, it should be remembered that their specificity may bias the inferences drawn. The increased frequency of a disease may be a result of its easier detectability in some professions (e.g., in physicians) or, it may be due to the fact that the practice of a particular occupation is conditioned by the possession of good health, as in the case of miners, metallurgists, etc. It should not be forgotten that for some occupations the health requirements are less strict and hence the health of persons employed in such occupations (e.g., doorkeepers, guards) may be significantly worse.

When studying the incidence of chronic diseases in different occupational groups, it is necessary to take into account the possibility that a worker formerly in an occupation responsible for a specific disease may have changed it for a safer one (*16*).

Time

Changes in health occur with time, which is an indispensable element in the definition of every epidemiological parameter and a basic variable in the concept of causation. Thus, it may be necessary to take into account the time that separates exposure to an etiological factor from the appearance of the first pathological symptom, e.g., the appearance of the first byssinosis symptoms is dependent on both the duration of exposure and the concentration of cotton dust. Moreover, in descriptive studies, the duration of a disease is frequently considered as a characteristic, although nowadays duration is no longer a specific feature of many diseases. It is significant, however, for the programming of medical care in a wider sense. For description of pathological phenomena in populations, time trends are also significant. Again and again analysis of time trends provides significant clues to the formulation of etiological hypotheses. The interpretation of trends in mortality, prevalence, or incidence is difficult, especially when crude rates are considered, and therefore the trends in age-specific rates or age-standardized rates (direct standardization) should be analysed (Fig. 4).

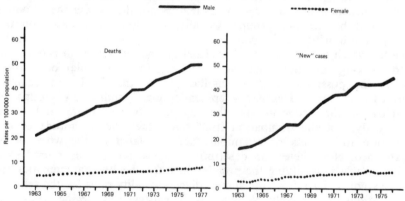

Fig. 4. Deaths from lung cancer and registered cases in 1963–1977 according to sex

Reproduced from: Olakowski, T. et al. [Epidemiology of lung cancer in Poland up to 1979]. *Pneumonologia polska,* **47**: 553 (1979).

In Fig. 4, the mortality from and incidence of lung cancer in men and women in the years 1963-1977 are placed side by side. From the distribution of the sex-related rates, it follows that during 15 years, the incidence of and mortality from lung cancer increased to a greater extent in men than in women. These crude data provide insufficient information, however, on which to base solid conclusions. When the data are broken down by age groups, it appears that the observed increase in lung cancer incidence in men in these years did not apply to all age groups to the same extent. The greatest increase was found in men aged 35–54 years, while in persons aged over 55 it had decreased by half. This observation shows the exceptional advance of the disease. From the natural history of chronic diseases it is known that a shift in the intensity of incidence (or mortality) towards younger age groups is evidence of the epidemiological progression of the disease. Analysis of lung cancer mortality in women did not reveal a similar phenomenon.

To elucidate time trends based on age-specific rates, it is necessary to take into account a number of factors that may lead to erroneous conclusions, especially when analysing longitudinal observations. These include changes in diagnostic criteria and classification, new record keeping procedures, and sometimes also progress in therapy.

Place
Variations in the frequency of occurrence of diseases depending on the geographical situation have already been given due consideration in descriptive studies. The World Health Organization has been collecting and publishing data on diseases, prevalence, and causes of death in many countries for a long time (*17*).

Although differences in diagnostic criteria and in methods of reporting and recording in different countries lead to lack of comparability of such statistics and sometimes make it difficult to draw conclusions from them, they are still a valuable source of information on the health of the populations and frequently provide a basis for formulating etiological hypotheses for further studies. Examples are mortality from cancer of the stomach in males in Japan or from lung cancer in England or a high death rate from heart diseases in Finland in comparison with very low rates in Japan (*17, 18*).

The study of the frequency of diseases according to the degree of urbanization and development of particular industries calls for comparison in many countries. Distinct differences between mortality in urban agglomerations and that in rural districts can be documented. The nature of the work, occupational hazards, and microclimatic conditions are among the main factors differentiating the two areas. The occurrence of a disease simultaneously in two different districts, or a change in the intensity of its occurrence in the same area, would suggest the need to look for causes of these phenomena among factors characteristic of each region.

References

1. **Barker, D. J. P.** *Practical epidemiology.* Edinburgh, Churchill Livingstone, 1976.
2. **Lilienfeld, A. M. & Lilienfeld, D. E.** *Foundations of epidemiology*, 2nd ed. London, Oxford University Press, 1980.

3. **Kostrzewski, J.** [*Contagious diseases in Poland and fight against them in the years 1961–1970*]. Warsaw, PZWL, 1973 (in Polish).
4. **Indulski, J. et al.** [Principles of determination of tasks in the planned health service activity]. *Zdrowie publiczne*, Suppl. 2 (1968) (in Polish).
5. **Jedrychowski, W.** [*Methods of epidemiological investigation in industrial medicine*]. Warsaw, PZWL, 1978 (in Polish).
6. **Kleczkowski, B. M.** [The use of epidemiological studies in programming health care in social diseases]. *Polski tygodnik lekarski*, **25**: 428(1970) (in Polish).
7. **Kostrzewski, J. et al.** [Objectives, methods and organization of epidemiological studies in social diseases]. [*Polish medical weekly magazine*], **25**: 421 (1970).
8. **Monson, R. R.** *Occupational epidemiology*. Boca Raton, FL, CRC Press, 1980.
9. *International classification of impairments, disabilities, and handicaps. A manual of classification relating to the consequences of disease.* Geneva, World Health Organization, 1980.
10. **Chiang C. L. & Cohen, R. D.** How to measure health: a stochastic model for an index of health. *International journal of epidemiology*, **2**: 1,7 (1973).
11. **Kostrzewski, J.** [*Health of Polish population in the light of data on diseases and deaths*]. Warsaw, PZWL, 1977 (in Polish).
12. **Miettinen, O.** *Principles of epidemiologic research*. Boston, Harvard School of Public Health, 1978.
13. **Armitage, P.** *Statistical methods in medical research*. Oxford, Blackwell Scientific Publications, 1971.
14. **Newhouse, M. L. & Thompson, H.** Mesothelioma of the pleura and peritoneum following exposure to asbestos in the London area. *British journal of industrial medicine*, **33**: 261 (1965).
15. **Noro, L.** Occupational and "non-occupational" asbestosis in Finland. *American Industrial Hygiene Association journal*, **29**: 195 (1968).
16. **Schilling, R. S. F.** *Occupational health practice*. London, Butterworths, (1981).
17. *World health statistics annual. Vol. I: Vital statistics and causes of death.* Geneva, World Health Organization (published annually).
18. **Staszewski, J.** *Epidemiology of cancer of selected sites in Poland and Polish migrants.* Cambridge, Ballinger, 1976.

8

Cross-sectional studies

G. Enderlein[a]

Cross-sectional epidemiological studies in occupational health record the prevalence of certain diseases or symptoms with regard to work-related factors.

The prevalence rate means the frequency of defined health impairments related to the total number of persons in a group. The medical findings are obtained mainly by medical examinations or, more rarely, by individual questionnaires or screening procedures.

It is necessary to distinguish between descriptive and analytical cross-sectional studies. Descriptive cross-sectional studies are quantitative descriptions of states giving actual exposures and findings. Such studies are of little scientific value, as conclusions regarding prevalences based on the actual exposures or work loads involve numerous possibilities of error. These result, above all, from variations in the exposures, work loads, and other factors in the occupational history, as well as from the element of selection introduced by changing occupation on health grounds. Social and other non-occupational factors and different age and sex structures of exposed and non-exposed populations can distort considerably the prevalence rates. Therefore, the physician undertaking epidemiological investigations is not concerned with merely descriptive studies. This does not exclude the use by the occupational health physician of descriptive registers of absenteeism, disability, morbidity, and mortality.

Only analytical studies in which the essential confounding factors (cf. pp. 137–140) are controlled by design or subsequent computing match the demands of modern scientific epidemiology. In cross-sectional studies, the prerequisites are retrospective ascertainment of the occupational history, which includes the period and level of exposure to occupational factors.

Analytical cross-sectional studies of occupational health provide information on interesting questions relatively quickly and with an

---

[a] Central Institute of Occupational Health, Berlin – Lichtenberg, German Democratic Republic.

acceptable level of expenditure. Such studies may have an exploratory character or may be oriented to the verification of a special working hypothesis (*1*).

Etiological cross-sectional studies

The study of work-related diseases is generally concerned with etiology but not with pathogenesis. The problem is to determine which exposures to chemical or physical factors and which physical or neuropsychic loads give rise to health disturbances, and also in which occupations, in which activities, and under which additional conditions such disturbances occur. Etiological cross-sectional studies could also be used to investigate the chronic effects of long-term exposures.

By recording reliably the intensity and duration of exposure to noxious agents, as well as by controlling the main confounders, for instance by matching or stratification, cross-sectional studies even permit the assessment of quantitative exposure–response relationships. In general, however, only extensive longitudinal studies satisfy the requirements for determining exposure–response curves.

The possibilities of etiological research in cross-sectional studies are very limited, owing to the relatively few patterns displayed by the reactions of the organism. The so-called specific effects of long-term exposure to chemical agents are an exception rather than the rule. Typical constellations of reactions and impairments can, however, provide a better indication of long-term effects. Repeated confirmations by different cross-sectional studies, in different localities and by different investigators, can contribute valuable evidence on the causation of some diseases.

Cross-sectional studies of adequate quality may be used as a baseline for cohort studies (*2*).

There are a number of purposes, however, for which cross-sectional studies are not, as a rule, indicated. Of these, the most important is the testing of the acute and subacute effects of new chemical or other potentially noxious factors in the occupational environment. For this purpose, longitudinal studies are needed. Moreover, cross-sectional studies can be used only to a very limited extent for establishing individual tolerance.

Descriptive cross-sectional studies

In occupational health, these studies describe the exposures at work and compare them with other health data obtained during the same period (*3, 4*). The main use of descriptive studies is for planning health policy. Such studies may cover a certain occupational group in a plant, industry, or the entire country, as well as geographically or occupationally defined groups.

Descriptive cross-sectional studies give information on:

136

— exposures in excess of occupational exposure limits;
— human working potential with respect to age, sex, qualifications, state of health, and suitability for a particular occupation;
— deficiencies in occupational health surveillance;
— deficiencies in medical care, particularly as regards periodic examinations;
— need for changes in job placement and other medical care measures;
— prevalence and nature of existing occupational diseases (1, 5).

In the German Democratic Republic, descriptive cross-sectional examinations of more than 113 000 persons resulted in the disease prevalence rates and disability assessments shown in Table 1 (5).

Table 1. Results of cross-sectional study of occupationally active men (5)

Diagnosis	Prevalence per 100 persons examined	Reduction in ability per 100 sick persons	Reduction in ability per 100 persons examined
	I	II	$\dfrac{I \times II}{100}$
Degenerative diseases of the vertebral column	11.9	27.3	3.25
Hearing losses	7.1	41.3	2.93
Chronic nonspecific lung diseases	4.2	44.0	1.85
Diseases of the central nervous system	0.8	43.6	0.35

Mass surveys of absenteeism in industrial concerns or areas are of no value in solving questions of occupational health. Analyses of absenteeism among occupational groups in a factory or branch of industry may, however, yield instructive results and be used for characterizing occupations.

Principles of designing cross-sectional studies

Just as for all epidemiological studies, the designing of cross-sectional studies is of fundamental importance. The interdisciplinary teamwork is absolutely necessary including, above all, the cooperation of the occupational physician with a scientist experienced in biometry and methodology. The quality of the results depends mainly upon the quality of the study design. Proper formulation of the purpose of the study is a prerequisite for the approach to be used in the investigations, including methodology and statistical procedures. In general, cross-sectional studies of an exploratory character are designed on a larger scale. Etiological studies should, however, be more limited. Frequently, they

137

are overloaded with too many questions, effect parameters, and other data that do not correspond with sufficiently exact data relating to the actual exposure. The following sections refer in more detail to the recording of exposure and effect parameters.

Before starting the studies, the results reported in the relevant literature, including the results of animal experiments, must be carefully evaluated and given due consideration at the design stage. In comprehensive, large-scale cross-sectional studies it is useful to carry out in advance a pilot study with a small typical sample to obtain information on the rationality and effectiveness of the study design and on organizational questions.

In both cross-sectional and follow-up studies, exposed groups are compared with reference groups with regard to differences in prevalence rates and their degree. For this reason, the principles of planning and of statistical evaluation are very similar (4, 6). The control of confounding factors is of fundamental importance. An influencing factor in cross-sectional studies is a confounder only under the following conditions:

(1) it is an etiological factor for the health impairment considered;
(2) it is correlated with the exposure examined.

Age and sex are typical common confounders. With increasing age, the prevalence of most diseases increases and the duration of exposure is closely correlated with age. Frequently, there are also sex differences, both in prevalence and exposure. Factors like smoking, nutritional habits (overweight), alcohol consumption, leisure time activities, and sports modify the course of many diseases. However, they distort the results as confounding factors only if the second condition applies, i.e., if the distribution of the confounder differs in the groups of differently exposed workers being compared. There are two possible ways of controlling the confounding factors (6).

(1) When planning the study the control group is selected in such a way that the factors assumed to be confounders have the same distribution as in the exposed group (the same age, weight, smoking habits, etc.). The number of persons to be included may be considerably reduced by the control of confounders, thus reducing random variation in the statistical evaluation.

(2) The confounders that cannot be allowed for in the study design are recorded for each person and controlled when making the analysis. It is important to record all probable confounding factors. Otherwise, uncontrollable biases due to confounding effects may result.

The most difficult problems when designing and evaluating cross-sectional studies are caused by the following types of confounders:

— various scarcely measurable additional exposures and work loads present in the occupational history of the different groups or distributed among the main groups actually exposed;

138

— selection due to the "healthy-worker effect" (7).

The causes of the healthy-worker effect are as follows.

— On the whole, workers are in a better state of health than persons not working.
— Persons with a certain physical and mental predisposition tend to enter occupations with corresponding requirements and work loads (primary self-selection).
— Many workers drop out from demanding jobs because, subjectively and objectively, they do not reach a tolerable social and healthy steady state (secondary self-selection).
— Above all, occupational physicians advise management not to engage persons lacking in the health prerequisites for particular jobs, or they exempt them from such jobs during the course of periodic surveillance (active selection).

The process of selection and its long-term effects does not necessarily operate uniformly over years in one direction only. This applies especially to occupations with complex work load distributions, such as shift work (8).

In contrast to the healthy-worker effect, people with impaired health generally prefer occupations with lighter requirements and work loads. Such groups, if serving as comparison groups, increase the bias caused by the healthy-worker effect and may diminish the value of cross-sectional studies. Spontaneous or active selection affects groups of workers in the following ways:

— different states of health at the beginning of exposure;
— different drop-out rates with and without certain impairments of health — psychosocial and economic causes must be carefully differentiated from health complaints or somatic effects.

These two confounders can be properly controlled only in carefully designed longitudinal studies. In cross-sectional studies, sufficient precision can often be achieved by the following measures:

— the confounder "better state of health at the beginning of exposure" can be controlled by analysing groups subject to short exposures, assuming that there are no delayed effects;
— for controlling the confounder "different drop-out rates", a random sample of exempted persons is examined to ascertain the prevalence of the impairment considered.

The control group is very important with respect to confounding factors. Particular attention must be paid to the healthy-worker effect. For this reason, administrative staff are not suitable for use as a control group for workers. Apart from the exposure to be examined, the other work loads and conditions should conform with those of the exposed group. In choosing samples of the exposed and control groups, the factor of personal motivation should be excluded as far as possible. Mostly, the motivation to participate is higher in exposed than in control persons. In control persons, the risk of bias results from the fact that sick persons and neurotics tend to be more motivated to participate.

If several investigators in the same specialty (internists, psychologists, etc.) participate in a cross-sectional study, they should examine equal proportions of exposed and control subjects. Frequently, the bias introduced by unequal participation of several investigators has been neglected. In mass studies, including numerous investigations undertaken for technical reasons, the bias of the investigators and sometimes somewhat unreliable methods of examination have to be accepted. However, increasing the number of investigators is likely to reduce systematic bias.

Blind test techniques, in which the investigator has no possibility of differentiating between exposed and control persons, should be included in etiological studies, as far as possible. This blinding is possible for laboratory, X-ray, and functional diagnostic examinations, as well as for electrocardiograms and other electrodiagnostic methods but, unfortunately, not for most clinical procedures (7).

If a sufficiently large population is examined by the same methods and during the same period, it can be used for comparative purposes. If the number of exposed persons is large enough a control group may be dispensed with, provided that stratification according to the intensity and length of exposure results in comparable groups with regard to numbers, age, and other relevant factors.

In planning the study, a sufficiently large random sample must be ensured. The numbers required for statistically significant statements depend on the accuracy of the intended result, the intensity of exposure, the interindividual variability, and the ways in which the effects are recorded. Sometimes, valid statements are possible with groups of 20–30 subjects, if the effect parameters are determined in a reliably quantitative manner. As a rough guide both the exposed and non-exposed groups should comprise at least 200 persons if the impairments of health can be recorded only as bivariate alternatives.

Assessment of exposures

It is essential that the assessment of exposure to a noxious agent should be performed as exactly as possible, especially in etiological cross-sectional studies. The levels of exposure experienced in previous years and decades must be taken into consideration, if such studies concern long-term exposures. Often, no reliable and regularly recorded values for such exposures are available; the assessment has to be made with due regard to the availability of information permitting the computing of a concentration-time product for groups or even for individuals.

On the whole, in the industrial countries, the recording of exposures has improved during the last few years, corresponding to increased interest in epidemiological and ecological studies. It may therefore be expected that the prerequisites for epidemiological studies will be met more completely in the future.

In etiological cross-sectional studies, assessment of the exposure costs more than assessment of the health injuries. The first necessity is to determine the level and duration of the exposure to the presumed noxious factor or factors.

In mass studies, experts should check the validity of measurements and estimates of previous occupational exposures. Then, the average exposure to the noxious agent is estimated for periods of approximately equal exposure. As a rule, the duration of exposure can be assessed much more exactly than the level. The product of the duration (d_i) and concentration (c_i) serves as an estimate of the dose received during an exposure period, and the total exposure of each individual worker is indicated by the sum of the period products, $\sum c_i d_i$. The use of data obtained by personal dosimetry to indicate the accumulated dose of the inhaled noxious agent is ideal, but rarely possible.

As a rule, data are obtained by regular monitoring of the working environment. The frequency of peaks of exposure should be considered because of possible modifications in effects. Temporary overexposures may give rise to increased quantities of metabolites (e.g., benzene or trichloroethylene), which could evoke carcinogenic or immunological effects in addition to the typical toxic symptoms. The results of exposure tests should be used as far as available. Here, caution is indicated if the data originate from samples obtained under different conditions. Biological monitoring has the great advantage that it takes account of increased body burdens due to absorption through the skin or to simultaneous forced ventilation occasioned by heavy physical work. If monitoring is performed frequently enough, data on the so-called internal exposure are obtained.

In the present state of knowledge, the measured values for external and internal exposure cannot replace one another for characterizing individual exposures. The two values must be checked in parallel against clinical effect parameters.

Frequently, it is possible to make only a rough estimate of the level of previous harmful exposures. The subjects are then grouped according to the level and duration of exposure. In a cross-sectional study at least the following two groups are needed for comparative purposes:

(a) a truly exposed group with a sufficient duration of exposure (e.g. over 5 years);

(b) a control group with a maximum exposure below a certain level.

Every possible further differentiation of the exposed group improves the evaluation, provided there are sufficient persons in the subgroups. Persons who cannot be safely designated are not considered. Cigarette smoking and alcohol consumption are relatively important non-occupational exposures. In general, records of the number of cigarettes smoked per day, the number of years of smoking, and the number of ex-smokers (at least 1 year without cigarettes) and non-smokers are necessary for epidemiological studies in occupational health.

Data on the consumption of alcoholic beverages are much more difficult to obtain. It is necessary to know both the quantity of alcohol ingested (ml per day) and whether or not the intake is regular. Alcohol consumption has to be considered in all etiological studies concerning health impairments, especially diseases of the liver and nervous system. In dealing with solvents, competitive as well as synergistic mechanisms of action have to be taken into account, with due regard to toxicokinetics, biotransformation, and target organs. Similar considerations apply to steroid contraceptives and other medicaments.

Social and personal characteristics, such as education, qualifications, income, overtime work, leisure-time activities, sports, and nutritional habits (especially those linked with obesity) are other factors that need to be considered as possibly implicated in the causation of certain diseases. In some studies, ecologically important industrial pollutants detected by environmental or biological monitoring have to be taken into account in certain regions with heavy concentrations of industry (e.g., lead, carbon monoxide, benzopyrene, sulfur dioxide, DDT, asbestos). These toxic substances may be confounders. The cross-sectional study may fail in its purpose or furnish false conclusions if due attention is not paid to potential confounders from the outset.

Assessment of the health effects

In etiological cross-sectional studies, deviations from normal health caused by long-term exposure must be detected, if possible, in the preclinical state. It has to be presumed that there is considerable variation with regard to the dose (concentration and duration of exposure) as well as the susceptibility of the population examined. Therefore, the spectrum of hypothetical health effects has to be extended to include not only a variety of laboratory and behavioural signs but also clinical evidence of functional and possibly organic changes. For epidemiological studies in occupational health the *International classification of diseases* should be used, although this tool has its limitations. The diagnostic criteria must be separately defined for each study according to the purpose of the investigation and the methods used.

Syndromes may be used as effect parameters, and statistical evidence from cross-sectional epidemiological studies may help in identifying them as work-related phenomena. The etiology and pathogenesis of many diseases and syndromes are ambiguous. For instance, psycho-organic syndromes, so important in industrial neurology, can be caused either by direct toxic brain damage or by degenerative, inflammatory, or infectious cerebrovascular processes. There is an analogous situation for polyneuropathies, disturbances of autonomic regulation, chronic bronchitis, degenerative diseases of the locomotor system, etc. The possible causal role of relevant occupational factors must always be considered in the following diseases of multiple etiology:

— chronic nonspecific lung diseases;
— liver damage;
— polyneuropathies;
— hypertension;
— ischaemic heart disease;
— psychosomatic syndromes;
— degenerative diseases of the vertebral column and joints.

In cross-sectional studies, just as in longitudinal studies, it is important to concentrate on early signs of disease, since manifestly ill workers may leave the job. The methods used for assessing the effect parameters must have been shown to possess sufficient validity (sensitivity and specificity) and reliability by appropriate preliminary tests. The requirements are not so high in cross-sectional studies, however, as in longitudinal ones. The application of metric (interval-scaled) measures reliably indicating health disturbances is always advantageous.

The selection of effect parameters and their precise definition depend on the aims of the cross-sectional study. The selection can be very limited if, for instance, the only question to be answered is whether the free erythrocyte-protoporphyrin and/or the delta-aminolaevulinic acid excretion in the urine are more closely correlated than the blood lead level with deviations in nervous conduction velocity. A considerably larger number of effect parameters would be needed, however, to decide whether arteriosclerosis, hypertension, and liver damage are promoted by long-term exposure to lead and under which conditions.

Frequently, cross-sectional studies use different effect parameters from longitudinal studies, e.g., when investigating the vasculotoxic effect of carbon disulfide. According to systematic epidemiological studies, myocardial infarction may be considered as a long-term effect of exposure to carbon disulfide at a concentration in air of about $50 \, mg/m^3$. Hernberg and his colleagues selected mortality from infarction as a critical parameter of the supposed effect in their longitudinal studies (2). In a cross-sectional study, however, an acute infarction is unlikely to be encountered. In this case, coronary heart disease would be selected as an effect parameter, starting from the hypothesis that carbon disulfide has an atherogenic effect. The diagnosis of coronary heart disease may be rendered uncertain, however, by extracardiac influences or, above all, by functional chest pain. The administration of a validated questionnaire and performance of an exercise ECG would help to reduce mistakes as far as possible.

The methodological tool employed has to be standardized with regard to sex and age. This applies especially to questionnaire methods, e.g., the use of a psychological-neurological short questionnaire as a screening method for assessing psycho-autonomic disturbances caused by long-term exposure to harmful neurotropic agents (9).

As an example of a careful, large-scale cross-sectional study of the problems of the chronic effects of organic solvents on the nervous system, reference may be made to an investigation of spray painters (*10*); psychiatric interview, 18 psychological tests, neurological and neurophysiological examinations, including visual evoked responses, computerized EEG examination, computerized tomographic scanning (CT scanning), electroneurography, vibratory perception thresholds, and ophthalmological examination were all applied. When psychological tests are applied, it is important to pay due attention to intelligence and education, especially when selecting the control group.

When making cross-sectional studies, the *International classification of diseases* should be taken as a basis for diagnostic purposes. The designation of diagnoses according to the stage of development has proved to be useful in a mass study carried out in the German Democratic Republic (*5*). These stages are intended to make possible a more precise characterization of the morbidity of the chronic diseases present and the level of fitness of individuals suffering from them. A number of socioeconomic parameters and data on the frequency and quality of occupational health examinations and health care activities are also of interest in such studies.

Statistical evaluation

The data on exposure and on the various modifying factors and effect parameters are recorded individually for purposes of evaluation. The use of standardized codes facilitates comparison with other studies and increases the precision of diagnoses.

The targets of statistical evaluation of epidemiological cross-sectional studies are as follows:

— quantitative description of the relationships between noxious factors (single and combined) and health impairments;
— computational control of recorded confounders that have not been eliminated in the planning stage; and
— statistical confirmation of the relationships ascertained.

The choice of statistical method is determined, above all, by the scaling of the causal and modifying factors and effect parameters. A distinction has to be made between variables with nominal, ordinal and internal (metric) scales. Occupation is an example of a nominally scaled variable. When health is ranked as good, average or poor, an ordinal scale is employed; successive numbers can be used to indicate order, but they do not imply equal intervals. The internal scale as used for variables such as age, blood pressure and lung function, however, indicates both order and distance.

As mentioned above, the use of quantitative effect parameters is to be preferred. Often, the modifying factors may be evaluated metrically, as well as by category. Thus, age can be graded by rank-category in terms of age groups. In general, exposure effects are more clearly demonstrable if scaled by category. Computer programs for the usual methods belong to the statistical software of computers (*11*).

144

The quantitative expression of relationships between exposures and health impairments with respect to nominally or ordinally scaled effect parameters is derived from rates of prevalence. In the simplest case, involving an exposed and a control group, a two-by-two table is constructed for persons with and without health impairment.

	Exposed (E_1)	Control (E_0)	Examined
Persons with a health impairment	a	b	m_1
Persons without a health impairment	c	d	m_0
Number of persons	n_1	n_0	N

For the exposed group (E_1), the prevalence $R_1 = \dfrac{a}{n_1}$;

for the control group (E_0), the prevalence $R_0 = \dfrac{b}{n_0}$.

The effect of exposure is expressed by the following simple measures of relative risk (6).

$$\text{Risk difference } RD = R_1 - R_0$$

$$\text{Risk ratio } RR = R_1/R_0$$

Approximate confidence intervals for RD and RR may be computed relatively simply. The significance of both measures is usually tested by the chi square test. Another important measure is the etiological fraction

$$EF = \frac{R_1 - R_0}{R_1}$$

which expresses the reduction in relative prevalence for persons with a given finding, if the exposure were to be eliminated.

In the case of metrical effects, the rates of prevalence are replaced by averages; exposure effects are expressed as differences between mean values. These differences are tested by variance analysis and the t-test. Expressions of exposure–effect relationships and their precision can be evaluated by regression analysis, provided that quantitatively recorded health impairments are present in some groups in which exposure is graded according to intensity and duration. For ordinally scaled effects, this approach is appropriate only if the exposed group is large and if the effect function is to be transformed e.g., by logit-transformation. The methods employed are analogous to the use of bioassays for evaluating dose–effect relationships in testing the toxicity of drugs. In principle, the computation of regression functions is possible in all cases, but it cannot be recommended if the groups are too small and the effect parameters insufficiently reliable. There is a temptation to make speculative judgements, as the confidence intervals for the estimated regression functions are so large that a very broad range of relationships is compatible with the values observed.

The grouping of subjects by particular characteristics, e.g., according to age groups, is the basic method of eliminating a confounder in computing. This method is called stratification. For each stratum (e.g., age group), the measures of the effect of exposure are computed. Separate statements of the exposure–effect relationships in each stratum are desirable if they differ widely between strata. In the case of age, for instance, it is then necessary to compute age-related risks and test the results statistically, if the number of subjects in each age group is sufficient. Aggregated measures covering all strata can be computed if the measures of exposure–effect do not differ very much between different strata, or if the numbers of subjects are not sufficient for strata-specific statements. The most important forms of aggregation are standardization and partial measures of relationships. Standardization (6) can be applied to differences in prevalence rates (RD) and to risk ratios (RR), as well as to differences in averages. Suppose the question to be answered is whether males exposed to cadmium show a higher rate of functional impairment than a control group, using the data in Table 2. Age is considered as a confounder, because the groups have considerably different age structures.

Table 2. Percentage of functional impairment in subjects exposed to cadmium and in a control group, by age.

Age (years)	Exposed group			Control group		
	No. in group (n_{1j})	No. with impairment (a_j)	% with impairment (a_j/n_{1j})	No. in group (n_{oj})	No. with impairment (b_j)	% with impairment (b_j/n_{oj})
< 24	71	12	$R'_1 = 16.90$	236	13	$R'_0 = 5.51$
25 – 44	65	10	$R''_1 = 15.38$	503	44	$R''_0 = 8.75$
> 45	30	12	$R'''_1 = 40.00$	215	45	$R'''_0 = 20.93$
Total	166	34	(20.48)	954	102	(10.69)

Direct standardization relates the prevalence rates to the age distribution of the control group. In this instance, the method is applicable because the control group contains sufficient numbers in all age groups.

The age distribution of the control group can be expressed as:

$$w_1 = 236/954 = 0.247 \qquad w_2 = 503/954 = 0.527 \quad w_3 = 0.225.$$

The prevalence rate standardized for the exposed group is

$$R^{St}_1 = w_1 \times 16.90 + w_2 \times 15.38 + w_3 \times 40.00 = 21.28\,\%.$$

Hence, $RD = R^{St}_1 - R_0 = 10.59\,\%$ and

$$RR = R^{St}_1/R_0 = 1.99$$

which means that the percentage of impairment is much higher in the exposed group than in the control group. Without standardization,

146

slightly distorted values would be obtained for *RD* and *RR*, namely:

$$RD = 20.48 - 10.69 = 9.79\%; \quad RR = 20.48/10.69 = 1.91.$$

For greater precision, proportional standardization (*6*) has to be applied if the numbers in some age groups are small and if the age distributions in the exposed and control groups are very different. Approximate confidence intervals can also be calculated for the standardized measures of relative risk (*6*). The statistical testing of significance is identical for both risk difference and risk ratio. It is done by means of an approximate chi square test, the so-called Mantel-Haenszel (M-H) test, if the presumed exposure effects show the same trend in all age groups (*12*). In the example, χ^2 M-H = 16.46 with one degree of freedom; this is significant, compared with χ^2 5% = 3.84.

Partial measures of relationships, such as Kullback's partial MDI-statistic, can be calculated by multidimensional contingency table analyses with respect to qualitative effect parameters (*13, 14*).

Covariance analysis may be used to eliminate confounders when testing a quantitative effect parameter, qualitative modifying factors, and quantitative confounders. Multiple regression analysis is used if all the factors and effect parameters are quantitative. In such cases, partial regression coefficients and partial coefficients of determination indicate the influence of the factors considered when eliminating the confounders.

Exact control and recording of the confounding factors when designing the study are crucial for the application of statistical methods of evaluation. It is not appropriate to use sophisticated methods to demonstrate that certain constellations of factors produce particular effects if these effects result largely from distorted random sampling and are not reproducible.

Inference

The health of the individual is always a paramount consideration. Nevertheless, it is a concern that in no way influences the scientific outcome of a cross-sectional study. The following practical requirements directly or indirectly concerned with the health protection of the working population must always be satisfied upon completion of a cross-sectional study.

— Persons suffering from or suspected of having a disease caused by factors to which they are exposed at work must be subjected to appropriate diagnostic and therapeutic procedures and recognized as having an occupational disease.
— Persons with relevant diseases of all kinds must be advised to have diagnostic check-ups and appropriate health care.
— Proposals must be submitted to the works manager and works physician for improvements in working conditions and for medical surveillance, with concrete advice on disabilities when this is indicated by the results.

All pertinent aspects of the study design must be described clearly when evaluating and reporting on the study. All uncritical, exaggerated or biased interpretations must be avoided in the evaluation (7).

The significance of statistical tests of the results of cross-sectional studies must not be overemphasized, as it has always to be borne in mind that undetected confounding factors may provoke a bias. The extent to which the hypotheses resulting from the study have been verified should be pointed out clearly. The investigator must report the outcome of the study in a way that affords all experts ample opportunity for an open and fruitful discussion.

References

1. **Bräunlich, A. et al.** [Methodological basis of epidemiology concerned mainly with health impairments caused by exposure to noxious chemicals]. *Zeitschrift für die gesamte Hygiene*, **24**: 237–247 (1978).
2. **Hernberg, S. et al.** Eight-year follow-up of viscose rayon workers exposed to carbon disulfide. *Scandinavian journal of work, environment and health*, **2**: 27–30 (1976).
3. WHO Technical Report Series, No. 535, 1973 (*Environmental and health monitoring in occupational health*. Report of a WHO Expert Committee).
4. **Schaefer, H. & Blohmke, M.** [*Social medicine, introduction to results and problems in medical sociology and social medicine*], 2nd ed. Stuttgart, Thieme, 1978, pp. 147 et seq.
5. **Häublein, H. -G. et al.** *Epidemiological investigations on correlations between working conditions, work load and health of workers*. Geneva, World Health Organization, 1980 (unpublished document 02/181/51(A)).
6. **Enderlein, G.** [The planning and statistical evaluation of epidemiological cross-sectional and longitudinal studies]. *Zeitschrift für die gesamte Hygiene*, **27**: 246–252 (1981).
7. **Hernberg, S.** Epidemiologic methods in occupational health research. *Work-environment-health*, **11**: 59–68 (1974).
8. **Angersbach, D. et al.** A retrospective cohort study comparing complaints and diseases in day and shift workers. *International archives of occupational and environmental health*, **45**: 127–140 (1980).
9. **Seeber, A. et al.** [A psychological-neurological questionnaire as a screening method for recording complaints of workers exposed to neurotoxic agents]. *Probleme und Ergebnisse der Psychologie*, No. 65: 23–43 (1978).
10. **Elofsson, S. -A. et al.** Exposure to organic solvents: A cross-sectional epidemiologic investigation on occupationally exposed car and industrial spray painters with special reference to the nervous system. *Scandinavian journal of work, environment and health*, **6**: 239–273 (1980).
11. **Nie, N. N. et al.** *SPSS-statistical package for the social sciences*, 3rd ed. New York, McGraw-Hill, 1979.
12. **Mantel, N. & Haenszel, W.** Statistical aspects of the analysis of data from retrospective studies of disease. *Journal of the National Cancer Institute*, **22**: 719–748 (1958).
13. **Bishop, Y. M. M. et al.** *Discrete multivariate analysis*. London, MIT Press, 1975.
14. **Enderlein, G.** [Analysis of the influence of certain factors on one or two response items in multidimensional contingency tables]. *Biometrical journal*, **15**: 65–77 (1973).

9

The cohort study

W. Halperin et al.[a]

The goal of occupational epidemiology is to delineate the association between occupation, or some aspect thereof, such as a particular chemical exposure, and disease. There are two main ways of achieving this goal: the case–control study and the cohort study. In the case–control approach, described in Chapter 10, the epidemiologist must first define the disease of interest. He must then establish who are the cases of the particular disease and who will serve as the controls. The next task is to collect information on previous exposures through interview with the cases and controls, review of exposure records, or from some other source. In this approach, the epidemiologist seeks to uncover differences in exposure between cases and controls.

The cohort study is the alternative approach. The epidemiologist first identifies a population that has experienced a given exposure and then determines the subsequent health outcome. Rates of disease in the cohort are then compared with those in some less exposed section of the cohort, with those in another nonexposed cohort that is as similar as possible except for the exposure under study, or with the health experience of the general population. The disease rate in the exposed population divided by the disease rate in the nonexposed or comparison group is a measure of the risk ratio, known as the relative risk.

Cohorts may be established in two ways, prospectively and historically. Although cohorts may be established in the present and followed forward through time (the prospective approach), this is rarely done. The advantage of the prospective study is that it can be planned in detail by the investigator so that specific risk factors can be identified. However, this advantage is often outweighed by the cost and time involved waiting for an answer. More commonly, cohorts are established historically from existing records and followed through time to the present. This chapter is concerned primarily with historical cohort

---

[a] For a complete list of authors, see p. 180.

studies. It will discuss studies of mortality as they are conducted in highly industrialized countries, although studies of morbidity are possible as well. Within the limitations of the authors' experience, cohort studies will be discussed primarily from the perspective of their conduct in the United States.

This chapter deals with three issues, in addition to introducing the concept of the cohort. The first issue discussed is deciding whether a study is feasible. The second is explaining how a cohort study is conducted. These issues are pragmatic and include discussion of the adequacy of records, means of tracing study members, availability of comparison disease rates, computer software, etc. The third issue is describing some of the main defects in the conduct of cohort studies that lead to invalid and misleading results, such as absence of exposure and insufficient sample size.

This chapter makes no pretention to be either a recipe for conducting cohort studies for those previously untrained or a state-of-the-art methodological discussion. It aims merely to provide an introduction to cohort studies for those who would try to read reports of such studies critically, or for those who might desire to understand the scope of the available options for answering occupational epidemiological questions.

Example of a cohort study—a historical cohort mortality study of uranium miners

To illustrate the major steps and fundamental concepts of a cohort study, an investigation of uranium miners (1) will be briefly described.

Hypothesis
Most cohort studies are designed to test a specific, well formulated hypothesis. In the study of uranium miners, there had been numerous reports since the 1800s indicating long-term health effects, specifically lung disease. More recent studies suggested that the lung cancer excess involved exposure to radon gas and its radioactive decay daughters. These observations were the basis for the hypothesis that uranium miners are at a greater risk than the general public of developing certain chronic diseases, specifically lung cancer.

Study design and cohort definition
Once a hypothesis had been formulated it was decided, in view of the nature of the hypothesis and the available data, to design the uranium miners' investigation as a historical cohort mortality study.

In order to identify a cohort for the study, the data from a US Public Health Service (USPHS) medical survey conducted in the 1950s were used. Teams of USPHS investigators had systematically visited uranium mines and mills where volunteer miners were examined and questioned about symptoms and risk factors, such as smoking. From these data the cohort was established by the researchers in 1980. The cohort was defined as all males identified in the medical surveys who had worked underground in uranium mines located in the Colorado Plateau for at least one month at any time prior to 1 January 1964. Workers became eligible for inclusion in the cohort at the date of the first examination. Since no medical examinations were conducted prior to 1950, workers entered the study at the time of their first medical examination in or after 1950 and were followed up until 1977. For example, a worker who was examined in 1950, who had started working in the mines in 1910, and who had accumulated the one-month work exposure, would be included in the study cohort. Fig. 1 shows the distribution by year first employed for white male workers.

The data from the USPHS survey contained the critical information necessary to identify cohort members and trace them through time. This information included: name, social security number, race, sex, date of

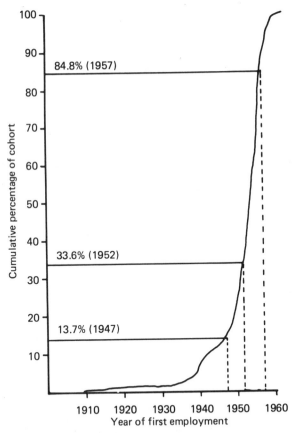

Fig. 1. Distribution of first year employed at uranium mining for white male miners

Source: Waxweiler et al. (1).

birth, and occupational history relating to the exposure or job being studied.

A medical survey such as the USPHS survey of uranium miners is not the most common source of data for identifying and defining historical cohorts. More often, records of inactive and active employees maintained by companies and occasionally by labour unions are used.

All the individuals who are eligible to enter the cohort are known as the actual population at risk. The cohort identified from the historical records should resemble the population at risk as closely as possible. Since historical cohort studies depend on existing records, which may be in poor condition or may be incomplete, it is advisable to verify that all individuals who could meet the definition of the study cohort have been identified.

Prior industrial hygiene records that can be used to estimate the exposures of interest either qualitatively or quantitatively are invaluable in a historical cohort study. In the uranium miners' study, an accumulated personal dose of radon daughter exposure was calculated in terms of working level months (2). This calculation was based on the miner's occupational history and historical records of radon daughter exposure measured in the individual mines during specific years.

Follow-up
An attempt to determine the vital status (alive or dead and cause of death) of at least 95% of the cohort is desirable. In the uranium miners' study, determination of vital status was attempted for all cohort members for the period between 1 January 1950 and 31 December 1977. The last date represents the closing date (also known as cut-off or study end-date) for the study. The analysis of the data effectively stopped at this point in time, even though the study was conducted in 1980 and individual cohort members were known to have died in the interim. The time lag allowed for completion of routine reporting and registration of deaths. Vital status was successfully ascertained for almost 100% of the uranium miners cohort. For all those miners who were known to have died during the study period (1950–1977), death certificates were requested and the underlying causes of death were coded by a professional nosologist according to the categories of the revision of the *International classification of diseases* (ICD) in effect at the time of death.

Analysis
The observed mortality experience of the cohort was compared with the expected mortality. Expected mortality was calculated by assuming that the cohort died at the same rate as a standard population, taking into account age, race, sex and calendar time. Only observed and expected deaths occurring during the study period were counted. The ratio of observed to expected deaths is a measure of relative risk known as the

Table 1. Cause-specific mortality among a cohort of white male uranium miners followed up until 1 December 1977

Cause of death	ICD category (7th rev.)	Observed	Expected	SMR
Tuberculosis	(001–109)	14	3.4	4.09
All malignant neoplasms	(140–205)	264	117.2	2.25
Liver and biliary tract	(155, 156A)	2	2.8	0.71
Pancreas	(157)	9	6.6	1.37
Larynx	(161)	0	1.9	0.00
Lung	(162, 163)	185	38.4	4.82
Breast	(170)	1	0.2	4.53
Prostate	(177)	7	5.9	1.18
Bladder	(181)	3	3.2	0.94
Skin	(190, 191)	5	2.3	2.16
Lymphatic and haematopoietic systems	(200–205)	9	12.0	0.75
Diabetes mellitus	(260)	4	8.5	0.47
Alcoholism	(322)	7	2.6	2.73
Accidents	(800–962)	155	46.8	3.31
Total		950	600.3	1.58

standardized mortality ratio (SMR). Table 1 illustrates the results obtained in the uranium miners' study.

The SMRs can be examined by exposure (measured by either length of employment or actual dose) and by time since first employed or since first exposed (latency). Examining risk by exposure is helpful in determining the existence of a positive dose–response. Examining the risk by latency is done to determine whether the risk is associated with a latency period, since most chronic occupational diseases, specifically occupational cancer, take several decades to develop after first exposure. Table 2 illustrates how lung cancer and nonmalignant respiratory disease rates are related to latency.

Of course, the decision as to how to examine the data is dependent on the hypothesis to be tested. In the uranium miners' study it was hypothesized that exposure to radiation from radon daughters is associated with an increased risk of mortality due to lung cancer. In order to test this hypothesis, the SMR for lung cancer was examined by exposure (length of employment and dose) and by latency. If the hypothesis were correct, the risk of lung cancer should increase with an increase in length or amount of exposure and the risk should become evident only after a substantial period between exposure and disease. After analysing the results in these two ways, the hypothesis was accepted as being correct.

In the following sections, more detailed discussions of important concepts, cohort selection, analysis of data, alternative approaches, and the advantages and disadvantages of cohort studies are presented.

Table 2. Observed and expected deaths due to respiratory diseases among white underground miners by years since starting uranium mining

Cause of death		Years after start of underground uranium mining (latency)						
		0–4	5–9	10–24	25–29	30–34	35	Total
Lung cancer (ICD 162–163)	Obs.	2	12	129	20	12	10	185
	Exp.	1.8	4.3	25.3	2.8	1.8	2.5	38.4
	SMR	1.11	2.79	5.10	7.14	6.67	4.00	4.82
Other non-malignant respiratory diseases (ICD 510–527)	Obs.	0	5	55	9	9	5	83
	Exp.	0.6	1.6	10.9	1.3	0.8	1.4	16.6
	SMR	—	3.12	5.05	6.92	11.25	3.57	5.00

Selection of an appropriate cohort

Once a hypothesis has been formulated, the epidemiologist must determine whether or not there is an appropriate cohort available for study. This is an extremely important step that will ultimately determine the validity of the study. Selecting an appropriate cohort is a very difficult task since the selection is based primarily on historical information.

Finding a cohort

The first step is to define the epidemiological characteristics of the cohort. Specific issues that need to be considered include:

1. The number of workers must be sufficiently large to permit detection of the hypothesized health outcome. Sample size estimates should be made before the search for the cohort begins.

2. The period between the initiation of exposure and the end-date of the study must be sufficiently long to allow development of disease. Exposures must have commenced early enough to allow development of disease in the study population.

3. The level of exposure must be adequate. For example, with a known carcinogen, the hypothesis may be that a low level of exposure is related to disease. However, to establish an exposure as a health risk, the study design should call for the highest available historical exposure. At no time should the epidemiologist accept a study population without evidence of actual exposure, as may be the case in a well engineered chemical production plant.

4. Exposures other than the one of interest complicate interpretation of the study results. It is desirable to find a cohort exposed only to the agent of interest.

5. Historical exposure data should be available in order to permit interpretation of the study findings in relation to the level of past exposure. In the absence of actual exposure data, other information must be used to estimate past exposure.

6. Records must be available to establish who was exposed and, consequently, who is to be studied, as well as to allow tracing of the individual through time to establish the health outcome. Standard items needed are social security number, date of birth, name, race, sex, and job history at the plant.

Clues to aid in finding the cohort

After defining the requirements for the cohort, the search begins. The first step is to learn as much as possible about the agent of interest, including how it is manufactured, what are its uses, and where in industry it can be found. There are many sources of information that can be helpful in answering these questions.

Several countries, including Finland, the Federal Republic of Germany, Hungary, Sweden, Switzerland, and Yugoslavia, are establishing registers of workers exposed to carcinogens. These registers are intended to identify workplaces where exposures exist. In the European Economic Community the European Inventory of Existing Chemical Substances (compiled in accordance with 67/548 Directive) could be used to locate exposures of interest. In Greece, the *Financial directory of greek companies* contains (*a*) an alphabetical list of 5 000 enterprises including for each the address, number of employees and financial data, and (*b*) an alphabetical list of products and the names of the companies that produce them. In Finland, the *Industrial statistics of the Central Statistical Office of Finland* includes data by industry on (*a*) the address, (*b*) the number of employees, and (*c*) the names and amounts of the chemicals used.

A number of sources that are useful in the United States are listed below, with a short description of the type of information each contains.

1. *The condensed chemical dictionary* (*3*) lists chemical name and trade name. It provides a brief description of chemical structure and physical properties, how the chemical is manufactured, hazards associated with the chemical, and major uses of the chemical.

2. *The Merck index (4)* is similar to *The chemical dictionary* as regards types of information provided. It also provides several cross-references for finding proper standardized (IUC/IUPAC) names from common names, trade names, or chemical structures.

3. The *National occupational hazard survey (NOHS)* (*5*) contains data on a sample of industry consisting of 4500 industrial sites throughout the United States surveyed during 1972–1974. The data available from the NOHS programme are extensive and include: estimates of total numbers of workers exposed to specific chemicals, estimates of numbers of workers exposed in specific Standard Industrial Codes (SIC) or occupational codes, and a distribution of industries or occupations in which exposure to specific chemicals has been observed.

4. The *Directory of chemical producers* (*6*) provides information on the products manufactured by 1500 companies in the United States and also an alphabetical list of products and their manufacturers. From this source, it is possible to find the names and addresses of all US companies producing a specific chemical (or product containing that chemical).

5. The *Chemical economics handbook (CEH) (7)* presents a wealth of information on the status of raw materials, primary and intermediate chemicals, and product groups (e.g., plastics and fertilizers). A typical CEH report includes: (1) status, (2) prospects, (3) manufacturing processes, (4) producing companies, plant locations, and capacities, (5) production and sales, (6) consumption, (7) price and unit sales values, (8) international prospects, and (9) other pertinent statistics and information. This is a valuable source for locating specific plants producing and using a specific chemical, including information on how much they use and thus how many workers may be exposed.

6. *The National Trade and Professional Associations and Labor Unions (of the United States and Canada) (8)* lists all trade associations

and labour unions, their addresses and phone numbers, the names of the current presidents, and any publications. These groups usually have information that can help in locating a cohort for study.

7. *The Dun and Bradstreet directories* (9) comprise the following. *Dun's census of American business* lists 4.5 million US businesses by SIC codes and includes information on the number of employees, annual sales, and a listing of the businesses by state. The *Million dollar directory* lists 120 000 US businesses with a net worth over US $500 000. This directory has three separate listings: alphabetically by business, geographically for all businesses, and a listing by SIC codes (products). The information on each business includes the address, number of employees, annual sales, SIC codes covered by the business, and the names of company officers.

These various sources should provide enough information to start contacting specific plants and unions.

Going to the field

The next step is to visit plants that are candidates for inclusion in the study in order to conduct walk-through surveys. During the course of the visits, the following information is gathered: availability of useful record systems; number of workers exposed; length and level of exposure; historical exposure data; make-up of the workforce (average age, sex, ethnicity); controls used to limit exposure; and major process changes over time. Even during this preliminary visit it is valuable to collect a few environmental air or bulk product samples to estimate current exposure.

At the end of several walk-through surveys, there should be enough information available to choose a cohort or decide if a cohort cannot be found.

Assessment of exposure

As mentioned earlier, critical questions that must be addressed in every cohort study are: has there been exposure and if so, what was the level of exposure? The assessment of exposure is typically the function of the industrial hygienist, who performs several tasks, including measurement of current levels of exposure. This sometimes requires the development of analytical methods for sample analysis. The industrial hygienist assesses historical measurements of exposure for adequacy of the methods employed and estimation of the level of exposure. In the absence of past measurements, the industrial hygienist reviews changes in process and control measures in order to arrive at a qualitative estimate of the probable level of exposure.

Even though the industrial hygienist establishes the fact that exposure has taken place, the task is not yet complete. Jobs within plants are not usually homogeneous with regard to exposure. If all employees are assumed to have been exposed, when in reality exposure was limited to a segment of the cohort, there is a possibility of obscuring a real excess in mortality. This is because the adverse

157

mortality experiences of the exposed population are diluted by the normal mortality experience of the nonexposed. The industrial hygienist therefore analyses the production process to identify locations of potential exposure and job titles with potential exposure. He also identifies extraneous exposures that may complicate the analysis.

The degree of success achieved in each of these steps varies considerably from one study to the next. In some studies, there is no consistent system of job titles and some form of classification must be developed by the study team. Corn & Esmen in *Workplace exposure zones for classification of employee exposures to physical and chemical agents (10)* offer a useful guide for assigning workers to exposure zones to yield a personal exposure value.

The search for historical exposure data can proceed in one of three ways. The ideal condition is when good quality historical exposure data are available. Even in this case, however, results can be compromised, since industrial hygiene sampling methods may have changed. For example, past data may have been collected using an antiquated sampling method, whereas a more modern (and accurate) method is used at the present time. This problem may be partially overcome by conducting sampling with the antiquated method and correlating the results with those obtained by a modern method; an alternative solution is simply to continue to sample using the antiquated method in order to gather sufficient present-day data to correlate with personnel records.

The second possibility is that good historical data do not exist but the investigator can resort to indirect reconstruction of past exposure from other correlatable exposure information. Such a reconstruction may be effected by identifying a parallel agent and comparing exposure data for the contaminant under study with available data for the parallel agent. Suppose, for example, that it is desired to determine individual workers' exposure to asbestos over an extended historical period but specific asbestos exposure data do not exist, although data on total dust exposure are available. Since exposure to asbestos could be correlated with total dust exposure, an estimate of past exposure to asbestos could be established.

The third possibility is that neither historical data nor correlatable data exist for reconstruction of past exposures. In this case, the investigator can only make a subjective estimate of exposure over a historical period. Certain strategies can be employed in this exercise. Some theoretical considerations are presented by Esmen (*11*). Influences that must be considered include: (1) changes in the process, (2) changes in the physical parameters of the agent, and (3) use of personal protective devices. In addition, the author suggests that changes in rate of production clearly influence estimated exposures.

Adequacy of personnel records
In a prospective cohort study the epidemiologist can define the detailed information to be collected. In a historical cohort study the epidemiologist is dependent on the scope and quality of information in existing

records. The adequacy of records determines not only whether a study can be done, but also the comprehensiveness of a study. At least two types of information must be available about a study population from a specific work site: that which identifies employees and that which defines employee exposure. Such information is usually found at the particular work site but can possibly be obtained from other sources, such as trade union, insurance, or government records.

The information must identify all employees who meet the requirements for entry into the cohort. The characteristics of the population to be studied, known as the population at risk, are set in the planning stage of the study by the epidemiologist. The population at risk does not necessarily include all exposed employees at the plant.

The epidemiologist may wish to include only some of those exposed. Typically, he will include all employees who worked during the time period when exposure is known to have occurred. A minimum length of employment prior to entry into the cohort is often specified. This is done to limit the number of cohort members to be traced, and hence, the cost of the study, and to eliminate those with short employment if their exposure is judged to be trivial. The epidemiologist may wish also to restrict the cohort to specific age or sex categories or sections of factories where exposures are of interest.

Inadvertent omission from the cohort study of employees who meet the definition of the population at risk may introduce bias. Although bias will probably not be introduced if the omissions are random, it may be introduced if the omissions are more or less systematic and related to the disease or exposure of interest.

Records must contain personal identifiers (items that identify the individuals being studied) specific to each country. This information is required so that the current health status (i.e., whether the study members are alive or dead and whether they have developed diseases) can be determined. In the United States, this information includes name and social security number. Descriptive information must also be included, such as birth date, ethnic group, sex, and any other factors known to influence the disease of interest. This is important so that the influences of these variables can be allowed for in the analyses. If such information is available from multiple record sources, the researcher is afforded the advantage of cross validating its accuracy and verifying completeness of the cohort.

Additional personal information that is frequently available at work sites can be very helpful in conducting cohort studies. This information includes: the names of next of kin, parents, spouse, children, references, and person to be contacted in case of accident; home address; doctor; insurance carrier; former employers; etc. This information helps to locate individuals so that their current health status can be ascertained.

Adequacy of study size and statistical power
In planning a cohort study, the investigator usually determines what study size is adequate to test the hypothesis. It is obvious that, in

159

general, the larger the study, the greater the sensitivity of the study to detect an excess in disease. In practice, a balance must be reached between the monetary cost of increasing the size of the study and the benefit of increased sensitivity. The size may also be limited by the number of exposed individuals available for study.

Adequate study size is closely related to the issue of statistical power. For a full understanding of statistical power and the estimation of an adequate sample size, a brief discussion of Type I and Type II errors is necessary.

Table 3 demonstrates four situations: two in which the results of a study are in concordance with reality, and two in which the results are discordant. The discordance that occurs when the study erroneously suggests a difference in disease rate between exposed and nonexposed individuals, is called Type I error, also known as alpha error or false positive. Translated to the occupational setting, an example of Type I error would be when the researcher concludes that agent x is significantly related to disease y when, in fact, the association in the study is due to chance. Type II, beta or false negative error, is actually more cause for concern in the occupational setting than is alpha error. In this situation, although it is concluded that the disease rate in the exposed population is not substantially different from that in the nonexposed, in reality the agent is related to the disease excess. The study is falsely negative. In some sense, the agent is exonerated and exposure will continue. This is a dangerous situation for the exposed worker.

Table 3. Comparison of real and observed situations involving Type I and Type II errors

Observed situation	Real situation	
	Difference	No difference
Difference	Concordance	Type I error (alpha or false positive)
No difference	Type II error (beta or false negative)	Concordance

The power of a study, defined as one minus beta, is the probability of avoiding a false negative result. In other words, power is the ability of the study to detect a difference in disease rate between the exposed and nonexposed population (12).

Study size can be estimated as a mathematical function of the Type I error level to be used in the significance test, the size of the relative risk (or SMR), and power. The Type I error level is chosen by the investigator and is usually set at either 0.05 or 0.01. For example, at the 0.05 level the researcher accepts as significant differences that may be due to chance alone 5% of the time. Power is also set by the investigator, usually at either 0.80 or 0.90.

The relative risk can occasionally be estimated from previous studies. However, often an arbitrary relative risk, such as 2.0, is used. It should be evident from the formula for power, as well as intuitively, that the power of a study increases with an increase in the strength of the association between exposure and effect reflected by the relative risk.

Several formulae exist for estimating adequate power in retrospective cohort studies (13–15). For studies in which rates from the national population or another large population are used for indirect standardization, formula 1(a) is appropriate (15):

$$Z_{1-\beta} = Z_a - 2(\sqrt{RR} - 1)(\sqrt{E}) \qquad 1(a)$$

where $Z_{1-\beta}$ = the standardized normal deviate corresponding to the power $(1-\beta)$, Z_a = the standardized normal deviate corresponding to the Type I error (a) level, RR = the relative risk (or SMR), and E = the expected number of deaths.

Formula 1(a) can be rearranged to formula 1(b) in order to estimate the sample size of a study when the power and relative risk are specified:

$$E = \left(\frac{Z_{1-\beta} - Z_a}{-2(\sqrt{RR} - 1)} \right)^2 \qquad 1(b)$$

The sample size in terms of person years can be estimated by dividing E by the mortality rate for the disease in the general population.

As an example, suppose an investigator is designing a cohort study to examine the possible relationship between a certain agent and the development of lung cancer. The investigator decides that the study must have a power of 0.8 $(Z_{1-\beta} = -0.84)$ for detecting a relative risk of 2.0 at a Type I error level of 0.05 $[Z_a(1 \text{ tail}) = 1.645]$. By using these numbers in formula 1(b), it is determined that $E = 9$ deaths, which, when divided by a lung cancer mortality rate of 10 per 100 000, results in a study size requirement of 90 000 person years.

It should be recognized that the largest, most powerful study will not avoid a false negative result if other conditions are not met. For example, the study population must have adequate exposure and the study must cover an adequate observation period to allow for development of disease.

Importance of sufficient latency

In occupational cohort studies, latency is commonly defined as the time from onset of exposure to measurable outcome. It is generally accepted that cohort mortality studies aimed at estimating relative risk for cancer should allow for 20 to 30 years of latency. This assumption is based on the latency distribution of known occupational carcinogens (16). The latency characteristics of the potential cohort should be examined when designing a cohort study. If the latency is insufficient, the study will have a high probability of underestimating the true risk because many cases of the associated disease will not have had time to develop.

The investigator may choose to identify a more suitable cohort or to postpone mortality follow-up until sufficient time has elapsed.

Insufficient latency is particularly problematic when the agent of concern has been in commercial use for 20 years or less. For example, the animal carcinogen (*17*), 4,4-methylenebis(2-chloroaniline) (MBOCA), a curing agent for isocyanate-containing polymers came into use in the United States in the 1960s. The most probable target organ for MBOCA in humans is the bladder. The reported latent period for known occupational bladder carcinogens varies from mean or median values as low as 18 years to as high as 44 years (*18*). Thus, a retrospective cohort mortality study of workers exposed to MBOCA, if conducted around 1980, would have had a high probability of giving a false negative result because less than twenty years had elapsed since the first exposure. Any relative risk estimated would be likely to underestimate the true risk. Because diagnosis of bladder cancer generally precedes death from bladder cancer by several years, and many patients diagnosed never die from the disease, an incidence study would be more sensitive to the adverse outcome earlier in the latency period.

Insufficient latency is an important problem in occupational health research because many new agents have been introduced into the work environment during the past 20 years. Although a conclusive assessment of the mortality risks associated with exposure to these agents must await sufficient latency, early indications of risk may be obtained by examining disease incidence or mortality early in the latency period. Caution should be exercised in the interpretation of these early studies to avoid a false sense of safety.

Length of exposure

In designing cohort studies, individuals with short lengths of employment may be excluded. The rationale for excluding workers with short employment, and hence, minimal exposure, is that their inclusion will dilute the overall relative risk. An underlying assumption, which is generally borne out by epidemiological studies, is that short exposures are less biologically significant or that their risk may be too small to be detected.

On the other hand, for agents with established overall relative risks, it is valuable to design studies that can estimate the risk associated with short exposures. For example, only one study of aromatic amine bladder carcinogens has had sufficient numbers of workers with short exposure to permit estimation of the relative risk for exposure of less than one year. This study demonstrated that even very short exposures to these agents substantially increased workers risk of death from bladder cancer (*19*).

Methodology of cohort studies

Once the cohort has been identified and defined, the data on individual workers are usually coded onto a computer file. Data on mortality outcome are then collected and added to the file.

Vital status follow-up

The next activity when conducting a retrospective cohort mortality study is determining the vital status of cohort members—in other words, verifying whether the cohort members are alive or deceased as of a specified date. In the United States, because of a delay in reporting death and filing certificates, follow-up efforts are usually directed to ascertaining vital status at a point in time several years prior to the date at which the study is undertaken. The date of vital status ascertainment is usually referred to as the "study cut-off date" or "the study end-date" for observation of observed and expected deaths. The means of verifying vital status differ, depending on the characteristics of the study cohort and the record systems available in the country where the study is being conducted.

Before an effective strategy for vital status follow-up can be planned, it is necessary to know the demographic characteristics of the study cohort. Important characteristics include such factors as race, sex, age, geographic location, calendar time of employment, and nature of the particular occupation being studied. A knowledge of these characteristics helps in choosing the most appropriate follow-up sources to use.

One strategy for vital status follow-up of a worker cohort in the United States is illustrated in Fig. 2. The United States has several systems of records (e.g., Social Security Administration, Internal Revenue Service, Division of Motor Vehicles) that are centralized for the entire country or for regions of the country. Access to these systems can be gained either manually or by computerized searches. Except for the National Death Index, these follow-up sources exist for reasons other than health research and are not ideal for vital status follow-up. Therefore, the data retrieved must be interpreted cautiously and sometimes a second source should be used for verification.

After the centralized record keeping systems have been exhausted, other sources of vital status information are used. These include various "decentralized" records, such as phone directories. These sources are more difficult and more costly to use since they are not computerized, and since they cover a relatively small geographical area (usually requiring personnel to visit the local area to do the actual tracing).

To obtain meaningful information from any of these record systems, certain identifying data are required to facilitate accurate matching. The nature of these data varies, depending on the record system being accessed. Most require, at a minimum, the social security number and the last name (often the date of birth is also needed). Therefore, it is extremely important that these identifying data be accurately coded before they are submitted for retrieval of vital status information.

In contrast to the United States, some countries, such as Italy, must rely primarily on information that is decentralized or maintained by cities and localities. An illustration of a follow-up strategy of this type is given in Fig. 3. Again, certain information on each cohort member is necessary, or at least desirable, for this strategy to be successful. In the case illustrated in Fig. 3, it is desirable to know the individual's name, date of birth, place of residence, or place of birth.

163

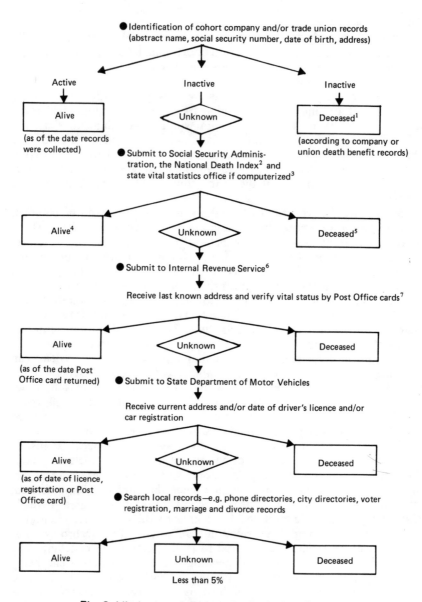

● Identification of cohort company and/or trade union records
(abstract name, social security number, date of birth, address)

Active

Alive

(as of the date records
were collected)

Inactive

Unknown

● Submit to Social Security Administration, the National Death Index[2] and
state vital statistics office if computerized[3]

Inactive

Deceased[1]

(according to company or
union death benefit records)

Alive[4]

Unknown

Deceased[5]

● Submit to Internal Revenue Service[6]

Receive last known address and verify vital status by Post Office cards[7]

Alive

Unknown

Deceased

(as of the date Post
Office card returned)

● Submit to State Department of Motor Vehicles

Receive current address and/or date of driver's licence and/or
car registration

Alive

Unknown

Deceased

(as of date of licence,
registration or Post
Office card)

● Search local records—e.g. phone directories, city directories, voter
registration, marriage and divorce records

Alive

Unknown

Less than 5%

Deceased

Fig. 2. Vital status follow-up in the United States
using centralized and decentralized records

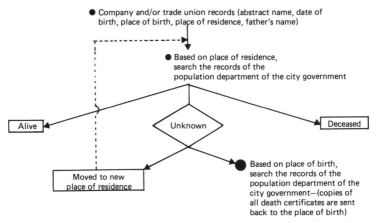

Identification of cohort

● Company and/or trade union records (abstract name, date of birth, place of birth, place of residence, father's name)

● Based on place of residence, search the records of the population department of the city government

Alive

Unknown

Deceased

Moved to new place of residence

● Based on place of birth, search the records of the population department of the city government—(copies of all death certificates are sent back to the place of birth)

Fig. 3. Vital status follow-up in Italy using decentralized records (local registries)

There are also several countries where strategies for vital status follow-up depend upon both centralized and decentralized systems of records. Of course, certain information on each cohort member is necessary for these strategies to be effective. In the United Kingdom, tracing members of a cohort who are not currently employed can start with searches of records at the medical unit, discharge records in the

---

Fig. 2 footnotes:

[1] Whenever there is an indication that an individual is deceased, this is followed by searching the records of appropriate state vital statistics office for the death certificate.

[2] The National Death Index (operated by the National Center for Health Statistics) maintains a computer file that includes identifying information on all deaths that have occurred in the USA since 1979. The index is updated yearly. The state in which death occurred, date of death, and death certificate number are given.

[3] This should be done only if the state vital statistics office in the state where the study is being conducted (and possibly contiguous states) maintains a computerized file of their death data.

[4] For the Social Security Administration data, the workers are assumed to have been alive approximately two years prior to the current date, based on earnings or on benefit payments.

[5] For the Social Security Administration, information on deceased individuals is divided into three group: (1) state where claim was filed, (2) state of last residence, and (3) funeral director's notice. For the first two groups, information on the date of death and state in which death occurred is given.

[6] Access to the Internal Revenue Service is available only to the National Institute for Occupational Safety and Health (NIOSH).

[7] In this activity, the worker's name and most recent address is submitted (on a postcard) to the regional post office to verify whether or not the individual is still living at that address. If he is not, the post office may be able to indicate whether the worker is alive or deceased or, if the worker has moved, to provide a forwarding address. The process is then repeated with the new forwarding address. One could also mail a registered letter with return receipt requested.

165

personnel department of the company, and trade union records. Subsequently, information from registers maintained by the Department of Health and Social Security is sought. In addition, vital status can be ascertained through the Office of Population Censuses and Surveys, where the causes of death are also coded. Presumed survivors may be contacted at addresses recorded in the company records or collected from electoral rolls. In Denmark, the study subjects can usually be traced in either the Danish National Central Person Register, in which all persons alive in 1968 or later are registered, or in the Danish Central Death Register, which is a manually operated microfilm register of all deaths since 1948. Vital status of the remaining subjects can be obtained from the municipal population registers.

Regardless of the methods used, the aim of vital status follow-up should be the same, namely, to verify the vital status, at the most recent date possible, for at least 95 % of the cohort. For those who are deceased, a copy of the death certificate must be obtained to verify the fact of death and to supply the date and the cause of death.

Death certificates

In most cohort mortality studies, the cause of death is defined in terms of information contained on the death certificate and coded according to the appropriate revision of the *International classification of diseases* (ICD). This coding of the cause of death follows a complex set of rules that change from time to time. Coding should be done by an experienced professional nosologist.

In determining the cause of death, examination of hospital records and autopsy reports may result in a more accurate determination than one based entirely on the examination of death certificates. Even when such detailed records are available for the exposed employee groups, comparable detailed information is not generally available for the comparison group. Hence, observed and expected mortality should be based on the cause of death as recorded on the death certificate.

Misclassification of the cause of death in an important issue. In a study of 48826 cancer deaths in the United States for which hospital records were available, there was only 65 % concurrence between cause of death on the certificate and that indicated in the medical record (20). However, this misclassification was not consistent for all cancer sites. Of the ten leading types of cancer, more than 80 % of the deaths for seven sites (lung, breast, prostate, pancreas, urinary bladder, haemopoietic tissue, and ovary) were confirmed by the hospital records. For others, the confirmation rate was much lower. For deaths due to bone cancer, only 50 % were confirmed.

There are two types of misclassification. First, the physician may improperly complete the death certificate or may ascribe the cause of death to some condition that does not accurately reflect the actual medical diagnosis (21). Second, despite accurate information regarding the cause of death, the underlying cause of death might not be coded properly by the nosologist.

166

The coding rules of the ICD are revised approximately every 10 years to take account of changes in disease classification and medical terminology. Therefore, if a study spans more than one revision of the ICD, the cause of death codes must be reclassified so that they are compatible over time. The nosologist must first accurately code the cause of death according to the rules in effect at the time of death. The epidemiologist must then establish rules for combining cause of death categories over the various ICD revisions.

Comparison disease rates
Comparison rates should be chosen according to the needs of the specific cohort analysis. Rates for large populations, such as the general population of the United States, are often preferred because they are large enough to provide stable rates even for rare diseases and causes of death. It might be necessary, however, to use regional, state, or local rates if the disease or cause of death in question is not distributed equally by region. Likewise, it might be necessary to use regional or state rates to obtain the correct rates by race or ethnic group. For example, in the United States, the nonwhite death rate for the general population is dominated by black Americans. If a researcher wishes to analyse the death rates for American Indians (e.g., Navajos), it would be necessary to use the nonwhite death rates from a state such as New Mexico, where there are many Navajos and relatively few blacks. One potential disadvantage of using local rates for comparison is that the rates may be elevated by the disease contribution of the undertaking being studied or of other industries causing the same disease.

Comparison rates from a large worker population are especially useful for occupational cohort studies because they reduce the problems in comparability created by the "healthy worker effect" (see pp. 174–175). In the analysis of lung cancer and other respiratory diseases, rates by smoking category would be very useful but, sadly, they are not usually obtainable. Within the realm of the comparison rates that are available, the researcher should use those rates that come closest to measuring meaningful differences in morbidity or mortality while remaining alert to the need for possible analytical adjustments occasioned by the use of less than ideal comparison rates.

Comparison rates have as their numerator the number of cases of disease or deaths that occurred over a specific interval of time. For example, the State of Connecticut maintains a statewide tumour registry. The numerator for the cancer incidence rate consists of the number of new cases of cancer diagnosed. Likewise, the United States maintains general population mortality data; the numerator for the mortality rate consists of the number of deaths that have occurred. Comparison rates have as their denominator the number of persons at risk over the specified interval of time. For example, the denominator for the cancer incidence rate of Connecticut consists of the number of persons in Connecticut at risk of acquiring cancer. The denominator for the United States mortality rate consists of the number of persons in the midyear population of the United States as interpolated or extrapolated from the census years. Comparison rates are usually expressed as the number of cases or deaths by cause per 100 000 persons at risk by year.

Calculation of observed and expected events
The results of occupational cohort studies are usually expressed as a comparison of the observed and the expected number of events (death or disease). The ratio of the observed to the expected number is called the mortality or morbidity ratio. To obtain the expected number of events, person-years at risk in the workers must be calculated.

Person-years at risk
Person-years at risk of developing disease or dying are the years of observation that each worker contributes to a cohort study. For example, if a worker entered a study at its beginning in January 1950 and survived until the end of the study in December 1979, he contributed 30 years. These 30 years are "at risk", since the death of a worker at any time in that interval would be counted. While the worker was alive prior to 1950 and in 1980 or later, those years were not "at risk" because a death occurring during those years would not have been counted.

There are two principal advantages to counting person-years rather than simply persons. First, person-years take into account differences in the amount of time workers contribute to a study. For example, a worker who is in a study for one year will obviously have much less chance of dying from an exposure than a worker who is in a study for forty years, and such workers should not be considered equally. Second, because workers' characteristics change while they are in a study, e.g., they get older and accumulate exposure, it is difficult to put workers into a single classification of these variables. Person-years, on the other hand, can be allocated to many different categories. While the worker is young his person-years are included with the person-years of other young workers, and as he ages his person-years are included with those of other older workers.

When person-years are calculated for the workers, expected deaths are easily derived from comparison death rates (e.g., national rates), because comparison death rates are also expressed in terms of person-years. For example, a death rate of 100 per 100 000 persons per year in the standard population is equivalent to 100 per 100 000 person-years. Thus, if there are 10 000 person-years in the cohort study, 10 deaths are expected.

Fig. 4 gives examples of workers contributing person-years to a study where the cohort is defined as "all workers employed in or after 1950" and where the study ends on 31 December 1979. (The model is simplified to a scale of full calendar years.) Worker A contributes 30 person-years and is alive at the end of the study. Worker B contributes 30 person-years, despite the fact that he was observed for 35 years, because only person-years in the 1950–1979 period are "at risk". If workers die prior to 1950, their deaths are not included in the study

168

because of the cohort definition (employed after 1950), thus person-years prior to 1950 are not "at risk" (see section on "Immortality", p. 175). Worker C contributes 20 person-years to the study and is counted as a death. Worker D contributes 20 person-years and is counted as alive, despite the fact that he died in 1981, because he was alive at the end of the study (in 1979). All four workers, then contribute a total of one

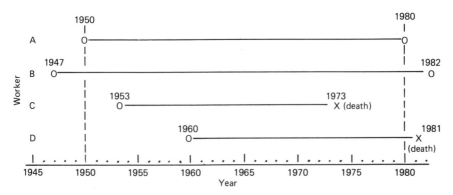

Fig. 4. Observation of four workers in a cohort study covering all workers employed in or after 1950 and observed to the end of 1979

death and 100 person-years. If the death rate in the standard population is 800 per 100 000 per year, then approximately 0.8 deaths are expected. The standardized mortality ratio is 1/0.8 or 1.25.

Standardization to avoid confounding
Comparison of crude mortality rates among populations that differ with respect to factors that significantly affect mortality (such as age, sex, and race) may be misleading. Standardization of rates in order to achieve comparability has been described in Chapters 5 and 7. Further information can be found in the relevant literature (e.g., refs. *22–24*).

The bias that can occur when mortality ratios are not standardized is called confounding. Confounding occurs when a disease rate is associated with an extraneous variable (i.e., one not of direct interest) and the exposed and nonexposed populations differ with respect to the extraneous variable. For example, lung cancer mortality rates increase with age. If the exposed population is younger than the nonexposed population, the expected number of lung cancer deaths (based upon the rates of the nonexposed) will be too large.

The data in Table 4 illustrate what can happen when the death rate increases with age. If only the total figures are used (no consideration of age) the overall mortality ratio is 1.50. However, the ratio for each age category is 2.00, and it is apparent that the correct overall ratio should

Table 4. Example of confounding when the exposed and nonexposed populations have different age distributions and death rates increase with age

| | Age (years) | | |
	Under 50	Over 50	Total
Nonexposed population			
Person-years	1000	2000	3000
Number of deaths	20	100	120
Death rate	0.02	0.05	0.04
Exposed population			
Person-years	2000	1000	3000
Observed deaths	80	100	180
Expected deaths[a]	40	50	120[b]/90[c]
Mortality ratio	2.00	2.00	1.59[b]/2.00[c]

[a] based upon nonexposed rates.
[b] if "total" data are used; 120 = 0.04 × 3000.
[c] if age-stratified data are used; 90 = 0.02 × 2000 + 0.05 × 1000.

also be 2.00. Thus, if stratification by age were not performed the mortality ratio, 1.50 in this example, would be biased (confounded) in that it would be much too low.

Examination of exposure and latency
In addition to calculating an overall standardized mortality ratio, it is often helpful to subdivide the results into categories of exposure (measured by length of employment or actual dose) and latency. Trends for the variables can then be examined. When mortality ratios increase with exposure or latency, this is further evidence that an exposure is harmful.

To create categories of exposure and latency, the person-years must be further stratified by these variables. Since the person-years are already stratified by age, calendar time, race, and sex (to avoid confounding), further stratification by exposure and latency means that a six-dimensional matrix of person-years is needed. Expected deaths are calculated for the cells in the matrix, and each observed death is allocated to the appropriate cell according to the exposure and latency interval when the death occurred. The observed and expected numbers can then be combined and displayed in any way that is interesting to the investigator. A common method of displaying exposure and latency is the two-way stratification illustrated in Table 5. With this method, exposure can be examined within latency strata and vice-versa. These are useful analyses because it is often required to examine the effect of one variable while controlling for the effect of the other.

Consideration of calendar time period
Ideally, the person-years at risk that are accumulated in a cohort should be only those contributed by persons who experienced the risk factor. Inclusion of any irrelevant person-years artificially inflates the

Table 5. Lung cancer in uranium miners: Examination of exposure and latency using two-way stratification of standardized mortality ratios

Latency (years)	Exposure (years of employment)				
	0–9	10–19	20–29	30+	Total
0–9	2.30				2.30
10–19	3.77	12.84			4.91
20–29	3.22	12.94	8.96		6.01
39+	0.00[a]	9,69	11.22	0.00[a]	5.15
Total	3.13	12.18	10.69	0.00[a]	4.82

[a] No deaths were observed.

expected number of disease events. This decreases the overall mortality ratio. It is best, then, to confine the study time period to one that allows both exposure experience and sufficient latency. For instance, in an evaluation of a plant that began operations in 1950 but introduced a particular chemical substance in 1960, it would be undesirable to count the person-years of individuals prior to 1960. It would also be undesirable to count person-years of individuals whose initial exposures occurred close to the study cut-off date, because insufficient time would have been allowed for the disease to manifest itself. The cohort could also then be limited to those persons who first experienced exposure 10 years prior to the study end-date. If the period of observation ended in 1980, the entire cohort would be defined as those persons who first experienced exposure between 1960 and 1970.

Consider the case of a study at a plant in operation since 1940. Unbeknown to anyone at the time of the study, 8 of 10 persons involved in an experimental process that began in 1952 and was abandoned in 1954 have subsequently succumbed to lung cancer. This "epidemic" could easily go unnoticed if the calendar time period studied was 1940 to 1980. Only by focusing the analysis on those exposed between 1952 and 1954 would this excess be identified.

Hypothesis testing

Once observed and expected numbers of disease have been determined, the null hypothesis of no association between exposure and disease can be statistically tested.

Often the comparison rates are based on very large numbers, as in the case of national rates, and the expected numbers of deaths derived from such rates are quite stable. In this situation, the Poisson distribution is applicable and an exact probability can be calculated (25). While the exact calculation is rather cumbersome, the following Byar-Bailar formula (26) provides a suitable approximation and is much easier to use:

$$z = \sqrt{9a} \times \left[1 - \frac{1}{9a} - \frac{E^{\frac{1}{3}}}{a} \right]$$

where $z =$ unit normal deviate that corresponds to the desired level of confidence; $a =$ observed events when the observed number is greater than the expected number, otherwise $a =$ observed events $+1$; and $E =$ expected events.

When the comparison rates are based on smaller numbers, such as when local rates are used, the expected events can no longer be considered invariant, and the above Poisson test should not be used. Other tests that take into account the variability in the expected events are available (26, 27).

Available computer software

Several computer programs are available for the analysis of occupational cohort mortality data. In general, these systems use the person-years approach in comparing the mortality experience of a group of exposed workers with that of a comparison population. Software for three systems is available in the United States.

The system developed by the National Institute for Occupational Safety and Health (NIOSH) (28) has the following features: (a) the program has 89 cause-of-death categories; (b) there is an option to use either the *International classification of diseases* (ICD) rules and codes in effect at the time of death, or the rules in effect at the time of death with conversion to codes of a single revision; (c) it will identify any combination of departments and/or jobs as "exposure" areas and provide a separate analysis of their mortality patterns; (d) it can assign a cumulative dose to person-years based on exposure data organized by person or by work area; (e) it can simultaneously examine any two of the following three variables: latency (time since first exposure), duration of exposure, and dose; (f) it allows comprehensive data editing and automatic deletion or modification of the data before analysis.

A similar system, developed by Dr Gary M. Marsh, has the advantage of performing analyses of standardized mortality ratios (SMRs) and proportionate mortality ratios (PMRs) (29). Unlike the NIOSH system, this system requires that death certificates be coded by the rules of a single revision of the ICD. One unique feature is its ability to analyse by exposure accumulated prior to specified intervals.

A third system, developed by Dr Richard Monson (30), requires that deaths be coded by the rules of the ICD in effect at the time of death, with subsequent conversion of these codes to 7th or 8th revision codes. An advantage of the Monson system is its ability to analyse by age and year first exposed. One limitation is that duration of employment is considered as the interval between the date first employed and the date last employed; gaps in employment are not allowed for.

Important issues

When conducting cohort studies, there are problems that are best considered at the design stage or, at least, during study analysis.

Comparison of SMRs

Standardized mortality ratios are often compared with one another. The comparison can be across different studies, as in a recent review of six studies concerned with brain cancer in vinyl chloride workers (15), or within a single study in order to examine the effect of a variable of interest, e.g., radiation dosage categories at a nuclear shipyard (31).

While such comparisons are useful, it is theoretically possible for them to be misleading. The problem occurs when the populations being compared differ with respect to a variable that modifies the effect of the exposure. The observation period after exposure to a carcinogen is an example of an effect-modifying variable in cancer studies. If in one study the risk is observed within the first 20 years from the initial exposure, while in another study the risk is observed after 20 years from the initial exposure, then the first study is likely to show a smaller relative risk, even if all the other study parameters are the same. This is because the time that has elapsed since the initial exposure (often termed latency) is important for cancer development—the first 10–20 years show very little effect relative to later years. Cancer studies with different observation periods must therefore be compared with caution.

Table 6 illustrates how a comparison between two studies can be misleading. As seen in the latency specific mortality ratios, the exposure affects only people who have had more than 20 years of exposure. Latency, therefore, modifies the effect. Since the population used for Study No. 1 has a much higher latency, in general, than the population used for Study No. 2, the overall standardized mortality ratios of the two studies differ substantially (4.33 versus 1.67), even though the latency category-specific mortality ratios are identical. Without seeing

Table 6. Lack of comparability of total SMRs when latency is an effect modifier and latency distributions are different

	Latency		Total
	Under 20 years	Over 20 years	
Standard death rate	0.02[a]	0.02[a]	0.02
Study No. 1			
Person-years	1000	5000	6000
Observed deaths	20	500	520
Expected deaths	20	100	120
Mortality ratio	1.00	5.00	4.33
Study No. 2			
Person-years	5000	1000	6000
Observed deaths	100	100	200
Expected deaths	100	20	120
Mortality ratio	1.00	5.00	1.67

[a] The same standard death rate is used for both the under 20 and over 20 years latency categories.

the latency-specific mortality ratios, one might mistakenly conclude that the studies yielded different results.

Similarly, comparison of SMRs within a single study can be misleading, as when cumulative exposure categories are compared. The problem is that higher cumulative exposure categories are generally based upon person-years for populations that are older in age, later in calendar time, and higher in latency. If age, calendar time, or latency is an effect modifier for the exposure in question, e.g., the exposure causes a greater effect in old people than young people, then the high dosage category might show the highest SMR simply because it is different with respect to the effect modifier.

The solution to the problem created by effect modification is to report stratum-specific results of offending variables such as latency. In practice, there are few examples of age or calendar time modifying the effect of an exposure. Latency is probably the most important effect-modifying variable in cohort studies of chronic disease; therefore, category-specific results for this variable should almost always be reported.

Healthy worker effect

In occupational cohort studies the overall morbidity or mortality among a worker population is usually compared with the overall morbidity or mortality among the national population. The national population is made up of persons both healthy and ill, while worker populations are made up only of persons who are healthy enough to be employed. Pre-employment screening is often used to eliminate the applicants "unfit" for work. The worker population, therefore, often appears quite healthy in contrast to the national population. This phenomenon is known as the "healthy worker effect".

McMichael (32) pointed out the following characteristics of the healthy worker effect. (1) It declines with increasing time since first employment. It is strongest among younger workers because they are closer to the time they were hired and, hence, the time when they were healthy enough to work. (2) It is greater for nonwhites than for whites. In general, the nonwhite population has higher rates of morbidity and mortality than the white population. Socioeconomic factors play a role in this difference. Thus, in comparison with white workers, the nonwhites selected for employment experience a relatively greater improvement in socioeconomic status and improved overall morbidity and mortality rates and, hence, a greater healthy worker effect. (3) The healthy worker effect does not operate equally for all diseases or causes of death. For example, the healthy worker effect is relatively strong for ischaemic heart disease. It is usually not as important for diseases such as cancer, where clinical manifestations are not detected at the time of first employment.

One method of minimizing the healthy worker effect is to choose a working population for comparison rather than the general population

174

(33). With this strategy, the initial selection for employment should apply equally to both the study group and the comparison group. However, comparable rates for nonexposed worker populations are often not available. If the general population must be used for comparison, the healthy worker effect may manifest itself differently for the various age, race, sex, work status, and calendar time period groups within the occupational cohort. Equally important, it will manifest itself differently for various diseases and causes of death.

Immortality
Immortality results when the person-years at risk of death are, by definition, based on time that has been survived. This problem usually occurs in studies that require a minimum duration of employment for entry of the worker into the study cohort (34, 35). Because of this requirement in the cohort definition, the study members have survived that minimum length of employment and, consequently, no deaths can possibly be observed for that time period. For example, if 1000 workers meet a study requirement of five years of employment for entry into the cohort, they would accrue 5000 person-years in the next 5 years. If the comparison disease rate is 10 per 1000, 50 cases would be expected. If, however, they are incorrectly considered at risk for the initial five-year employment period, person-years would be inflated to 10 000 and 100 cases would be expected. If 100 cases of disease were observed, the real mortality ratio would be 2, and the ratio based on immortal person-years would be· 1.

The problem of immortality may be more subtle, as in assigning person-years at risk of death to exposure or cumulative dose categories (36). In this situation, workers in a study may have been employed in jobs that involved exposure in two or more exposure categories. If an individual survived the time working in the first exposure category and went on to work in a job in the second exposure category, then it would not be permissible to include the time spent working in the first exposure category in the person-years at risk attributed to the second exposure category.

Retired worker studies
Cohort mortality studies of retired workers are an alternative to investigating the mortality experience of an entire industrial population, which would include all active, terminated, and retired workers. They are particularly desirable when trade union or company records of non-retired workers are inadequate for use in a cohort study. Retired worker studies have several advantages: (1) they are relatively quick and inexpensive to conduct; (2) the vital status of retired workers is often readily ascertainable owing to the financial incentives of the pension programme; (3) retired workers usually have long employment and time since first exposure (latency); (4) because the average age of retired workers is usually over 50, there is a large proportion of deaths relative to the size of the total study group; (5) death rates and corresponding

cause-specific SMRs can be calculated; and (6) the healthy worker effect is reduced for most cohorts of retired workers (37).

Retired worker studies may be biased because only those employees who survive until retirement are studied. As a result of the basic definition of the retired worker cohort, individuals who die prior to retirement age, or who live to retirement age but do not remain in employment long enough to retire, are excluded from study. The magnitude of this bias is dependent upon the difference in mortality between those included and those excluded from the study. An example of this bias was reported in a study of workers exposed to vinyl chloride (37). In this study, it was demonstrated that workers who left a job within the first 15 years of employment had an overall standardized mortality ratio 50% higher than those still employed after 15 years. On the other hand, some occupational cohort studies have demonstrated that significant findings among the entire work force are reflected among those who have retired (38–42). Since in the extreme, disease excess may occur solely before or after retirement, it is difficult to predict whether the mortality experience of the retired worker population adequately reflects the experience of the entire exposed population.

Smoking
The effects of smoking should be considered when the disease or diseases of interest are affected by the habit, and when there is an unequal distribution of smokers between the cohort and the comparison population. If these two conditions are met, smoking is likely to be a source of uncontrolled confounding and, if possible, appropriate adjustment should be made in order to evaluate the exposure of interest.

The experience of the national population is most often used as the basis for disease rate comparisons. However, there is evidence that smoking is more common among blue-collar workers than among the national population and that they smoke more (43, 44). Therefore, when studying an occupational exposure where the cohort is composed primarily of blue-collar workers, comparison with the general population will cause an underestimation of the expected number of deaths for diseases that are affected by smoking. This problem can be substantial, especially when evaluating the effect of an occupational exposure on the risk of lung cancer, since lung cancer rates are estimated to be 4–7 times greater for light smokers (1 pack or less daily) and 7–16 times greater for heavy smokers (more than one pack daily) when compared with nonsmokers (45, 46).

It is helpful to have data on smoking for all the study cohort members. These data permit an accurate assessment of the difference in smoking patterns between the cohort and the general population with which comparison is being made. Although this situation is ideal, it is the exception. Data on smoking are rarely collected on a routine basis in industrial populations and it is extremely difficult, if not impossible, to collect such data retrospectively for all the cohort members.

An alternative procedure is to collect smoking data for a representative sample of the cohort. The difference in the observed level of smoking in the sample compared with national smoking statistics can be used to adjust the expected number of deaths for such disease categories as lung cancer. This method, although both costly and time consuming, gives an estimate taken from the actual cohort members.

A third procedure uses hypothetical estimates of smoking to adjust the number of expected deaths (47). This method permits the evaluation of various hypothesized distributions of smoking for the cohort when compared with a national smoking survey. It is possible to evaluate various hypothetical percentages of light, medium, heavy, and non-smokers. If a significant excess risk of smoking-related disease is found (such as lung cancer), this method can determine what level of smoking would be necessary to account for the observed risk.

Regardless of the method used, when evaluating the effect of an occupational exposure on a disease category that is also affected by smoking, attention must be paid to the possible differences in smoking habits between the cohort and the comparison population.

Other methods of investigating mortality among occupational groups

In addition to cohort studies, there are other methods that can be used to determine whether or not occupations or exposures are associated with unusual mortality patterns. These include proportionate mortality studies, case–control studies, and variants of the cohort study.

In a proportionate mortality study, the proportion of deaths due to a specific cause among the deceased persons is compared with the proportion for the same cause in an appropriate comparison group, adjusted by age, sex, race, and calendar period. The comparison group is usually deceased persons in the national population. The results are expressed as a relative risk estimate, the proportionate mortality ratio (PMR). The advantage of the PMR as compared with the SMR is the ease and limited cost of data collection. The major disadvantage is uncertainty about whether all deaths have been ascertained, and if not, whether biases have greatly affected the conclusions (48). PMRs are most appropriate for initial inquiries into association between exposure and disease.

In a case–control study, the proportion exposed to a risk factor among those with a particular disease is compared with the proportion exposed in those without the disease. The association between the risk factor and the disease is determined by calculation of an odds ratio (12).

In general, there are two categories of occupational case–control study: (1) the geographical case–control, and (2) the plant-based case–control. They differ in terms of the population from which the study group is chosen, and the purpose of the investigation. In the geographical study, the study group is drawn from the general population in geographical proximity to the industry of interest. The purpose of such a geographical or community-based case–control study is to identify plants or industries with excess risks of certain diseases. The plant-based study, on the other hand, is usually conducted after an excess risk has been documented at a particular plant. Cases and controls are selected only from that plant population in order to determine whether the excess is associated with a particular production area or exposure. When the cases and controls are identified through a cohort study, this approach is called a "nested" case–control study.

The advantage of the nested case–control approach is that only the detailed work or exposure history of the cases and controls need be determined. The cohort approach, on the other hand, requires detailed work or exposure history on all cohort members, a much more onerous task.

In addition to PMR and case–control studies, variants of the cohort study have been developed and used, primarily to reduce costs. If the cohort is sufficiently large and resources are limited, a cohort study of a properly chosen sample of the entire population at risk can be a reasonable alternative. An example of a sample cohort study is a study of mortality among a 10% sample of coal miners (49).

References

1. **Waxweiler, R. J. et al.** Mortality follow-up through 1977 of the white underground uranium miners cohort examined by the United States Public Health Service. In: Gomez, M., ed. *Radiation hazards in mining*. New York, American Institute of Mining, Metallurgical and Petroleum Engineers, 1981.
2. *General Industry Standards, Code of Federal Regulations*, 29 CFR 1910.1000, November 7, 1978.
3. *The condensed chemical dictionary*, 9th ed. New York, Von Nostrand Reinhold Company, 1977.
4. *The Merck Index*, 8th ed. Rahway, NJ, Merck & Co., 1968.
5. *National Occupational Hazard Survey*, Vol. I: *Survey Manual*. Washington, DC, US Department of Health, Education and Welfare, 1974 (NIOSH Publication No. 74–127).
6. *1980 Directory of chemical producers, United States of America*. Menlo Park, CA, SRI International, 1980.
7. *Chemical economics handbook*. Menlo Park, CA, SRI International.
8. *National Trade and Professional Associations of The United States and Canada and Labor Unions*. Washington, DC, Columbia Books (published annually).
9. *The Dun and Bradstreet Directories*. Parsippany, NJ, The Dun and Bradstreet Corporation (published annually).
10. **Corn, M. & Esmen, N. A.** Workplace exposure zones for classification of employee exposures to physical and chemical agents. *American Industrial Hygiene Association journal*, **40**: 47–57 (1979).
11. **Esmen, N.** Retrospective industrial hygiene surveys. *American Industrial Hygiene Association journal*, **40**: 58–65 (1979).
12. **Fleiss, J. L.** *Statistical methods for rates and proportions*. New York, John Wiley & Sons, 1973.
13. **Schlesselman, J. J.** Sample size requirements in cohort and case-control studies of disease. *American journal of epidemiology*, **99**: 381–384 (1974).
14. **Miettenin, O. S.** Individual matching with multiple controls in the case of all or none responses. *Biometrics*, **25**: 339–354 (1969).
15. **Beaumont, J. J. & Breslow, N. E.** Power considerations in epidemiologic studies of vinyl chloride workers. *American journal of epidemiology*, **114**: 725–734 (1981).

16. **Cole P. & Goldman, M. B.** Occupation. In: Fraumeni, J. F., ed. *Persons at high risk of cancer: an approach to cancer etiology and control.* New York, Academic Press, 1975, pp. 167–184.
17. **National Institute for Occupational Safety and Health, Centers for Disease Control.** Special hazard review with control recommendations for 4,4'-methylenebis (2-chloroaniline). Cincinnati, OH, US Department of Health, Education and Welfare, 1978.
18. **Matanoski, G. M. & Elliott, E. A.** Bladder cancer epidemiology. In: Nathanson, N. & Gordis, L., ed. *Epidemiology reviews,* 3: 203–229 (1981).
19. **Case, R. A. M. et al.** Tumours of the urinary bladder in workmen engaged in the manufacture and use of certain dyestuff intermediates in the British chemical industry. *British journal of industrial medicine,* 11: 75–104 (1954).
20. **Percy, C. et al.** Accuracy of cancer death certificates and its effect on cancer mortality studies. *American journal of public health,* 71: 242–250 (1981).
21. **Glasser, J. H.** The quality and utility of death certificates. *American journal of public health,* 71: 231–233 (1981).
22. **Hill, A. B.** *Principles of medical statistics,* 8th ed. London, Lancet, 1966.
23. **Liddell, F. D. K.** The measurement of occupational mortality. *British journal of industrial medicine,* 17: 228–233 (1960).
24. **Kalton, G.** Standardization: a technique to control for extraneous variables. *Applied statistics,* 17: 118–136 (1968).
25. **Armitage, P.** *Statistical methods in medical research.* Oxford, Blackwell Scientific Publications, 1971.
26. **Rothman, K. J. & Boice, J. D.** *Epidemiologic analysis with a programmable calculator.* Bethesda, MD, US National Institutes of Health, 1979 (Publication No. 79–1649).
27. **Hakulinen, T.** A Mantel-Haenszel statistic for testing the association between a polychotomous exposure and a rare outcome. *American journal of epidemiology,* 113: 192–197 (1981).
28. **Waxweiler, R. J. et al.** A modified life table analysis system for cohort studies. *Journal of occupational medicine,* in press.
29. **Marsh, G. M. & Preininger, M.** OCMAP: A user-oriented occupational cohort mortality analysis program. *American statistician,* 34: 245–246 (1980).
30. **Monson, R. R.** Analysis of relative survival and proportional mortality. *Computers and biomedical research,* 7: 325–332 (1974).
31. **Rinsky, R. A. et al.** Cancer mortality at a naval nuclear shipyard. *Lancet,* 1: 231–235 (1981).
32. **McMichael, A. J.** Standardized mortality ratios and the "healthy worker effect": scratching beneath the surface. *Journal of occupational medicine,* 18: 165–168 (1976).
33. **Monson R. R.** *Occupational epidemiology.* Boca Raton, FL, CRC Press, 1980, p. 57.
34. **Duck, B. W. et al.** Mortality study of workers in a polyvinyl-chloride production plant. *Lancet,* 2: 1197–1199 (1976).
35. **Wagoner, J. K. & Infante, P. F.** Vinyl chloride and mortality. *Lancet,* 2: 194–195 (1976).
36. **Beaumont, J. J.** Dose response and immortality in SMR studies. *Journal of occupational medicine,* 22: 574–576 (1980).
37. **Fox, A. J. & Collier, P. F.** Low mortality rates in industrial cohort studies due to selection for work and survival in the industry. *British journal of preventive and social medicine,* 30: 225–230 (1976).
38. **Collins, J. F. & Redmond, C. K.** The use of retirees to evaluate occupational hazards. *Journal of occupational medicine,* 18: 595–602 (1976).
39. **Doll, R.** The causes of death among gas-workers with special reference to cancer of the lung. *British journal of industrial medicine,* 9: 180–185 (1952).
40. **Olsen, J. & Sabroe, S.** A follow-up study of non-retired and retired members of the Danish Carpenter/Cabinet Makers Trade Union. *International journal of epidemiology,* 8: 375–382 (1979).
41. **Enterline, P. et al.** Respiratory cancer in relation to occupational exposures among retired asbestos workers. *British journal of industrial medicine,* 30: 162–166 (1973).
42. **Pinto, S. S. et al.** Mortality experience of arsenic-exposed workers. *Archives of environmental health,* 33: 325–330 (1978).

43. **Sterling, T. D. & Weinkam, J. J.** Smoking characteristics by type of employment. *Journal of occupational medicine,* **18**: 743 (1976).
44. **Covey, L. S. & Wynder, E. L.** Smoking habits and occupational status. *Journal of occupational medicine,* **23**: 537 (1981).
45. **Hammond, E. C.** Smoking in relation to the death rates of one million men and women. In: Haenszel, W. ed. *Epidemiological approaches to the study of cancer and other chronic diseases.* Washington, DC, US Department of Health, Education and Welfare, 1966 (National Cancer Institute Monograph No. 19), pp. 127–204.
46. **Haenszel, W. et al.** Lung-cancer mortality as related to residence and smoking histories. 1. White males. *Journal of the National Cancer Institute,* **28**: 947 (1962).
47. **Axelson, O.** Aspects on confounding in occupational health epidemiology. *Scandinavian journal of work, environment and health,* **4**: 85 (1978).
48. **Beaumont, J. J. et. al.** Occupational data sets appropriate for proportionate mortality ratio analysis. In: *Banbury report 9: Quantification of occupational cancer,* 1981, pp. 391–411.
49. **Rockette, H. E.** Cause-specific mortality of coal miners. *Journal of occupational medicine,* **19**: 745–801 (1977).

Authors

William Halperin
James Beaumont
David Brown
David Clapp
Marie Haring
Charles McCammon
Theodore J. Meinhardt
Andrea Okun

Gordon Reeve
Robert Rinsky
Robert Roscoe
Alexander Blair Smith
Leslie Stayner
Frank Stern
Elizabeth Ward
Theodore Bazas

Dr Bazas is with the University of London, United Kingdom. All other authors are with the Industrywide Studies Branch, National Institute for Occupational Safety and Health, Cincinnati, OH, USA.

10

Case–control studies, with a note on proportional mortality evaluation

O. Axelson[a]

Occupational health hazards have usually been evaluated epidemiologically through cohort studies. The cohort design tends to be rather cumbersome and expensive, however, since several hundreds or thousands of people have to be enrolled and studied with regard to exposure and outcome (i.e., disease or death). Frequently, therefore, the case–control or case–referent study is an attractive alternative, demanding fewer resources. This type of epidemiological approach has also become more popular during recent years and important methodological improvements took place during the 1970s. There is still some controversy, however, about the validity of case–control studies (*1*) but most concerns seem to originate from ignorance or misunderstanding of the general principles underlying this particular study design. As will be further discussed later, there is also a rather close relationship between the so-called proportional mortality study and a case–control study involving deceased subjects only.

The principle of the case–control or case–referent approach

Before going into a more detailed discussion of such aspects of the case–control study as its specific structure and related problems of validity, its general character may be recalled by citing the example of the relation between smoking and lung cancer. If 100 people with lung cancer are enrolled and interviewed with regard to their smoking habits, 95 of them might be found to be smokers. The question then is, whether there would also be some 95 % of smokers in the population from which the

---

[a] Department of Occupational Medicine, University Hospital, Linköping, Sweden.

cases were drawn or whether lung cancer patients tend to have a higher frequency of smoking than the source population. In order to find out, "controls" (or preferably "referents", since nothing is "controlled") might be chosen from the particular population that provided the cases, or from among individuals with another disease (which bears no relation, positive or negative, to smoking) representing the same population. It might then be found that, say, only 60% of these referents were smokers. In other words, there would be a difference in exposure frequency between cases and referents. This example illustrates that the primary information to be obtained from a case–referent study is the exposure frequency among cases and referents. As will be shown later, this information on exposure frequency can be further utilized in the calculation of various epidemiological parameters, particularly the rate ratio ("relative risk")— in the case of lung cancer, for example, for smokers versus nonsmokers. It is also possible, however, to make an indirect assessment of the absolute rates of mortality or morbidity among the exposed and the nonexposed, if the overall rate is known for the study population (2).

The case–referent study requires the exposure to be reasonably common to be effective, since otherwise the study might involve very few (or even no) exposed individuals and would then not provide informative data on the epidemiological problem at issue. Specific industrial exposures tend to be relatively rare from the viewpoint of the general population, however, but this problem can be overcome by optimizing the particular population for the study, e.g., by trying to find a setting in which a fairly large group of workers subject to the exposure in question live in a parish, a small town, etc. The study population might also be taken as those employed in a specific trade or industry in which the exposure is known to occur (it should be noted, however, that the case–referent approach is also ineffective if the exposure is extremely common in the population under study and the design fails if everyone— cases and referents—is equally exposed).

Local population registers of smaller administrative units exist in some countries and may include registration of causes of death. Such registers are valuable sources of cases and can usually also provide referents as well. Another possibility is to utilize the records of a (local) hospital for case-finding; referents might also be taken from the same source, but can perhaps be drawn from other sources, e.g., employment or trade union records (if these sources are used, the number of cases has also to be reduced to include only those with employment or trade union affiliation, so that cases and referents represent the same population). Trade union registers should not be forgotten in this context, since the membership might be covered by disability and life insurance and therefore the registers would also yield information on causes of disability and death.

The structural relationship of case–referent and cohort studies

In a more detailed consideration of the structure of the case–referent study, three different types might be distinguished (2). These correspond to the various measures of disease occurrence: (1) incidence rate or incidence density, i.e., the number of new cases (or deaths) per person-years of observation; (2) cumulative incidence or cumulative risk, i.e., the proportion of individuals falling ill (or dying) during a defined period of time; (3) prevalence rate, i.e., the proportion of individuals who are ill at a certain point in time. These various types of case–referent study are discussed in somewhat more detail in the following sections, particularly in relation to the common and more easily understandable cohort approach.

The incidence rate (or incidence density) type of study

The new cases of the illness of interest that occur in the population during the study period may be either exposed (a) or nonexposed (b). From a cohort point of view, the calculation of incidence rates among exposed and nonexposed would require that the person-years at risk could be ascertained as the denominators. However, by using a sample of the healthy population, or taking individuals with another disease as representative of this population in terms of exposure, (c) being the exposed and (d) the nonexposed individuals drawn from the population in the course of the study period ($t' - t''$), it is possible to develop a formula for the incidence rate (or incidence density) ratio. If the incidence rate among the exposed is denoted by IR_1 and the incidence rate among the nonexposed by IR_0, then

$$IR_1/IR_0 = \frac{a}{kc(t'' - t')} \left/ \frac{b}{kd(t'' - t')} \right. ; \tag{1}$$

The constant k is the reciprocal of the (unknown) fraction of the healthy population represented in the study by the referents c and d; consequently the denominators represent the number of person-years with and without exposure, omitting the exposed and nonexposed cases a and b, respectively.

Formula (1) may be simplified to:

$$IR_1/IR_0 = \frac{ad}{cb} \tag{2}$$

Thus, the incidence rate ratio is calculable if the exposure status is known for the cases and for a sample of the study population, i.e., the referents in the case–referent study, even though the number of person-years at risk remains unknown. It is obvious, also, that although it is not necessary for all cases to be enrolled in the study, the proportion of exposed and nonexposed cases must be the same.

As c and d must adequately reflect the exposure of the healthy individuals (healthy with regard to the disease under study) throughout the study period, this also means that, in principle, a non-case individual enrolled as a referent in the earlier part of the study period might later become a case and thus be represented twice in the same study. If only deceased persons are used, whether diagnosed as cases or as free from the disease in question (i.e., referents) this still constitutes a case–referent study of the incidence rate type, since dead non-cases apparently represent the exposure situation of the population over the period of the study (cf. discussion of reference entity, however). In this alternative approach one avoids the intellectual frustration of having an individual who was initially enrolled as a referent later developing the disease under study and then being designated a case.

It is worth pointing out, perhaps, that this type of case–referent study, referring to the incidence rate, has been commonly applied in epidemiological research, but its specific structure has not been generally recognized, i.e., that it relates to an open (or dynamic) population in which dead or "immune"[a] individuals are being continuously replaced and that the referents (or "controls") are there to provide (relative) information on the fraction of person-years relating to current or earlier exposure in the population.

The cumulative incidence (cumulative risk) type of study

If cases are accumulated throughout the study period and referents are chosen from among those members of an original population who remain healthy at the end of the study period, this results in the cumulative incidence type of case–referent study, which should be thought of as nested within a cohort or a closed population, i.e., no turnover takes place through replacement of dead or "immune" individuals.

Again, let a and b represent exposed and nonexposed cases originating from the exposed and nonexposed sections of the population, M_1 and M_0, respectively. Now, since there is no turnover in the population, at the end of the study period there will remain $M_1 - a$ and $M_0 - b$ individuals as the source of the referents, respectively. These expressions approximate to M_1 and M_0 only as long as the disease is rare, i.e., if a and b are small fractions of M_1 and M_0. The risk ratio or the ratio of the cumulative incidences, CI_1 and CI_0 might be taken as:

$$CI_1/CI_0 = \frac{a}{M_1} \left/ \frac{b}{M_0} \right. \approx \frac{a}{M_1 - a} \left/ \frac{b}{M_0 - b} \right. \tag{3}$$

or

$$CI_1/CI_0 \approx \left[\frac{a}{b}\right]\left[\frac{(M_0 - b)}{(M_1 - a)}\right] = \frac{ad}{bc} \tag{4}$$

since the quotient $\frac{(M_0 - b)}{(M_1 - a)}$ corresponds to the quotient of nonexposed to exposed referents, d/c.

This approximation to the risk ratio might preferably be referred to as an "odds ratio", and furthermore, as already indicated, this type of case–referent design requires the disease under study to be rare in order to ensure that the calculated odds ratio comes reasonably close to the cumulative incidence rate ratio. Traditionally, also, a rare disease

---

[a] "Immune" is here taken in the broad sense of no longer at risk, e.g., an individual surviving a heart attack might be considered "immune" since he is now suffering from that particular type of disease.

condition has been taken as a prerequisite for the "case–control" design to be applicable, but the incidence rate type of study has been more common than this "classical", rare disease study design.

A variant of the cumulative incidence type of study that does not involve the rare disease condition is also possible if the referents, c and d, are chosen as a sample from the entire population rather than from persons not diagnosed as cases, i.e., some cases would also be included in c and d. There might not always be suitable registers available for this approach, which has been rarely applied in spite of its theoretical advantages (cf. also the corresponding approach in the context of the prevalence type of study discussed below).

The prevalence type of study

A case–referent study can also be cross-sectional, with prevalent cases and referents drawn from the study population, either open or closed. There are two options, however, with regard to the choice of referents. If the referents are taken as individuals not diagnosed as cases, which might seem the "natural" thing to do in a case–referent study, one sets the incidence rate (or incidence density) ratio and not the prevalence rate ratio, as perhaps might be expected. For an unbiased estimate, however, it is necessary that the duration of the disease should be uninfluenced by the exposure (which probably is the usual situation). The reason for this and other considerations are further discussed in Appendix 1.

The other option in selecting referents would be just to take a cross-sectional sample, T_p, of individuals out of the population, irrespective of whether or not they are healthy with regard to the disease under study. Such a sample would provide the ratio of exposed to nonexposed individuals in the population, and the ratio of the prevalence rates (in which cases also figure in the denominators) is obtained by multiplying the quotient of exposed to nonexposed cases by the ratio of nonexposed to exposed individuals in the population sample (d_p and c_p, respectively, being the components of T_p), or using the usual notation:

$$\text{prevalence rate ratio} = \frac{a}{b} \times \frac{d_p}{c_p}$$

It should be remembered, however, that d_p and c_p in this instance may include cases that are also included in a and b, since a population sample was drawn regardless of health status. Apparently, this approach can be an efficient and economic variant of the cross-sectional study, since not every person free from the disease in question has to be dealt with in data acquisition, e.g., when checking background parameters such as smoking. Furthermore, if the fraction of the population represented by the sample T_p is known, then the prevalence rates can also be obtained by multiplying c_p and d_p by a corresponding factor (i.e., by 20 for a 5% sample, 10 for a 10% sample, 5 for a 20% sample, etc). Finally, it should be noted that, if the disease is rare, the incidence rate ratio comes close to the prevalence rate ratio, which is also reflected in the fact that

185

individuals with a rare disease are unlikely to be caught in a population sample, i.e., to be included in c_p or d_p.

It is unusual for reports of epidemiological studies to mention which type of case–referent study has been made; nor is this necessary in most cases, since a distinction between the various types of study is of little practical importance so long as the disease is relatively rare, which is usually the situation. When dealing with diseases of frequent occurrence, however, such as cardiovascular disease, bronchitis, or certain common symptoms, it is necessary to be more careful in designing the case–referent study, which should preferably be of the incidence rate type to provide good estimates of the epidemiological parameters. Sometimes mosaic types are created and these, again, are acceptable so long as the disease is rare. Nevertheless, it is useful to distinguish between the various types of case–referent study in order to gain a better understanding of the principles behind the case–referent approach in epidemiological research, especially with regard to validity questions and because of the specific relationships between the various measures of occurrence of disease and the more readily accepted cohort designs.

A note on proportional mortality evaluation

Many occupational health studies have been based on so-called proportional mortality. If the number of case deaths in the exposed population is a and the number of other deaths (i.e., not due to the disease in question) is taken as c, the proportional mortality for the exposed part of the population is $a/(a + c)$. Similarly, the proportional mortality for those not exposed is given by $b/(b + d)$, where b denotes the number of cases and d the number of other deaths. The proportional mortality ratio is consequently $[a/(a + c)]/[b/(b + d)]$ or, if a and b are small in comparison with c and d, the ratio may be simplified to ad/bc.

As with the incidence rate type of case–referent study, the use of only dead subjects leads to a situation where the referents reflect the exposure situation of the source population during the study period. Thus, an exact estimation of the rate ratio can be conveniently based on the same information as is necessary for carrying out a proportional mortality study (in fact, even less information can be used, since not all deaths other than those taken as cases necessarily have to be enrolled in the case–referent approach, although they are required for the proportional mortality evaluation; it might even be preferable to select only particular individuals with specific diagnoses as referents, as further discussed below under validity aspects of the reference entity).

It is suggested, therefore, that there is not necessarily a need for the proportional mortality study, a viewpoint that has been expanded upon elsewhere (3, 4). Sometimes, however, it might be useful to compute the percentage distribution of various causes of death in a population, i.e., the "proportional mortality rate", in order to obtain an overview or for descriptive purposes, since this rate can be easily understood even by people who have no insight into epidemiological questions.

Validity aspects of the case–referent study

The various sources of systematic error in an epidemiological study might be referred to as the "validity aspects"; they derive basically from the fact that there is no possibility of randomly distributing exposure and nonexposure in the population studied so as to ensure comparability between the groups to be evaluated with regard to health effects. The validity of case–referent studies tends to be even more vigorously debated than that of cohort studies, which are simpler and similar in structure to studies in which groups with and without exposure are compared. These problems of validity are discussed from a general point of view in Chapter 14. Certain aspects that apply especially to case–referent studies are examined in more detail below.

The various aspects of validity may be discussed with reference to (a) the possible role of exposure status in the selection of cases and/or referents, (b) difficulties in obtaining adequate information on exposure among cases and referents, (c) the possible relation of the reference entities (when using persons with diagnoses other than the disease in question) to the exposure, and (d) the effects of confounding.

Selection of subjects

With regard to case–referent studies in general, the suspicion may arise that the exposure status might have influenced the selection of either cases or referents. This would constitute a most serious bias. The possibility of such a bias is closely connected with the nature of the source from which the subjects used in the study were drawn. In investigations on the side-effects of drugs, for instance, there is often reason to believe that treated subjects may have been more actively cared for than the population at large, since they would usually have had better access to medical aid, once having entered the health care system. Similarly, certain disorders might be more easily diagnosed in that subsection of the population that is particularly liable to utilize the health care system; asymptomatic gallstones might be thought of as an example of such a selective diagnosis.

In occupational health epidemiology, selection bias of this kind is less likely to be of major importance, at least as long as hospital registers or other general registers are used for finding cases and referents, since medical care and diagnostic procedures in general are not usually influenced by a particular industrial exposure. However, the situation may be different if a register from an industrial health care unit or a clinic of occupational medicine is used as the source of subjects, since the occupational health physician(s) might have been particularly interested in health disturbances among individuals with a specific type of exposure, e.g., blood pressure among workers exposed to lead. The possibility of such relationships between the assessment of the case diagnosis (as well as diagnoses qualifying for enrolment as a referent) and the exposure should be carefully considered; a potential source of subjects for a study might then turn out to be unusable, or certain restrictions might need to be placed on the use of this source of subjects.

An example from a study of neuropsychiatric disorders and solvent exposure in Sweden (5) might help to explain somewhat further the problem of selecting subjects. In this study, it was judged necessary to restrict use of the source of subjects, a pension fund register, to a period in time when no particular consideration was being given to the possible relationships between solvent exposure and neuropsychiatric disorders. Subsequently, interest in this possible connection grew considerably among physicians, in consequence of which it might be suspected that painters and similar workers would be more likely to be diagnosed as suffering from neuropsychiatric disorders than other, nonexposed workers. In fact, it is even doubtful whether it would be possible to conduct such a study in Sweden at the present time, since great attention is now paid to neuropsychiatric symptoms, especially among workers exposed to solvents.

Although exceptions may occur, diagnoses related primarily to exposure seem to be a fairly rare problem in occupational health epidemiology as far as the study of new areas of possibly work-induced disorders is concerned. In the future, however, the situation may be different when improvements in the work environment are evaluated epidemiologically with regard to known hazards, and when awareness of relationships between industrial exposures and certain disorders becomes more common among physicians in general and might therefore influence diagnostic thinking. Even then, the diagnosis of a serious and relatively well defined disease, such as cancer or cardiovascular disease, will hardly be influenced by the physician's knowledge of the exposure status of the individual. However, there may perhaps be exceptions, e.g., in the diagnosis of conditions such as mesothelioma and liver haemangiosarcoma, should an exposure to asbestos or vinyl chloride, respectively, be known to the pathologist when having difficulty with the classification of a suggestive neoplasm.

Information on exposure
Assessing exposure is usually difficult in all types of epidemiological study, whereas diseases or causes of death can be more easily characterized. Occupational titles can be used as acceptable indicators of exposure with regard to some of the more specific jobs, e.g., painter with regard to solvent exposure, uranium miner with regard to radon daughter exposure, chimney sweep with regard to soot, etc., whereas such titles as metal worker or wood worker tend to be too unspecific for the characterization of exposure. In this context it should be emphasized that just to associate a particular job with a certain health hazard is not the aim of occupational epidemiology; instead, it is desirable to pinpoint the responsible agent or process and to prevent health impairment by eliminating or changing hazardous materials or manufacturing procedures.

188

If the case–referent study can be applied locally within a small town, a parish, etc., as was discussed at the beginning of this chapter, it is often possible and beneficial to cooperate with an employer and/or a trade union to obtain information about exposure of the subjects, at least in qualitative terms, i.e., a certain case or referent should preferably, if not necessarily, be assigned blindly to an exposure category. For this purpose, lists of cases and referents mixed together, without diagnoses, might be handed over to the factory and/or the trade union representatives for identification of the individuals as employees and specification of the job tasks, as far as possible, over the years. Measurements of various workplace pollutants might exist and make it possible also to obtain, directly or indirectly, some quantified information about exposure.

However, many types of exposure tend to be scattered and widespread in the community and therefore usually need to be assessed by questionnaires and interviews, especially if the exposure of interest took place in small firms without any effective registration of the employees. In such cases, the validity of the observation of exposure becomes open to question, since there could be a tendency, conscious or not, for those suffering from a particular disease (the cases) to exaggerate the exposure in comparison with the referents. Should the disease be severe, resulting in deaths, it might be preferable to utilize dead individuals as referents, corresponding to dead cases, to ensure a similar inquiry situation for cases and referents. Also, for the same reason, it can be beneficial to use individuals with other disorders as referents. However, as will be further discussed below when considering the reference entity, certain diagnoses should be avoided among the referents, namely those that might have a positive or negative relationship with the exposure. Not only is blinding of the interviewers desirable, but the questions should be standardized and cover various extraneous matters in addition to the specific exposure(s) at issue. The same considerations apply to self-administered questionnaires.

Although various measures can be taken in the design of the study to ensure good quality of information, there is nevertheless always the suspicion that the observation of exposure might be inadequate if reliance has to be placed on questionnaires or interviews. However, some evaluation of the accuracy of the information is obtainable from the data, assuming that information is available not only about the specific exposure but also about other characteristics to which the exposure is related. A prerequisite is, however, that these other characteristics can be more precisely assessed than the exposure itself. For example, an exposure that has occurred among workers in a certain type of job, trade, etc. might be more or less adequately assessed by questionnaires, but it is usually possible to achieve less controversial and more exact information about the occupational titles, at least in general terms, such as farmer, forestry worker or metal worker. Then, if an increased rate ratio is obtained with regard to the specific exposure, owing to a tendency on the part of the cases to exaggerate and/or the referents to underestimate their exposure, the consequence will be a rate ratio less than unity among the nonexposed individuals within the particular job or trade in which the exposure occurs. This aspect is particularly relevant in situations with fairly high rate ratios, such as are often obtained in occupational health studies, whereas less credence can be given to such an evaluation if the rate ratio is only slightly or moderately increased, i.e., if it is about 2–3 or less. Moreover, this evaluation procedure also becomes uncertain if the actual exposure is relatively rare, say 10% or less, even within the particular job, trade, etc., and if the majority of the referents are apt to recall the exposure poorly (6).

The principle discussed here is described more fully in algebraic terms in Appendix 2. An example is provided by a study on mesenchymal tumours and exposure to phenoxy acids in farming and forestry work, the data for which are shown in Table 1 (7). It might still be argued that the association between exposure to phenoxy acids and mesenchymal tumours, as presented in the table, is not a true one by suggesting that an increased risk of mesenchymal tumours exists (for some reason) among farming and forestry workers. The appearance of Table 1 would then be the result of this background risk combined with an observational bias, as this could theoretically result in the estimates obtained for the various subpopulations as defined in the table. However, that suggestion would require another study to show that these particular types of tumour occur with an increased frequency among people in farming and forestry work (in which case, exposure to phenoxy acids would be considered to be a confounding factor; cf. below).

Table 1. Exposure to phenoxy acids among cases and referents with regard to occupational group[a]

	Farming/forestry		Other occupations	
	Exposed	Nonexposed	Exposed	Nonexposed
Cases	13	18	1	78
Referents	5	42	0	172
Rate ratio	5.7	0.9	—	(1.0)

[a] Adapted from Eriksson et al. (7).

When making observations of exposure, it is important to allow for the possibility of an induction-latency period, which is usually pertinent to studies of chronic disease. So far as cases are concerned, it is quite simple to allow for the induction-latency time by disregarding the last, say, five or ten years of exposure prior to diagnosis. If dead referents are used, it is also easy to disregard the most recent exposure in much the same way as for cases, since the "endpoint" is given. In matched series, the exposure of the referent(s) might be considered as effective during the same period of time as for the case, but it becomes more difficult to handle this problem in stratified series when living referents are used who are not matched with specific case individuals. In this event, one might use either the time of the diagnosis of the referent or the time of his or her enrolment in the study as the temporal reference point; otherwise an artificial reference point in time has to be created, a procedure that might be less suitable if the study encompasses a period of a decade or more.

The reference entity
As follows from the earlier discussion about the relationship between case–referent and cohort studies, the referents are expected to reflect the exposure frequency in the study population providing the cases. To achieve similar conditions for the observation of exposure among cases and referents, it was suggested that diseased or dead individuals might be used as referents, but if that is done it is necessary to consider carefully what diagnoses can be accepted, since the exposure might also cause other disorders or result in other causes of death than the one taken as the case entity of the study. Should individuals with other disorders that might also have been caused (or prevented) by the exposure(s) under study appear in the referent series, then the exposure frequency in the population from which the cases were drawn would not be properly reflected among the referents.

It follows, therefore, that on the basis of *a priori* judgements, adequate reference entities, i.e., diagnoses for singling out referent individuals, have to be selected as having no (known) relationships (positive or negative) with the exposure or exposures under study. It is worth noticing, in this context, that although the case–referent study permits a multiple comparison between cases and referents with regard to the frequency of a number of exposures, there is a possibility that the chosen reference entities might have primary relationships with some of the exposures but not with others. This apparently means that some of the comparisons of the exposure frequencies could be biased, whereas others would be valid (8).

Since there is often a lack of knowledge about possible relationships between an exposure and various potential reference entities, it is preferable to use a variety of diagnoses in the reference series, rather than a single diagnosis (or closely interrelated diagnoses), in order to decrease the influence of unknown connections between the exposure(s) and particular disorders. Should the exposure frequency be "falsely" increased among the referents through an unknown relationship between the exposure and the referent diagnoses selected, the result would be an underestimation of the effect of the exposure. Quite often it is possible to use two or more separate series of referents, and an intercomparison between the series with regard to exposure frequency may help to reveal an existing bias.

A few examples may further clarify the principles of choosing suitable reference entities. Thus, in a study of cardiovascular deaths and a possible causal role of exposure to nitroglycerine and nitroglycol in the explosives industry (9), it was obvious that individuals who had died from explosives accidents had to be excluded from the reference series, since there was clearly a primary relationship between the dynamite exposure and this particular cause of death. Similarly, in a study of the relationship between arsenic exposure and lung cancer (*10*), there was evidence from earlier studies (*11*) that cardiovascular deaths could have a causal relationship to arsenic exposure and would not, therefore, be an appropriate reference entity for the study of lung cancer with regard to arsenic exposure.

If the subjects of the study come from a general register covering the entire population from which the cases are drawn, there will be no particular apprehensions about reference entities, not even if a referent individual appears to have a disease with a known relationship to the exposure. The reason this situation is not of concern is that when referents are selected from the population register, the exposure frequency of the population is directly reflected among the referents, whereas if diseased or dead individuals were used as referents, they would provide only an indirect indication of the exposure frequency of the population. It that event, "distortions" might easily arise owing to primary relationships between the exposure and various disorders. Consequently, it would be necessary to exclude (potentially) exposure-related diagnoses. Therefore, if the exposure can be adequately observed (preferably through entirely objective means) there is good reason for selecting referents from population registers (in countries where these are available) rather than using registers of diseased or dead individuals. However, sometimes selective mechanisms might also operate within the health care system, causing certain cases to enter a particular (hospital) register; should such a mechanism be influenced, for example, by social status, there might also be an inherent relationship with occupational exposures, in which event similarly selected individuals from the same (hospital) register but not suffering from the disease in question might be the referents of choice.

Finally, it might be pointed out, in this context, that in the incidence rate type of case–referent study, the referents should represent the exposure situation over the whole of the study period. This may not always be achievable in retrospect since it may be very difficult to find a source of healthy referent individuals that will cover the entire study period, population registers usually being continuously updated. The difficulties of getting individuals to cooperate in answering questionnaires and interviews have already been mentioned.

Confounding

Confounding factors (confounders) are risk factors for the particular disease, which are either direct (e.g., smoking or a chemical exposure) or indirect (e.g., sex or social status) in character,[a] and which also tend to

---

[a] "Risk factor" is therefore a less suitable term although used here; risk indicator or determinant would be a more appropriate concept when indirectly acting or poorly understood determinants of disease are involved.

occur together with the exposure(s) under consideration. Because individuals exposed to the particular agent under study may also be exposed to these other disease-causing factor(s), any association found between exposure and disease might be mistaken for a causal one, if such confounding is not identified and accounted for either in the design or in the analysis of the study. It is important, however, that factors linking the effect to the exposure should not be confused with confounders; e.g., if an industrial exposure may cause hypercholesterolaemia, which then results in cardiovascular morbidity, hypercholesterolaemia is a link in the chain of causation and not a confounder (cf. Ref. *12*). However, in a study to examine the question of whether there is also another, additional mechanism of causation, not involving hypercholesterolaemia, then hypercholesterolaemia acquires the property of a confounding factor.

Since a confounding factor is related to both exposure and disease, it is rather obvious that such risk indicators as smoking or alcohol abuse do not usually have any strong confounding properties in occupational health epidemiology, since they are not closely related to a specific occupational exposure. Exceptionally a closer association may occur, but it should be observed that an association is also sometimes negative, e.g., workers subject to a specific exposure suspected of causing lung cancer or liver disease might turn out to have rather low levels of smoking or alcohol use. For example, painters were found to smoke somewhat less than the average (or reference) population in a region of Sweden (*13*) and, in general, a stable labour force, accumulating long-time exposure to a certain agent or process, would be expected to represent a sounder "life-style" than the average population. On the other hand, dirty jobs might recruit workers with an unhappy social background, in which case alcohol abuse and heavy smoking might well appear to be associated with a less attractive work task, besides being suspected of implying a health hazard. It is always wise, therefore, to consider various "general risk factors", but their influence should not be exaggerated.

Sometimes it might be necessary or even reasonable to try to evaluate theoretically the possible influence on the epidemiological estimates (such as the rate ratio) of an uncontrolled confounding factor, e.g., when smoking data are not available. A model for such evaluation of confounding due to smoking with regard to lung cancer has been presented elsewhere (*14*; cf. also Chapter 14) and seems to indicate that the confounding effect of smoking should be expected to be rather moderate in occupational health epidemiology. A similar approach may be taken to the evaluation of other potentially confounding effects in studies of various exposures causing disease, since it is important to obtain some quantitative information instead of just maintaining a general idea of uncontrolled confounding being present. Such evaluation is always difficult, however, since it is inherent in the situation that the study itself does not provide any information in this respect, and little might be known in general about the magnitude of the health effects imputable to the presence of a certain factor suspected of confounding properties.

Some general comments on the nature of confounding factors might also be mentioned in this context. In the first place, confounding factors should be either true risk factors for the disease or associated with the true risk factors within the particular study population. It is not possible to obtain definite information in this respect from case–referent data, but nevertheless it seems unjustified to invent confounders *ad hoc* simply to avoid having to accept a certain exposure as a probable cause of a disease when studies have shown it to be associated with that particular disease. It is reasonable, also, to require that evidence (from outside the study) with regard to causality for a suggested confounding factor is of about the same firmness as the evidence for accepting a connection between exposure and outcome in a particular study.

Attributing an effect to confounding necessitates suggesting an alternative, more or less generally accepted route of causation to explain an association between exposure and disease. Interestingly, the development of knowledge about causal relationships in medicine provides for a sort of hierarchical order for the concerns about confounding, sometimes even with moral overtones. Thus, knowledge about the unequal distribution of diseases with regard to social background and life-style, or a well established, long-known relationship between a common exposure and disease, e.g., lung cancer and smoking, calls for attention to the possibility of confounding whenever other exposures are studied, both in occupational health and in epidemiology in general. Apparently, however, confounding tends to be reciprocal, and it can perhaps sometimes be worth while to reconsider established relationships in the light of new information. Thus, it is interesting that there seems to be some evidence of a higher prevalence of smoking among blue-collar workers, especially when those engaged in dirty jobs are compared with professionals, managers, and technically trained individuals, at least according to observations made in the United States (*15*). Thus, blue-collar work might be looked upon as a confounding factor because of its association with potentially carcinogenic exposures, and it ought perhaps to have been taken into account in the context of establishing the widely accepted relationship between smoking and lung cancer. Taking a quantitative view, it is worth noticing, however, that the difference between various occupational groups in the frequency of persons who have never smoked is less than 19 % (*16*), which means that the impact of uncontrolled confounding from carcinogenic industrial exposures would, in fact, be rather small in comparison with the tenfold, or more, increase in lung cancer risk among smokers.

Even though the possible reciprocal nature of confounding is worth some consideration with regard to generally occurring risk factors for disease, it deserves special attention when two or more occupational exposures tend to be mutually confounding, i.e., when they could each be the cause of the illness under study. For example, exposure to volatile anaesthetics seems to cause miscarriages and birth defects (cf. Ref. *17*)

194

but some years ago it was also suggested that birth defects might be due to the use of hexachlorophene soap (18). That suggestion also meant there was a possibility of confounding in studies on volatile anaesthetics and miscarriages, since personnel exposed to anaesthetics might also have used hexachlorophene soap for surgical washing. In other words, the exposures to anaesthetics and hexachlorophene, both related to surgery, might confound one another with regard to teratogenic effects. In trying to establish the possible effect of either of these exposures, the other has to be taken as a confounding factor. Similarly, farming and forestry workers tend to be exposed to a variety of pesticides, several of which might have carcinogenic properties. This makes it difficult to evaluate the effect of the specific compounds, especially as impurities in the various preparations might also imply a hazard (for some practical examples, see Ref. 7 and 19). Another example is provided by the extremely complex chemical environment in the rubber industry (20). Although it would be desirable to elucidate the effects of the specific chemicals to permit successful prevention, in many of the complex situations encountered in practice epidemiological evaluation would necessarily be confined to considering the whole industrial process as the exposure rather than the mutually confounding chemical exposures of which it is made up.

In the present context of confounding, the phenomenon of effect-modification (12) might be mentioned, since there is a tendency in general thinking to mix up confounding with effect-modification. Whereas confounding has to do with an alternative, non-causal explanation of an association between exposure and a disease, namely, one that depends on an unequal distribution of another risk factor, the confounder, among exposed and nonexposed subjects (e.g., a difference in age determining the duration of exposure), effect-modification has merely to do with the pathogenesis and the nature of the manifestation of a disorder. An example of effect-modification would be a higher frequency of lung cancers among asbestos workers who smoke. Even if there were the same frequency of smoking among individuals with and without asbestos exposure (no confounding) or even fewer smokers among asbestos workers than in the reference population (negative confounding), lung cancer would have struck the smokers. If there had been more smokers among the asbestos workers than in the reference population, there would have been both an effect from confounding, smoking in itself causing some extra cases, and from effect-modification since the majority of cases again would have been created through the combined effect of asbestos and (the effect-modifying) smoking. In all three situations, also, a few cases might have been expected among the exposed non-smokers (cf. Ref. 21).

Confounding in the data might be dealt with in various ways, one being restriction of the study to encompass only a narrow range (or a certain level or category) of the factor of concern. Usually, however, confounding has been dealt with either by matching cases and referents at the data acquisition stage of the study, or by stratification and/or multivariate analysis of the data, although these procedures have no influence on a confounded situation in the study population itself.

Matching means that for each case one or more referents are selected from the same category of one or several confounding factors of which age and sex will almost always be included. The matched pairs or sets should be maintained in the analysis of the data, if the matching is relevant, i.e., when there tends to be a correlation in the exposure pattern within the pairs or sets (22, 23). In consequence of this tendency to correlation in the exposure pattern of cases and referents, it is important to recognize what the result will be if matching is undertaken on the basis of an exposure-related factor that is not a risk indicator for the disease (i.e., does not fulfil the criteria of a confounding factor). Obviously such matching would lead to a similar exposure pattern among cases and referents, and the study would become insensitive and fail to reveal (fully) a true association between an exposure and a disease. Matching on a confounding factor obviously brings about a similar situation by decreasing the difference in exposure frequency between cases and referents, especially if the correlation with the exposure is strong.

Further aspects of matching have been discussed by Miettenen (24). In general, however, it is desirable to avoid matching since, if other confounding factors appear than those matched for, analysis of the data tends to become more difficult as the pairs or sets have to be maintained in considering the additional confounders (25, 26). Should the original matching result in little or no correlation in the exposure pattern between cases and referents (which is quite common when matching is based on disease-related factors only), the pairs or sets might be dispensed with in considering additional confounders, e.g., by stratification. In stratification, the individuals in the particular categories of the confounding factor(s) are grouped together in the corresponding confounder-specific strata. Other traditional approaches are to apply multivariate methods (26–28) or a combination of multivariate analysis and stratification by means of multivariate confounder scoring (29).

Finally, it should again be emphasized that confounding must be taken into account both when planning the study and when evaluating the results. In the prestudy situation, judgements about possible confounding factors must be made with regard to data collection, and in the poststudy situation, confounding is re-examined in the light of the data yielded by the study (but random variation in the referents may obscure the true situation in the study population from which the cases and referents derive). When an evaluation is made, the primarily suspected confounding factors sometimes turn out to be weak or negligible as they appear in the data (cf. Ref. 30) or even negative and masking, as in the case of age for cardiovascular disease and exposure to explosives in a study on dynamite workers (9). For judgements whether or not confounding is present in the data, the tendency to correlation of the exposure among cases and referents might be considered in matched studies, as mentioned above, and a comparison made between the rate

196

ratios obtained with and without maintaining the matched pairs or sets. In stratified data, equal ratios of exposed to nonexposed referents in the various strata means absence of confounding (no relation between the exposure and the suspected confounding factor meaning no confounding); the same conclusion may be drawn from equal ratios of cases to referents among the nonexposed in the various strata (i.e., the suspected confounding factor does not appear as a risk factor for the disease; cf. also equation (1) where a constant ratio of b/d would obtain in the various strata if no effect were attributable to the stratification factor). The strength and the direction of confounding as affecting the data yielded by the study may be measured as the quotient of the crude rate ratio to the SMR (standardized morbidity/mortality ratio, i.e., the observed number of exposed cases, a, over the expected number, a', the expected numbers from each stratum being obtainable from $a'd/bc = 1$; $a' = bc/d$; cf. Ref. 31). A quotient greater than unity means positive (or exaggerating) confounding and a value smaller than unity means negative (or masking) confounding when risks are studied (the situation is reversed in studies on prevention).

Examples of case–referent studies

It is not possible in this context to exemplify fully all aspects of a case–referent study, from considerations of design and data acquisition through various aspects of validity and statistical analyses to the interpretation of the results. Usually, the various problems are taken into account in the presentation of a study and it may suffice here to refer to some publications on methodology and to provide the non-statistical reader with a few examples of calculations of measures of effect, such as the SMR (standardized morbidity/mortality ratio) and the SRR (standardized rate or risk ratio; 31, 32), and to illustrate how to compute the chi-square statistic by means of the Mantel-Haenszel test (33) (but testing of dose–response trends has been omitted; cf. Ref. 34) and how to use the McNemar test for matched pairs (for the analysis of data with multiple controls see references 22 and 23). Approximative confidence limits of the rate ratio can be calculated according to the principles given by Miettinen in 1976 in a most important paper on estimability and estimation in case–referent studies (2), which also explains how to derive the incidence rates for exposed and nonexposed individuals from case–referent data by use of the so-called etiological and preventive fractions, respectively, together with information from the study itself, or from other sources, about the combined incidence rate for exposed and non-exposed. These various calculations are shown in Appendix 3.

Epilogue

This review of the case–referent design has hopefully shown that this approach is often an effective and convenient alternative to other designs in occupational health epidemiology. It has often been suggested that the validity of case–referent studies are difficult to control and to evaluate, but recent developments and increasing insight into the nature of such studies have rendered the benefits of this design more and more obvious. It is also quite easy, as a rule, for interested readers to re-analyse case–referent data, since these data can be easily supplied in the presentation of the material, whereas background data for cohort studies are much more extensive and therefore difficult to include in any detail in a paper. A general conclusion might be, therefore, that there is little reason for considering the case–referent study design as totally inferior to the cohort approach from the methodological point of view, nor is the information obtained from adequately designed and executed case–referent studies less reliable than that stemming from cohort studies. It even seems likely that the case–referent study might predominate in epidemiological research in occupational health in the future.

References

1. **Ibrahim, M. A. & Spitzer, W. O., ed.** *The case control study. Consensus and controversy.* Oxford, Pergamon Press, 1979.
2. **Miettinen, O. S.** Estimability and estimation in case-referent studies. *American journal of epidemiology,* **103**: 226–235 (1976).
3. **Axelson, O.** The case-referent (case-control) study in occupational health epidemiology. *Scandinavian journal of work, environment and health,* **5**: 91–99 (1979).
4. **Miettinen, O. S. & Wang, J-D.** An alternative to the proportionate mortality ratio. *American journal of epidemiology,* **114**: 144–148 (1981).
5. **Axelson, O. et al.** A case-referent study on neuro-psychiatric disorders among workers exposed to solvents. *Scandinavian journal of work, environment and health,* **2**: 14–20 (1976).
6. **Axelson, O.** A note on observational bias in case-referent studies in occupational health epidemiology. *Scandinavian journal of work, environment and health,* **6**: 80–82 (1980).
7. **Eriksson, M. et al.** Soft-tissue sarcomas and exposure to chemical substances; a case-referent study. *British journal of industrial medicine,* **38**: 27–33 (1981).
8. **Axelson, O. et al.** A comment on the reference series with regard to multiple exposure evaluations in a case-referent study. *Scandinavian journal of work, environment and health,* **8** (suppl. 1): 15–19 (1982).
9. **Hogstedt, C. & Axelson, O.** Nitroglycerine-nitroglycol exposure and the mortality in cardio-cerebrovascular diseases among dynamite workers. *Journal of occupational medicine,* **19**: 675–678 (1977).
10. **Axelson, O. et al.** Arsenic exposure and mortality: A case-referent study from a Swedish copper smelter. *British journal of industrial medicine,* **35**: 8–15 (1978).
11. **Lee, A. M. & Fraumeni, J. F.** Arsenic and respiratory cancer in man: an occupational study. *Journal of the National Cancer Institute,* **42**: 1045–1052 (1969).
12. **Miettinen, O. S.** Confounding and effect-modification. *American journal of epidemiology,* **100**: 350–353 (1974).
13. **Hane, M. et al.** Psychological function changes among house painters. *Scandinavian journal of work, environment and health,* **3**: 91–99 (1977).
14. **Axelson, O.** Aspects on confounding in occupational health epidemiology. *Scandinavian journal of work, environment and health,* **4**: 85–89 (1978).
15. **Sterling, T. D.** Does smoking kill workers or working kill smokers. *International journal of health services,* **8**: 437–452 (1978).

16. **Sterling, T. D. & Weinkam, J. J.** Smoking patterns by occupation, industry, sex and race. *Archives of environmental health*, **33**: 313–317 (1978).
17. International Agency for Research on Cancer. *Monographs on the evaluation of carconogenic risk of chemicals to man*, vol. 11. Lyon, 1976.
18. **Halling, H.** Misstänkt samband mellan hexaklorofenexposition och missbildningsbörd [Suspected link between exposure to hexachlorophene and congenital malformations]. *Läkartidningen*, **74**: 542–546 (1977).
19. **Axelson, O. et al.** Herbicide exposure and tumor mortality: An updated epidemiologic investigation on Swedish railroad workers. *Scandinavian journal of work, environment and health*, **6**: 73–79 (1980).
20. **Gamble, J. & Spirtas, R.** Job classification and utilization of complete work histories in occupational epidemiology. *Journal of occupational medicine*, **18**: 399–404 (1976).
21. **Saracci, R.** Personal-environmental interactions in occupational epidemiology. In: McDonald, J. C., ed. *Recent advances in occupational health*. London, Churchill Livingstone, 1981, pp. 119–128.
22. **Miettinen, O. S.** Estimation of relative risk from individually matched series. *Biometrics*, **26**: 75–86 (1970).
23. **Miettinen, O. S.** Individual matching with multiple controls in the case of all-or-none response. *Biometrics*, **25**: 339–355 (1969).
24. **Miettinen, O. S.** Matching and design efficiency in retrospective studies. *American journal of epidemiology*, **91**: 111–118 (1970).
25. **Holford, R. T. et al.** Multivariate analysis for matched case-control studies. *American journal of epidemiology*, **107**: 245–256 (1978).
26. **Breslow, N. E. & Day, N. E.** *Statistical methods in cancer research*, vol. 1. Lyon, International Agency for Research on Cancer, 1980.
27. **Anderson, S. et al.** *Statistical methods for comparative studies*. New York, Wiley, 1980, pp. 122–127.
28. **Siegel, D. G. & Greenhouse, S. W.** Multiple relative risk functions in case-control studies. *American journal of epidemiology*, **97**: 324–331 (1973).
29. **Miettinen, O. S.** Stratification by a multivariate confounder score. *American journal of epidemiology*, **104**: 609–620 (1976).
30. **McMichael, A. J. et al.** Solvent exposure and leukemia among rubber workers: An epidemiologic study. *Journal of occupational medicine*, **17**: 234–239 (1975).
31. **Miettinen, O. S.** Components of the crude risk ratio. *American journal of epidemiology*, **96**: 168–172 (1972).
32. **Miettinen, O. S.** Standardization of risk ratios. *American journal of epidemiology*, **96**: 383–388 (1972).
33. **Mantel, N. & Haenszel, W.** Statistical aspects of the analysis of data from retrospective studies of disease. *Journal of the National Cancer Institute*, **23**: 719–748 (1959).
34. **Mantel, N.** Chi-square tests with one degree of freedom: Extensions of the Mantel-Haenszel procedure. *American Statistical Association journal*, **58**: 690–700 (1963).

Appendix 1

Algebraic calculations referring to the prevalence type of case–referent study

The relationship between the incidence rate, IR, and the prevalence rate, PR, is usually taken as

$$PR = IR \times D \tag{1}$$

where D is the duration of the disease. This relationship requires that the denominator of the incidence rate includes the person-years of the individuals suffering from the disease. If a distinction is made between the contribution of person-years from healthy and diseased individuals, relationship (1) becomes:

$$PR = \left(\frac{a}{Mt + aD} \right) \times D \tag{2}$$

where t denotes the observation time of healthy individuals, M; thus Mt and aD represent person-years at observation. This relationship may be rewritten as:

$$PR = \frac{a}{Mt} \times D \Big/ \left(1 + \frac{a}{Mt} \times D \right) \tag{3}$$

or

$$PR = IR \times D / (1 + IR \times D) \tag{4}$$

which may be further simplified to:

$$IR \times D = PR / (1 - PR) \tag{5}$$

However, the denominator of IR now only includes person-years of healthy individuals.

If the incidence rates among the exposed are denoted as IR_1 and among the nonexposed as IR_0 the incidence rate ratio, according to (5), may be written as

$$IR_1 / IR_0 = [PR_1 / (1 - PR_1)] / [PR_0 / (1 - PR_0)] \tag{6}$$

assuming that the exposure does not influence the duration of the disease. Equation (6) may finally be rearranged as

$$IR_1 / IR_0 = (PR_1 / PR_0) / [(1 - PR_1) / (1 - PR_0)] \tag{7}$$

For case–referent study, a and b may denote the exposed and nonexposed cases, i.e., a/b represents PR_1 / PR_0, and similarly, if c and d denote exposed and nonexposed referents (non-cases), c/d equals $(1 - PR_1) / (1 - PR_0)$ or

$$IR_1 / IR_0 = (a/b) / (c/d) = \frac{ad}{bc} \tag{8}$$

Thus, the incidence rate ratio is obtainable from the prevalent cases and non-cases (referents) in the population, assuming that the duration of the disease, D, is not influenced by the status of exposure. Should the prevalence be low, the prevalence rate ratio comes close to the incidence rate ratio, as follows from (7).

Appendix 2

Algebraic calculations in relation to observational bias

The issue of observational bias in a case–referent situation might be illustrated more exactly as follows:

Job/trade category	Case/Ref.	Exposed	Nonexposed
Within job/trade	Cases	a_1	b_1
	Referents	c_1	d_1
Outside job/trade	Cases	—	b_0
	Referents	—	d_0

On the assumption that no hazard is associated with the exposure, i.e., if the rate or risk ratio (odds ratio) is unity, the following relationship applies for the study as a whole:

$$a_1(d_1 + d_0)/c_1(b_1 + b_0) = 1 \tag{1}$$

while for a given job or trade, the relationship is:

$$a_1 d_1/c_1 b_1 = 1 \tag{2}$$

Relationships (1) and (2) may therefore be combined and rearranged to give:

$$(b_1 + b_0)/b_1 = (d_1 + d_0)/d_1 \tag{3}$$

and therefore:

$$b_1/b_0 = d_1/d_0 \tag{4}$$

Now, if the exposure is inadequately observed so that cases tend to be classified as exposed more often than they really are (by gaining a quantity Δb_1 from b_1) and/or the exposure of the referents is underestimated (by losing a quantity Δc_1 from c_1, this being gained by d_1) then equation (1) becomes:

$$\frac{(a_1 + \Delta b_1)(d_0 + d_1 + \Delta c_1)}{(b_0 + b_1 - \Delta b_1)(c_1 - \Delta c_1)} > 1 \tag{5}$$

This expression reflects the increased risk ratio through biased observation. However, for the relationship shown in (4) the consequence will be:

$$(b_1 - \Delta b_1)/b_0 < (d_1 + \Delta c_1)/d_0 \tag{6}$$

which corresponds to "prevention" among the nonexposed in the given trade, since

$$(b_1 - \Delta b_1)d_0/b_0(d_1 + \Delta c_1) < 1 \tag{7}$$

where $(b_1 - \Delta b_1)$ represents the (inadequately observed) nonexposed cases and $(d_1 + \Delta c_1)$ the (also inadequately observed) nonexposed referents in the trade.

Should the risk or rate ratio be relatively low—say, less than 2 or 3—or the exposure quite rare in the particular job, trade, etc., this method of evaluating a possible observation bias will be less reliable, especially if the exposed referents are apt to have forgotten their exposure.

Appendix 3

Illustration of some statistical methods

This appendix is intended to assist the non-statistician in applying some statistical methods commonly used in epidemiology. These methods are all of an approximative character and may not always give optimum results, especially for small sample sizes. In this respect, objections could be raised even about the examples given. On the other hand, the limitations of epidemiological research are occasioned mainly by other circumstances relating to the validity issues discussed in the main text, rather than by approximations in the statistical evaluations. When carrying out epidemiological studies, it is wise to cooperate with epidemiologically trained statisticians or to acquire the relevant knowledge from the methodological literature dealing with the border area of statistics and epidemiology (to which some references have already been given). The reader is especially referred to the book by Rothman & Boice, *Epidemiologic analysis with a programmable calculator (1)*.

Matched case–referent data
The data in Table 1 are taken from a paper by Hardell & Sandström (2) on exposure to phenoxyacetic acids (Ph) and chlorophenols (Ch), excluding 30 pairs where neither the cases nor their referents were subject to exposure. Only the first set of referents are considered in this context (for the complete data see the original paper and for the analysis with multiple referents see references 3 and 4).

Considering both types of exposure together (there is some rationale for doing so since phenoxyacids are derivatives of chlorophenols) it is possible to construct Table 2, which also includes the 30 pairs not shown in Table 1.

The odds ratio or risk ratio,[a] RR is obtained as

$$RR = \frac{b}{c} \qquad (1)$$

i.e., the ratio of exposure-discordant pairs. Thus,

$$RR = \frac{16}{3} = 5.3.$$

---

[a] Also known as rate ratio; the study is a mosaic of the incidence rate and cumulative incidence types of study, the latter providing a true "risk ratio".

Table 1. Exposures of matched pairs of cases and referents to phenoxyacetic acids or chlorophenols

Pair No.	Exposure	
	Case	Referent
01	Ph	—
02	—	Ph
10	Ch	—
12	Ch	—
16	Ph	Ph
18	Ph	—
19	Ph	—
20	Ch	Ph
22	Ph	—
23	Ph	—
27	Ph	—
35	Ph	—
39	Ph + Ch	—
43	Ph	—
49	Ph	—
50	—	Ph
51	Ph	—
52	Ph	—
54	Ch	Ph
56	Ch	—
57	Ch	—
58	—	Ph

The McNemar chi-square statistic with one degree of freedom is

$$\chi^2(1) = \frac{(b-c)^2}{b+c} \tag{2}$$

or

$$\chi^2(1) = \frac{(16-3)^2}{16+3} = 8.89$$

and hence the two-tailed P-value (from a chi-square table) is obtained as $P \doteq 0.003$.

The approximate confidence interval of the risk ratio (5; cf. also Ref. 6) may be taken as

$$\underline{RR}, \overline{RR} = (RR)^{1 \pm z_{1-\alpha/2}/\chi} \tag{3}$$

where $\underline{RR}$, $\overline{RR}$ are the lower and upper confidence limits and $z_{(1-\alpha/2)}$ is the $100(1-\alpha/2)$-percentile of the standardized normal distribution (for the 90 % confidence interval, $z_{(1-\alpha/2)} = 1.645$ and for the 95 % confidence interval $z_{(1-\alpha/2)} = 1.960$).

Thus, in the above example, the 95 % approximate confidence interval is

$$\underline{RR}, \overline{RR} = (5.3)^{1 \pm 1.960/\sqrt{8.89}}$$

whence $\overline{RR} = 5.3^{1+0.66} = 5.3^{1.66} = 15.9$

and $\underline{RR} = 5.3^{1-0.66} = 5.3^{0.34} = 1.7$

A more exact value, together with confidence limits, might be obtained using the binomial distribution, which would give a somewhat more conservative result. The so-called Yates correction was not used in the above calculations as it is no longer generally recommended (cf. Ref. 7).

Now, consider the data in Table 2 after dispensing the pairs, as shown in Table 3.

Table 2. Concordance and discordance of exposures and matched pairs

	Referents	
Cases	Exposed	Nonexposed
Exposed	$a = 3$	$b = 16$
Nonexposed	$c = 3$	$d = 30$

Table 3. Exposure among cases and referents after dispensing from matching

	Exposed (Ph and/or Ch)	Nonexposed
Cases	$a = 19$	$b = 33$
Referents	$c = 6$	$d = 46$

The risk ratio is given by

$$RR = \frac{ad}{bc} \tag{4}$$

and thus

$$RR = \frac{19 \times 46}{33 \times 6} = 4.4$$

which is a lower estimate than was obtained when the pairs were maintained. This is an indication of negative confounding by the matching factors (in this study, the matching factors were sex, age, and place of residence); also there is a negative correlation of exposure among cases and referents in Table 1, i.e., when the case is exposed the referent is nonexposed and vice versa. For further analyses in the study by Hardell & Sandström it was considered reasonable, therefore, to work with dispensed sets, although this implied some loss of validity with regard to negative (or masking) confounding.

Stratified case–referent data

The data presented in Table 4 are from a study of (among other effects) cardiovascular disease and arsenic exposure (8). In assessing exposure, induction-latency time was allowed for, disregarding exposures more recent than 17 years prior to death. A register of deaths from the parish around the

Table 4. Stratified data from a study of the relationship between arsenic exposure and cardiovascular disease

Stratum	Age (years)	Case/ Ref.	Nonexposed All	Never employed	Employed	Exposed	Total
1	30–54	C	$6 = b_1$	4	2	$7 = a_1$	$13 = N_{11}$
		R	$14 = d_1$	8	6	$2 = c_1$	$16 = N_{01}$
			$20 = M_{01}$			$9 = M_{11}$	$29 = T_1$
2	55–64	C	$25 = b_2$	17	8	$19 = a_2$	$44 = N_{12}$
		R	$16 = d_2$	14	2	$6 = c_2$	$22 = N_{02}$
			$41 = M_{02}$			$25 = M_{12}$	$66 = T_2$
3	65–74	C	$45 = b_3$	34	11	$27 = a_3$	$72 = N_{13}$
		R	$26 = d_3$	20	6	$10 = c_3$	$36 = N_{03}$
			$71 = M_{03}$			$37 = M_{13}$	$108 = T_3$
1–3	Total	C	76	55	21	53	
		R	56	42	14	18	
Crude rate ratio			(1)	1.0	1.1	2.2	
SMR			(1)	0.9	1.3	1.9	
SRR[a]			(1)	0.9	1.5	2.2	
$\chi^1(1)$ Mantel-Haenszel						5.52	
Rate ratio (Mantel-Haenszel)							
–point estimate						2.1	
–90% confidence interval						1.2–3.5	

[a] with the nonexposed as the standard.

copper smelter was used as the source of subjects, both cases and referents being dead individuals (for further details see the original study). In addition to the stratified data, the table also includes the symbols used in the formulae that follow.

The Mantel-Haenszel chi-square statistic (with one degree of freedom) has the structure[a]

$$\chi^2(1) = \frac{(\Sigma_j a_j - \Sigma_j N_{1j} M_{1j}/T_j)^2}{\Sigma_j N_{1j} N_{0j} M_{0j} M_{1j}/T_j^2 (T_j - 1)} \tag{5}$$

with N_{1j} etc. denoting the exposed and N_{0j} etc. the nonexposed, as shown in the table.

The Mantel-Haenszel estimator of the rate ratio (or odds ratio) is

$$RR = \Sigma_j (a_j d_j/T_j)/\Sigma_j (b_j c_j/T_j) \tag{6}$$

and might be looked upon as "weighing together" the ratios of the various strata to give a reasonable overall estimate of the rate ratio.

---

[a] Σ_j means summation over the strata, j; for example $\Sigma_j a_j$ means $a_1 + a_2 + a_3 = a$, or in the example $7 + 19 + 27 = 53$.

The crude rate ratio, CRR, the standard morbidity/mortality ratio, SMR, and the standardized rate ratio, SRR, are obtained (9) as

$$CRR = \Sigma_j a_j \Sigma_j d_j / \Sigma_j b_j \Sigma_j c_j \qquad (7)$$

$$SMR = \Sigma_j a_j / \Sigma_j (b_j c_j / d_j) \qquad (8)$$

$$SRR = \Sigma_j (a_j d_j / c_j) / \Sigma_j b_j \qquad (9)$$

The SRR is useful for comparing the rate ratios of various exposure categories (the SRR always has the same standard, e.g., a defined age-distribution if the data are age-stratified, whereas the SMR standard is derived from the various exposure categories; therefore the SMRs of various exposure categories are not comparable, although such comparisons are commonly made, especially in occupational health epidemiology).

The calculations can now be made with the aid of the following scheme:

Stratum	a_j	$N_{1j} M_{1j} / T_j$	$N_{1j} N_{0j} M_{1j} M_{0j} / T_j^2 (T_j - 1)$	$a_j d_j / T_j$	$b_j c_j / T_j$
1	7	4.034	1.590	3.379	0.414
2	19	16.667	3.504	4.606	2.273
3	27	24.667	5.456	6.500	4.167
Σ_j	53	45.368	10.550	14.485	6.854

Equation (5) then becomes

$$\chi^2(1) = \frac{(53 - 45.368)^2}{10.550} = 5.52$$

and equation (6)

$$RR_{M-H} = \frac{14.485}{6.854} = 2.11$$

From equation (3) it is then possible to calculate the 90 % confidence interval:

$$\underline{RR}, \overline{RR} = (2.1)^{1 \pm 1.645/\sqrt{5.52}}$$

or

$$\underline{RR}, \overline{RR} = (2.1)^{1 \pm 0.70}$$

$$\underline{RR}, \overline{RR} = 1.2 \text{ to } 3.5$$

The SMR is given by

$$SMR = (7 + 19 + 27) / \left(\frac{6 \times 2}{14} + \frac{25 \times 6}{16} + \frac{45 \times 10}{26} \right)$$

or SMR $= 53/(0.86 + 9.38 + 17.31) = 1.9$.

The SRR is obtained analogously. Since the crude rate ratio as obtained from equation (7), i.e., the rate ratio for the totals for strata 1 − 3, is

$$CRR = \frac{53 \times 56}{76 \times 18} = 2.2,$$

some confounding is present, because the confounding rate ratio, Conf RR (*10*) becomes

$$Conf\ RR = CRR/SMR \tag{10}$$

or   Conf RR = 2.2/1.9 = 1.2

This value represents the magnitude of the confounding that was controlled through the stratification.

If the case/referent ratios for the nonexposed are compared between the various strata, the increase in incidence of cardiovascular disease with age is apparent, i.e., age fulfils the first criterion of a confounding factor as far as being a risk factor for the disease. The ratio of exposed to nonexposed referents also increases with age, so apparently older individuals tend to be more often exposed than younger ones, i.e., age is associated with exposure and therefore also fulfils the second criterion of being a confounding factor, although the association is not very strong as shown by the confounding rate ratio.

The stratum-specific rate ratio is 8.2 for the youngest stratum, 2.0 for the middle one, and 1.6 for the oldest stratum, i.e., there is a negative effect-modification with age in terms of the rate ratio. From the calculations made for the SMR it can be seen that the excess number of cases in the youngest stratum is 7 − 0.86 = 6.1, in the middle one 19 − 9.38 = 9.6, and in the last 27 − 17.3 = 9.7. The number of person-years for exposed individuals is not available, but since there are probably more young than old people, the rate difference or excess rate apparently increases with age, i.e., on a formal level, effect-modification by age works in opposite directions, depending on whether the rate ratio or the rate difference is taken as a yardstick. In absolute terms, the risk for arsenic-related cardiovascular disease is high in older ages (and presumably with long exposure); in relative terms, the risk is high among younger exposed individuals, although their absolute risk would still be relatively low. (For further considerations of dose-response relationships and for a discussion of the validity aspects of this specific study the reader is referred to the original paper).

References

1. **Rothman, K. J. & Boice, J. D., Jr.** *Epidemiologic analysis with a programmable calculator.* Washington, DC, US Department of Health, Education and Welfare, 1979 (NIH Publication No. 79-1649).
2. **Hardell, L. & Sandström, A.** Case-control study: soft tissue sarcomas and exposure to phenoxyacetic acids or chlorophenols. *British journal of cancer,* **39**: 711–717 (1979).
3. **Miettinen, O. S.** Individual matching with multiple controls in the case of all-or-none response. *Biometrics,* **25**: 339–355 (1969).

4. **Miettinen, O. S.** Estimation of relative risk from individually matched series. *Biometrics*, **26**: 75–86 (1970).
5. **Miettinen, O. S.** Estimability and estimation in case-referent studies. *American journal of epidemiology*, **103**: 226–235 (1976).
6. **Fleiss, J. L.** Confidence intervals for the odds ratio in case-control studies: the state of the art. *Journal of chronic diseases*, **32**: 69–77 (1979).
7. **Miettinen, O. S.** Comment. *Journal of chronic diseases*, **32**: 80–92 (1979).
8. **Axelson, O. et al.** Arsenic exposure and mortality: a case-referent study from a Swedish copper smelter. *British journal of industrial medicine*, **35**: 8–15 (1978).
9. **Miettinen, O. S.** Standardization of risk ratios. *American journal of epidemiology*, **96**: 383–388 (1972).
10. **Miettinen, O. S.** Components of the crude risk ratio. *American journal of epidemiology*, **96**: 168–172 (1972).

11

Study of combined effects

J. I. Kundiev[a] *& A. O. Navakatikyan*[a]

Investigation of the combined effects of various factors on the human organism is crucial for medical science. To establish a combined effect the following items must be elucidated:

1. The effects of a given group of factors on health, work capacity, or individual functions.
2. The proportion of observed variation explained by all the factors considered as compared with those not covered in the study.
3. The direction in which individual factors affect the variables under study.
4. The proportion of observed variation explained by each factor.
5. The ways in which the factors interact in their effects.
6. The mathematical modelling of the quantitative interrelations between factors and the outcome variables.

The interaction of factors (item 5) is considered to result in an additive effect when the overall effect equals the sum of the individual effects. The effect may also be larger or smaller than additive. Thus, interactions may either increase or decrease the overall effect (antagonism or synergism) (*1*).

Some authors claim that it is necessary to recognize an independent effect as well, but this may be considered to be simply a special type of additive effect. The analysis of all possible interactions becomes more complicated with an increase in the number of factors considered.

Many methods are used for investigating combined effects, of which the simultaneous action of two chemical substances has been most comprehensively studied. However, methods for assessing the combined impact of factors of different types (chemical, physical, biological) are of great interest.

---

[a] Scientific Research Institute of Labour Hygiene and Occupational Disease, Kiev, USSR.

Planning of studies and selection of variables

One of the most important tasks in multifactorial analysis of epidemiological data is the selection of variables to express the effect of a number of factors on one or more indicators of health. The variables selected should be such that they enable the investigator to adopt the most effective approach to the problems under study and reduce research time. There are no formal rules for selecting the requisite variables, so adoption of the right approach depends entirely on the knowledge, competence, and intuition of the investigator. The selection of variables must be based on comprehensive knowledge of a particular field.

Initial evaluation of the suitability of a variable can be made logically or intuitively. The first consideration is how amenable it is to formal description and this is one of the difficulties. A variable may often be easy to identify but hard to quantify. For example, it is very difficult to express numerically the state of health of a human being, the degree of fatigue at work, and other similar variables.

Data may be divided into two groups: quantitative and qualitative. Body temperature, blood pressure, body weight, and some other data are quantitative, continuous variables and are obviously easy to use. Qualitative data include different kinds of higher nervous activity, state of health, etc. A scale may be developed for quantifying discrete data. It is sometimes advisable to combine several variables in order to cut down the number of calculations and simplify the algorithms used in the study (2).

The set of values of each particular variable determines the nature of the sample, which should be representative, i.e., its size (n) must be sufficiently large. As a rule, to achieve this aim several biological variables are examined simultaneously in epidemiological studies. A method of evaluating the sample size is to indicate margins of error for the variables considered to be most important in regard to the study. The requisite sample size is first determined separately for each of the most important variables (3, 4). After finding the n-values for individual variables, the situation as a whole may be examined. It may happen that all the n-values obtained are close to one another. If the greatest value of n is acceptable for the study budget, it is usually the one selected. However, more often the values of n vary widely and selection of the largest value is not the method of choice in view of the time it would take to follow-up, or because of the need for greater accuracy than had at first been planned. In this case, in order to justify use of a smaller n value, a lower level of accuracy may be accepted for certain variables. In some cases, the values of n required vary excessively, thus making it impossible to study some variables because the accuracy that would be attained is insufficient (5).

Methods of sequential analysis may be used to facilitate the compilation of data for study. Initially, only a small number of observations are made. If the data obtained confirm or refute the hypothesis with a sufficient degree of significance, the observations are

210

then terminated. If the degree of significance is insufficient, the observations have to be continued. After some time, the working hypothesis is again subjected to statistical evaluation. The procedure is repeated until the hypothesis is confirmed or refuted. This approach demands considerably fewer observations that is usual with other research strategies.

With sequential analysis, after each observation one of the following three conclusions may be arrived at:

(a) confirmation of the working hypothesis;
(b) confirmation of an alternative hypothesis;
(c) observations must be continued.

Depending on the basic goals of the study, provision must be made to draw up future study plans. At present, considerable progress is being made in the design of studies, allowing the required data to be obtained with fewer observations and greater accuracy than when research has been less well planned.

Assessment of health impairment due to the impact of two determinants

The combined effects of two determinants have been more fully investigated than more complex situations. Usually, health impairment is studied in two or more groups of people.

In some situations, it is sufficient to demonstrate the direction of the effects of the occupational factors studied. For example, Tartakovskaya (6) has shown that local vibration and mercury have a synergistic action. When studying the occurrence of vibration syndrome among drill operators in mercury mines, she found that persons with a mercury content of $10\,\mu g$/litre or more in their urine suffered from vibration syndrome twice as frequently as workers who had lower urinary mercury levels (44.4% as against 27.7%). On the other hand, the risk of toxic effects of mercury was increased in the case of simultaneous exposure to vibration. Thus, in this investigation, it is possible to answer only the first and third of the questions implicit in the six items listed at the beginning of the chapter and only a synergism between the effect of vibration and exposure to mercury can be demonstrated. If it later transpires that additive effects also occur at higher exposure levels, the occupational exposure limits will have to be reconsidered for situations in which workers are exposed to the two factors simultaneously.

For the assessment of the combined effects of two factors it would be informative to compare incidence rates in four study groups: nonexposed referents (O), groups A and B exposed to factors A and B respectively, and group AB simultaneously exposed to both factors. Fig. 1 gives an example of such exposures according to Breslow & Day (7) and Hammond et al. (8). The figure shows the increase in the incidence of death from lung cancer for simultaneous exposures to asbestos dust (A) and smoking (B).

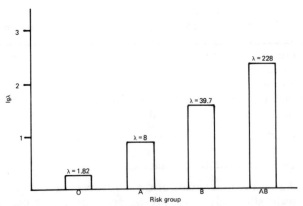

Fig. 1. Joint impact of asbestos dust (A) and smoking (B)
on lung cancer risk. Mortality from lung cancer in workers (λ)
exposed to asbestos dust (groups A and AB) and in referrents of the same age
who were smoking the same number of cigarettes per day.
For convenience the data are given on a logarithmic scale.

To assess the interaction of factors A and B, it is necessary to find out whether the combined effect of exposure to both factors (AB) equals the sum of the separate effects. If no interaction occurs, the mortality in workers exposed to AB should be $(8 - 1.82) + (39.7 - 1.82) = 44.06$. The actual mortality is higher than expected, i.e., an evident synergism has taken place.

In this example, it may be concluded that simultaneous exposure to factors A and B multiplies their effects. This multiplication can be recognized more easily if the logarithms of the incidence rates ($\lg\lambda$) are treated in the same way as the actual rates above. The expected effect of multiplication (AB_m) would be given by $\lg\lambda_A - \lg\lambda_0 + \lg\lambda_B - \lg\lambda_0$. For the example, $AB_m = 0.9 - 0.26 + 1.6 - 0.26 = 1.98$. The observed mortality rate for lung cancer was even higher ($\lg 228 = 2.36$), i.e., simultaneous exposure to asbestos and smoking had a multiplying effect.

Thus, the above type of data analysis makes it possible to answer the first, third, and fifth of the questions listed at the beginning of the chapter. Multivariate statistical methods must be used in order to answer the remaining questions.

Multiple correlation – regression analysis

The combined effects of several factors are often best characterized by multiple regression analysis. Many phenomena of interest in medical epidemiology are interrelated. Very often, but not always, the link is a causal one, i.e., the causes affect the outcomes, changing the direction of the outcome indicators. Such links are often of interest to medical research, but are hard to quantify. It is practically impossible to take into account all the factors affecting the complex processes taking place in a

212

living organism; only those factors producing the largest effects, or which are otherwise of special interest, are considered in studying their relationships. In such a study, one is dealing not with functional relationships but rather with correlations, where a change in one variable is not always associated with a strictly defined change in another.

The study of such correlations implies, first, defining how close they come to functional dependence. Various correlation indices are calculated for this purpose. The linear index of correlation – the correlation coefficient (r) – ranges from zero to one. If $r = 0$, there is no linear relation between the phenomena under study. If $r = 1$, this implies the strongest possible relation – a functional one. Intermediate values indicate stronger or weaker associations. The coefficient may have a positive or a negative sign. If it is positive, there is a direct relation, if negative, the relation is inverse.

Having determined the value of the correlation coefficient, it has to be established whether there is a cause-and-effect relationship, or possibly some other link, or whether the association between the variables is merely coincidental (9). However strong a statistical association might be, it can never actually prove a causal relation. Hypotheses as to cause and effect must ultimately be proved by some other means (10).

The methods of correlation analysis require that the variables under study should satisfy a number of conditions: they should not show a tendency to change with time, and they should be free from intermittent and seasonal variations. However, most functions of the organism and their indicators do undergo changes with the passage of time. In view of this, the investigator must have means of allowing for such changes. In some situations, spurious correlations appear as a result of the non-uniformity of the material under study. In all raw data some anomalies occur, but if the values of the variables under study have standard deviations that differ considerably from those of the data basis as a whole, a correlation coefficient does not reflect the true interrelation objectively, and this distorts the closeness of the association. Finally, spurious correlations may be observed owing to errors in the recording of the data (11).

Calculation of pair correlation coefficients is usually carried out by computer, for which programs are given in several publications (12–15). An absence of statistical significance in the pair correlation coefficients does not in itself prove that there is no association between the variables studied. Increasing the number of observations might establish the existence of an association. It is possible to estimate the size of sample needed for a desired level of significance for different values of the correlation coefficient (16).

If two variables are correlated with each other, this may also be due to each of them being correlated with a third variable or a set of variables. Such a situation has dictated the need to consider theoretical correlations between two variables while others are kept constant. The partial correlation coefficient $(r_{0i.n...21})$ is the indicator of such a need. It

213

shows that one is dealing with the correlation between an outcome variable (X_0 or Y_0) and an i-variable (X_i), with constant values of other, $n-1$ (X_1, X_2 ... X_n except X_i). If the correlation between two variables decreases when others are kept constant, it means that their association is due partly to the effect of the latter variables. If the partial correlation coefficient approaches zero, it may be concluded that the association is due entirely to this effect. In contrast, when a partial correlation coefficient is greater than the corresponding pair correlation coefficient, this shows that the other variables have weakened or "masked" the correlation. Even such a finding does not constitute complete proof of a causal relation because some extraneous variable, quite different from those considered in the analysis, may be the source of the observed correlation. Causality cannot be established only on the basis of statistical correlation observed in epidemiological studies; one has to consider other exposure data such as those from experimental toxicological studies (*10, 17*).

In practice, epidemiological research is often concerned with ascertaining the effect of several determinants on a single outcome variable. The multiple correlation method makes it possible to determine the intensity of such an effect by calculating the value of the multiple correlation coefficient. This depends on the ratio between the dispersion of outcome values $\hat{y}$ established by means of the multiple regression equation:

$$\hat{y} = a_0 + a_1 x_1 + a_2 x_2 + \ldots + a_n x_n$$

and the dispersion of the values giving the determinant index y ($\hat{y}$ = theoretical value, y = values based on individual observations, n = number of variables under study). The smaller the scatter of the determinant index values around the multiple regression line, the higher is the value of the multiple correlation coefficient, the absolute value of which may range from zero to one (*11*). FORTRAN programmes are available for calculating the matrix of pair and partial correlation coefficients, multiple correlation coefficients, and regression coefficients (*14, 15*).

The following equation is used as a criterion for checking the null hypothesis for the significance of the pair correlation coefficient:

$$t = r \sqrt{\frac{N-2}{1-r^2}} \geqslant t_{st}$$

where N = sample size, $N-2$ = number of degrees of freedom, t_{st} = the standard t-value, when $P = 0.05$ or $P = 0.01$. If $r < 0.20$ or the sample size is small, a Fisher transformation must be made by recalculating r as a z value (*10, 13, 14, 16*).

The sample size for each value of the pair correlation coefficient can be determined and the null hypothesis tested at two levels of significance using Table 1 (*13*). For example, if $r = 0.30$ the correlation may be considered significant when $N = 43$.

214

To check the significance of an individual partial correlation coefficient, the following expression is used:

$$t = r_{\text{oi.n.} \ldots 21} \sqrt{\frac{N - k - 2}{1 - r^2_{\text{oi.n.} \ldots 21}}} \geq t_{\text{st}}$$

where $k =$ the number of variables kept constant. There are $v = N - k - 2$ degrees of freedom. Instead of calculating t, Table 1 may be used to determine whether the sample size is sufficient for the coefficient to be significant for the values of $r_{\text{oi.n.} \ldots 21}$. In doing so, the value k should be added to the value N estimated by Table 1.

Table 1. Number of pairs of values (N) needed for significance testing of the correlation coefficient (r)[a]

r	N ($P < 0.05$)	N ($P < 0.01$)	r	N ($P < 0.05$)	N ($P < 0.01$)
0.02	9604	16640	0.46	19	30
0.04	2401	4159	0.48	17	27
0.06	1068	1848	0.50	16	25
0.08	601	1039	0.52	15	23
0.10	385	664	0.54	14	21
0.12	267	461	0.56	13	20
0.14	196	338	0.58	12	18
0.16	150	259	0.60	11	17
0.18	119	204	0.62	10	16
0.20	96	165	0.64	10	15
0.22	80	136	0.66	9	14
0.24	67	114	0.68	9	13
0.26	57	97	0.70	8	12
0.28	49	83	0.72	8	11
0.30	43	72	0.74	7	10
0.32	38	64	0.76	7	10
0.34	34	56	0.78	7	9
0.36	30	50	0.80	6	9
0.38	27	45	0.82	6	8
0.40	24	40	0.84	6	7
0.42	22	36	0.86	5	7
0.44	20	33	0.88	5	7
			0.90	5	6

[a] After Bogach et al. (13).

The significance of a multiple correlation coefficient is quite easy to test for values $P \leq 0.05$ and $P \leq 0.01$ with the aid of Tables 2 and 3. These show v_1 as the number of independent variables, v_2 as the number of degrees of freedom ($v_2 = N_1 - v_1 - 1$, where N is the sample size). If the multiple correlation coefficient calculated exceeds the value indicated in the table, its actual value will differ from zero at either $P \leq 0.05$ or $P \leq 0.01$.

To determine the effects of the factors under study, the coefficients of determination are calculated. They show the proportional effect of the

Table 2. Critical values of multiple correlation coefficients, $P = 0.05$[a,b]

v_2	v_1 = 1	2	3	4	5	6	7	8	9	10	12	14	16	18	20	24	30	40	50
10	0.58	0.67	0.73	0.76	0.79	0.81	0.83	0.84	0.86	0.87	0.88	0.90	0.91	0.92	0.93	0.94	0.96	0.96	0.96
12	0.53	0.63	0.68	0.72	0.75	0.76	0.79	0.81	0.82	0.83	0.85	0.87	0.88	0.89	0.90	0.91	0.93	0.94	0.95
14	0.50	0.59	0.65	0.69	0.72	0.74	0.76	0.78	0.79	0.81	0.83	0.84	0.86	0.87	0.88	0.90	0.92	0.93	0.94
16	0.47	0.56	0.62	0.66	0.69	0.71	0.73	0.75	0.77	0.78	0.80	0.82	0.84	0.85	0.86	0.88	0.90	0.92	0.93
18	0.44	0.53	0.59	0.63	0.66	0.69	0.71	0.73	0.74	0.76	0.78	0.80	0.82	0.83	0.84	0.86	0.88	0.91	0.92
20	0.42	0.51	0.56	0.60	0.64	0.66	0.68	0.70	0.72	0.74	0.76	0.77	0.80	0.81	0.82	0.85	0.87	0.89	0.91
22	0.40	0.49	0.54	0.58	0.61	0.64	0.66	0.68	0.70	0.72	0.74	0.76	0.78	0.80	0.81	0.83	0.85	0.88	0.90
24	0.39	0.47	0.52	0.56	0.59	0.62	0.64	0.66	0.68	0.70	0.73	0.74	0.76	0.78	0.79	0.82	0.84	0.87	0.89
26	0.37	0.45	0.51	0.55	0.58	0.60	0.63	0.65	0.66	0.68	0.71	0.73	0.75	0.76	0.77	0.80	0.83	0.86	0.88
28	0.36	0.44	0.49	0.53	0.56	0.59	0.61	0.63	0.64	0.66	0.69	0.71	0.73	0.75	0.76	0.79	0.82	0.85	0.87
30	0.35	0.43	0.48	0.51	0.55	0.57	0.59	0.61	0.63	0.65	0.68	0.70	0.71	0.73	0.75	0.78	0.81	0.84	0.86
34	0.33	0.40	0.45	0.49	0.50	0.54	0.57	0.59	0.60	0.62	0.65	0.67	0.69	0.71	0.73	0.75	0.78	0.82	0.85
40	0.30	0.37	0.42	0.46	0.48	0.51	0.53	0.55	0.57	0.59	0.61	0.64	0.66	0.68	0.69	0.72	0.75	0.79	0.82
44	0.29	0.36	0.40	0.44	0.47	0.49	0.51	0.53	0.55	0.56	0.59	0.60	0.63	0.66	0.67	0.70	0.74	0.78	0.81
50	0.27	0.34	0.38	0.41	0.44	0.46	0.49	0.50	0.52	0.54	0.57	0.59	0.61	0.63	0.65	0.68	0.71	0.75	0.78
60	0.25	0.31	0.35	0.38	0.41	0.43	0.45	0.47	0.48	0.50	0.53	0.55	0.57	0.59	0.61	0.64	0.67	0.72	0.75
80	0.22	0.27	0.30	0.33	0.36	0.38	0.40	0.41	0.43	0.44	0.47	0.49	0.51	0.53	0.56	0.58	0.61	0.66	0.70
100	0.20	0.24	0.27	0.30	0.33	0.34	0.36	0.37	0.39	0.40	0.43	0.45	0.47	0.49	0.50	0.53	0.57	0.62	0.65
150	0.16	0.20	0.23	0.25	0.27	0.28	0.30	0.31	0.32	0.34	0.36	0.38	0.39	0.41	0.42	0.45	0.48	0.53	0.57
200	0.14	0.17	0.20	0.22	0.23	0.25	0.26	0.27	0.28	0.29	0.31	0.33	0.35	0.36	0.37	0.40	0.43	0.48	0.51
300	0.11	0.14	0.16	0.18	0.19	0.20	0.21	0.22	0.23	0.24	0.26	0.27	0.29	0.30	0.31	0.33	0.36	0.40	0.43
500	0.09	0.11	0.12	0.14	0.15	0.16	0.17	0.17	0.18	0.19	0.20	0.21	0.23	0.24	0.25	0.26	0.29	0.32	0.35
1000	0.06	0.08	0.09	0.10	0.11	0.11	0.12	0.12	0.13	0.13	0.14	0.15	0.16	0.17	0.18	0.19	0.21	0.23	0.25

[a] After Überla (28).

[b] v_1 = number of independent variables; v_2 = number of degrees of freedom ($v_2 = N - v_1 - 1$, where N = size of sample). If the calculated coefficient of multiple correlation exceeds the value specified, its true value is different from zero at $P < 0.05$.

Table 3. Critical values of multiple correlation coefficients, $P = 0.01$ [a, b]

v_2															v_1					
	1	2	3	4	5	6	7	8	9	10	12	14	16	18	20	24	30	40	50
10	0.71	0.78	0.81	0.84	0.86	0.87	0.89	0.90	0.90	0.91	0.92	0.93	0.94	0.94	0.95	0.96	0.96	0.97	0.98
12	0.66	0.73	0.77	0.80	0.82	0.84	0.86	0.87	0.88	0.88	0.90	0.91	0.92	0.92	0.93	0.94	0.95	0.96	0.97
14	0.62	0.69	0.74	0.77	0.79	0.81	0.83	0.84	0.85	0.86	0.88	0.89	0.90	0.91	0.91	0.92	0.94	0.95	0.96
16	0.59	0.66	0.71	0.74	0.76	0.78	0.80	0.81	0.83	0.84	0.86	0.87	0.88	0.89	0.90	0.91	0.92	0.94	0.95
18	0.56	0.63	0.68	0.71	0.74	0.76	0.77	0.79	0.80	0.81	0.83	0.85	0.86	0.87	0.88	0.89	0.91	0.93	0.94
20	0.54	0.61	0.65	0.69	0.71	0.73	0.75	0.77	0.78	0.79	0.81	0.83	0.84	0.85	0.86	0.88	0.90	0.92	0.93
22	0.52	0.59	0.63	0.66	0.69	0.71	0.73	0.75	0.76	0.77	0.79	0.81	0.83	0.84	0.85	0.87	0.89	0.91	0.92
24	0.50	0.56	0.61	0.64	0.67	0.69	0.71	0.73	0.74	0.75	0.78	0.79	0.81	0.82	0.83	0.85	0.87	0.90	0.91
26	0.48	0.55	0.59	0.62	0.63	0.67	0.69	0.71	0.72	0.74	0.76	0.78	0.79	0.81	0.83	0.84	0.86	0.89	0.91
28	0.46	0.53	0.57	0.61	0.63	0.66	0.68	0.69	0.71	0.72	0.74	0.76	0.78	0.79	0.81	0.83	0.85	0.88	0.90
30	0.45	0.51	0.56	0.59	0.62	0.64	0.66	0.68	0.69	0.71	0.73	0.75	0.77	0.78	0.79	0.82	0.84	0.87	0.89
34	0.42	0.49	0.53	0.56	0.59	0.61	0.63	0.65	0.66	0.68	0.70	0.72	0.74	0.76	0.77	0.79	0.82	0.85	0.87
40	0.39	0.45	0.49	0.53	0.55	0.58	0.59	0.61	0.63	0.64	0.67	0.69	0.71	0.72	0.74	0.76	0.79	0.82	0.85
44	0.38	0.44	0.47	0.51	0.53	0.55	0.57	0.59	0.61	0.62	0.65	0.67	0.69	0.70	0.72	0.74	0.77	0.81	0.83
50	0.35	0.41	0.45	0.48	0.50	0.53	0.55	0.56	0.58	0.59	0.62	0.64	0.66	0.68	0.69	0.72	0.75	0.79	0.81
55	0.34	0.39	0.43	0.46	0.48	0.51	0.52	0.54	0.56	0.57	0.60	0.62	0.64	0.65	0.67	0.70	0.73	0.77	0.80
60	0.33	0.38	0.41	0.44	0.47	0.49	0.51	0.52	0.54	0.55	0.58	0.60	0.62	0.64	0.65	0.68	0.71	0.75	0.78
80	0.28	0.33	0.36	0.39	0.41	0.44	0.45	0.46	0.48	0.49	0.51	0.54	0.56	0.57	0.59	0.62	0.65	0.69	0.73
100	0.25	0.30	0.33	0.35	0.37	0.39	0.41	0.42	0.44	0.45	0.47	0.49	0.51	0.53	0.55	0.57	0.60	0.65	0.68
150	0.21	0.24	0.27	0.29	0.31	0.32	0.34	0.35	0.36	0.37	0.40	0.41	0.43	0.45	0.46	0.49	0.52	0.56	0.60
200	0.18	0.21	0.24	0.25	0.27	0.28	0.29	0.31	0.32	0.33	0.35	0.36	0.38	0.39	0.41	0.43	0.46	0.50	0.54
300	0.15	0.17	0.19	0.20	0.22	0.23	0.24	0.25	0.26	0.27	0.29	0.30	0.32	0.33	0.34	0.36	0.39	0.43	0.46
500	0.12	0.14	0.15	0.16	0.17	0.18	0.19	0.20	0.21	0.21	0.23	0.24	0.25	0.26	0.27	0.28	0.31	0.34	0.37
1000	0.08	0.10	0.11	0.12	0.12	0.13	0.13	0.14	0.15	0.15	0.16	0.17	0.18	0.18	0.19	0.20	0.22	0.25	0.27

[a] After Uberla (28).

[b] v_1 = number of independent variables; v_2 = number of degrees of freedom ($v_2 = N_1 - v_1 - 1$, where N = size of sample). If the calculated coefficient of multiple correlation exceeds the value specified, its true value is different from zero at $P < 0.01$.

various factors as fractions of 1. The multiple determination coefficient (D) is equal to the square of the multiple correlation coefficient, $D = R^2$. The partial determination coefficient (d) for the i-factor is calculated by using the formula:

$$d_i = \frac{a_i \sum (X_i - \overline{X}_i)(X_0 - \overline{X}_0)}{\sum (X_0 - \overline{X}_0)^2}$$

where $\overline{X}_i$ and $\overline{X}_0$ are mean values of the i-factor and the outcome variable X_0, respectively (18).

Some examples are given below of multiple correlation–regression analysis as applied to the study of the effect of conditions of life on morbidity.

Morbidity causing temporary loss of work capacity (X_0) was studied in 10 occupations at six thermal power stations (19, 20) in relation to seven factors: (1) physical load (X_1), (2) mental stress (X_2'), (3) industrial noise exposure (X_3), (4) coal-dust air pollution (X_4), (5) climatic zone (X_5), (6) degree of hazard (X_6), and (7) level of medical care (X_7).

All the factors were evaluated quantitatively so that a mathematical analysis could be made. Physical load and mental stress were each assessed on a 4-point scale developed at the Kiev Institute of Labour Hygiene and Occupational Disease, while the degree of hazard was classified using a 7-point scale developed by the Institute of Labour Hygiene and Occupational Disease of the USSR Academy of Medical Sciences. The noise level in some occupations exceeded the occupational exposure limit by as much as 16 dB(A), while the dust content varied from zero to more than the occupational exposure limit. Climatic zones varied from 1 (central Asia) to 5 (north European part of the USSR), and medical care varied from 5 (excellent) to 3 (fair). The study thus covered persons living in a wide variety of conditions and doing work involving different degrees of physical and mental strain. Altogether, 60 different combinations of the determinants (X_1 to X_7) and of the outcome variable, morbidity (X_0), were available for analysis ($N = 60$).

The results obtained indicate a statistically significant association between general morbidity (excluding accidents) and the factors studied ($R = 0.45 \pm 0.10$). The multiple determination coefficient D was 0.2, indicating that 20% of all morbidity was explained by variations in the basic occupational factors studied. Airborne dust and industrial noise were the main harmful factors incriminated, indicating how overall morbidity is related to working conditions. This was shown by the relatively high values of their partial and pair correlation coefficients ($r_{03.765521} = 0.34$; $r_{03} = 0.30$; $r_{04.765321} = 0.24$; $r_{04} = 0.24$).

General morbidity, being an aggregate index, does not reflect accurately the effect of working conditions on the organism of a worker. A consideration of multiple and partial correlation coefficients by diagnostic groups (Table 4) shows that different diseases vary in the degree and nature of their dependence on working conditions and that the effect of working conditions on some disease entities may be quite

Table 4. Relationship of certain diseases causing temporary incapacity for work to the determinants studied

Classification of disease	Multiple coefficient of correlation	Multiple coefficient of determination	Basic factors determining the value of R	Partial coefficients of correlation between determinants and morbidity
Influenza, catarrh of respiratory tract	0.65 + 0.08	0.42	Noise Hazard class Physical load Dust content	0.41 0.35 −0.30 0.24
Tonsillitis	0.53 + 0.09	0.28	Climate zone Dust content	0.28 0.26
Cardiac diseases	0.46 + 0.11	0.21	Mental stress Physical load	0.17 −0.17
Gastroduodenal ulcer	0.47 + 0.10	0.22	Hazard class Noise	0.25 0.21
Neuralgia, root syndrome	0.40 + 0.10	0.16	Noise Physical load	0.25 0.11
Bronchitis	0.58 + 0.09	0.33	Noise Dust content Physical load	0.33 0.30 0.27
Acute gastrointestinal diseases	0.72 + 0.07	0.56	Climate zone Medical care	0.32 −0.19
Vascular changes of autonomic origin	0.42 + 0.10	0.18	Shift work Mental stress	0.32 0.16

[a] After Navakatikyan & Nagornaya (20).

strong. High values for multiple and partial correlation coefficients were obtained when studying the effects of various factors on sickness absenteeism caused by acute gastrointestinal disease ($R = 0.72$) This applied mainly to climatic zone and high temperatures at workplaces, as well as to the efficiency of medical care. The multiple correlation coefficient, reflecting the association between morbidity of influenza and acute respiratory infections and the factors under study, was 0.65. This coefficient was mainly determined by the effects of temperature (a major item for the hazard class in this industry) and of dust and industrial noise at the workplaces. A negative correlation ($r = -0.30$) with the physical load was another interesting finding. This may be interpreted as indicating that physical work has a greater protective value than inactivity. However, the influence of spontaneous job selection should not be excluded, for persons in a good state of health and resistant to acute respiratory diseases may tend to select jobs involving hard physical work. The multiple correlation coefficient reflecting the dependence on certain factors of cardiovascular morbidity leading to absenteeism was 0.46. The factors studied thus explain only 21 % of the total variation.

The data collected thus made it possible to calculate multiple regression equations for different disease entities and for persons in different occupations at a number of power stations. From the regression equations it is possible to predict with a high degree of probability the extent of the reduction in both general morbidity rates and those for individual disease entities that might be achieved by reducing the harmful effects of poor working conditions at the place of work.

Another example is the study by Izmerov (21) of the effect of air pollution on child and adult morbidity in 37 cities. Morbidity indicators were determined for 29 separate diseases and groups of diseases based on the incidence of initial visits to clinics and dispensaries per 10 000 population for three age-groups: children (up to 14), adolescents (14–17) and adults. Calculation of the pair correlation coefficients did not, in most cases, reveal any significant association with mean concentrations of airborne dust, carbon black, sulfur dioxide, nitrogen dioxide or carbon monoxide. Only a few associations were found to be of any importance, e.g., the dependence of the frequency of gastric and peptic ulcers on the atmospheric concentrations of sulfur dioxide ($r = 0.52 \pm 0.16$). Multiple correlation coefficients between morbidity rates and various pairs of pollutants revealed that combined exposure to sulfur dioxide and nitrogen dioxide was associated with acute infections of the upper respiratory tract in adolescents and with bronchiectasis in children; associations were also discovered between chronic gastritis in children and combined exposures to sulfur dioxide and carbon black in air ($R = 0.59 \pm 0.18$) and between peptic ulcer in adults and combined exposures to sulfur and nitrogen dioxides ($R = 0.57 \pm 0.16$) or to sulfur dioxide and carbon monoxide ($R = 0.52 \pm 0.17$). The multiple correlation coefficients obtained for other disease entities did not reach statistical significance.

Thus, multiple correlation analysis permits a quantitative determination of the closeness of the association between the outcome variables and the various determinants included in the study, whether considered individually or combined. The direction in which the determinants act is shown by a positive or negative correlation coefficient.

A linear model was used in the above examples. This may be sufficient, if R turns out to be high enough. Moreover, some results suggest that if the various determinants act over a long time an additive effect is likely to be observed (22) and in such situations, too, the linear model is adequate. But this model cannot be employed to determine the interactions between separate factors when the combined effects differ from the additive ones. To determine the effects of non-additive interactions more sophisticated models have to be used, the correlation coefficients of the outcome variables being calculated using the squares or products of the levels of the determinants. With these models, the calculation methods remain unchanged in principle but the computations need to be increased considerably in number.

Canonical correlation

It has already been pointed out that simple correlation between two variables X and Y is expressed by a pair correlation coefficient. Multiple correlation is an aggregate concept derived from simple correlation where the variable Y is correlated with several variables X. Thus, multiple correlation may also be treated as a pair correlation between variable Y and variable $\hat{Y}$, where the latter is a linear combination of several variables X:

$$\hat{Y} = a_1 X_1 + a_2 X_2 + \ldots + a_n X_n$$

(a_i being coefficients of the multiple regression equation).

As multiple correlation is a generalized form of pair correlation, where several X variables are dealt with in a similar way, canonical correlation is a generalized form of multiple correlation when dealing with a number of variables X and Y. Thus, canonical correlation is a correlation between a linear combination of a number of variables X and a linear combination of several variables Y (*10*).

The linear combination P of variables X and q of variables Y may be determined as follows:

$$\hat{X} = a_1 X_1 + a_2 X_2 + \ldots + a_p X_p;$$
$$\hat{Y} = b_1 Y_1 + b_2 Y_2 + \ldots + b_q Y_q.$$

The basic task in calculating canonical correlations is to find the unknown coefficients a_i and b_i in order to obtain the maximum correlation between $\hat{Y}$ and $\hat{X}$. Let the maximum canonical correlation coefficient R'_c be designated as the first canonical coefficient of correlation between the two sets of variables and let the corresponding linear combinations $\hat{X}_1$ and $\hat{Y}_1$ be called the first canonical variables. The second canonical correlation will be determined by linear combinations $\hat{X}_2$ and $\hat{Y}_2$, such that the second canonical variables will yield, of all linear combinations not correlated with $\hat{X}_1$ and $\hat{Y}_1$, the second largest coefficient of correlation R''_c. The third canonical correlation will be determined by linear combinations $\hat{X}_3$ and $\hat{Y}_3$, such that the third canonical variables will give, of the various linear combinations not correlated with the first two variables, the third largest canonical correlation coefficient. Each subsequent set of canonical variables is determined in a similar manner and, if $q \leqslant p$, q canonical correlations and q sets of canonical variables are obtained.

This method may be especially useful in epidemiological studies when two large sets of variables are available for an investigator and he is interested in the relationship between them. If the sets are large, he can confine himself to investigating a small number of linear combinations from each set. The researcher may find it interesting to determine those linear combinations that are most highly correlated. For example, one set of variables may be related to working conditions and daily life, while a second one may be related to physiological functions or state of health. It may also happen that the interrelations are almost completely covered by the correlation coefficient between the first canonical variables (*23*). Programs for computing the canonical correlation coefficients are described in the literature (*14, 15*).

The entities under study, and in particular the individual components that make up the entities, seldom vary in respect of the value for one single variable. Usually, there is a difference in the values for several variables. If the significance of such a difference is evaluated only for the values of individual variables, characterizing the individual components of the entity under study, wrong and biased information may be obtained. The differences in the values of any one of the variables compared may not be significant, although a significant difference exists when they are considered as a whole. Therefore, multivariate statistical methods need to be used in order to make a comprehensive evaluation of the correlations being studied.

Analysis of variance

In epidemiological research it is often important to determine the effects of several different factors on the variable under study, both when acting together and when each is considered in isolation. To solve such problems, the use of the method known as analysis of variance is recommended (*12–15*). Variance analysis may be performed on both large and small samples, whether homogeneous or heterogeneous. The method has proved to be one of the most efficient and popular ones for analysing medical and biological data. Two examples are offered for consideration.

The effect of teaching work on heart rate and regulation of the endocrine system was studied by Navakatikyan & Karpenko (*24*) in female secondary school teachers. Three independent variables (*A, B, C*) were examined. *A* was the academic year (second, third, and fourth terms of the school year, coinciding with autumn, winter, and spring). This variable was included in order to determine the possible accumulation of fatigue and its chronic development. Variable *B* related to type of work (mathematics and language teachers as compared with teachers of other subjects and teachers in junior forms). Variable *C* was the amount of teaching per week (more or less than the mean for each of the two groups). Table 5 shows that the secretion of epinephrine and norepinephrine and variations in heart rate registered over a period of 100 minutes were mainly connected with the amount of teaching during the week. Secretion of 17-oxycorticosteroids (17-OCS) and heart rate variations over 50 minutes depended on the season and only heart rate variations over 100 minutes appeared to be sensitive to the type of work ($P < 0.01$). It can be seen that the associations between the occupational factors and the physiological outcome variables can be very complex and still yield significant results in the analysis.

The significance and relative impact of different factors acting in isolation are determined by using equations and tables found in a number of manuals (*9, 14, 16*). In the example described in the preceding paragraph, the interaction of the variables *ABC* proved to be highly

222

Table 5. Effect of teaching on the secretion of epinephrine, norepinephrine, 17-hydroxycorticosteroid (17-HCS) and fluctuations in heart rate in periods of 100 and 50 minutes, according to variance analysis[a]

Teaching work characteristics (combinations included)[c]	Significance of effect (P) and Fisher criterion[b]				
				Fluctuations in heart rate	
	Epinephrine	Norepinephrine	17–HCS	100 min	50 min
A	—	—	0.01	—	0.05
	—	—	5.33	—	3.47
B	—	—	—	0.001	—
	—	—	—	10.4	—
C	0.01	0.01	—	0.01	—
	7.4	7.0	—	7.3	—
AB	—	—	0.01	0.01	0.05
	—	—	6.0	7.3	3.5
AC	0.05	0.01	0.01	0.01	0.01
	4.1	5.5	5.8	6.8	5.2
BC	0.01	0.001	0.01	0.0001	0.05
	8.5	13.29	5.6	22.3	3.8
ABC	0.001	0.0001	0.0001	0.0001	0.001
	11.8	18.5	20.6	29.3	13.9
Intensity of combined effect ABC	29%	39%	27%	46%	36%

[a] After Navakatikyan & Karpenko (24).
[b] P, upper line; Fisher criterion, lower line
[c] A = period of academic year (terms 2, 3, 4); B = type of work; C = teaching hours per week (higher or lower than average per group).

significant. This means that the combined effect of variable *A* and combination *BC* or combination *AB* and variable *C* differed considerably from the algebraic sum of their separate effects.

Thus, variance analysis makes it possible to reveal whether two or more factors interact. But when dealing with four or more factors the value of this method decreases because it is difficult to make a specific evaluation of the direction of such an interaction.

Classification of data—discriminant analysis

In the methods discussed above, the morbidity of a disease or the rate of functional changes was used as an index of the state of health of occupational groups. The problem of causal association between diseases and occupational factors may be approached by means of epidemiological studies, omitting the calculation of rates but using instead associations of personal characteristics. Such an opportunity is provided by various methods of pattern recognition and by discriminant analysis in particular. Subjects, situations, and phenomena encountered in practical research frequently have to be divided into groups or classes. Those included in a given class possess certain peculiarities not found in other classes. Successful classification systems presuppose the establishment of rules governing the allocation of an object to one or more groups (categories, classes) by measuring some of its characteristics (variables).

In situations where the presence or absence of significant differences among the objects under study has to be estimated on the basis of several variables, the method of discriminant analysis is used (9). Discriminant analysis differs from other methods in that assignment of an object to one of the classes is made by an algebraic discriminant function whose parameters, in their turn, are algebraic functions of the distribution parameters of the corresponding sums.

The two classes may be discriminated by Fisher's linear discriminant function. One such linear function is:

$$X_0 = \beta_1 X_1 + \beta_2 X_2 + \ldots + \beta_p X_p$$

which maximizes the standardized distance between the centres of the sets (25).

The value of the discriminant function in general is a number that, using the statistical approach, is adjusted in line with the boundary dividing the two groups of homogeneous objects. Such a boundary placed on a line is a point; on a surface, a line; while in space it is a surface (or a hypersurface in multidimensional space).

In principle, selection of the discriminant function means selection of the code (program) that will allow digital-to-analogue conversion of information, replacement of the probability description of an object by a deterministic one, and of the continuous function by a discriminant one.

If a sample is split into two parts, it means that the discriminant function is determined. If a sample has to be split into several classes, it can be split first into a pair of classes, and then each of the discriminated classes can be checked for homogeneity before being split into a further pair, and so on. Since the splitting of the objects into two classes and the assignment of the objects to one of them is the most characteristic method of pattern recognition, the discriminant function method is also a case of pattern recognition.

Unlike other methods, discriminant analysis is more comprehensive in the account it takes of the correlation of variables, which increases the "resolution" of the diagnostic system, thus decreasing the percentage of errors in assigning individuals (or objects) to their corresponding classes. In particular, this means that discriminant analysis should be applied to solving problems of epidemiological studies where alternative classes are especially difficult to distinguish.

The problem of checking the significance of the discriminant function obtained is of major importance. If it is based on variables of general sets, errors of assignment occur less frequently the larger the number of variables used in the discriminant function. However, increases in the number of variables entail additional computation, both in establishing the discriminant function and in applying it. Meanwhile, certain variables contribute unequally to Mahalanobis generalized distance (D^2), the indicator of remoteness among groups of objects for classification, which accounts not only for averages and variances for each variable, but also for the interaction between any pair of the measured variables. It would thus seem desirable to leave out those variables that fail to contribute significantly to the quality of discrimination. However, the situation is aggravated by the fact that after omitting the least valuable variable from a given set, the relative value of the remaining ones in the set $p-1$ differs from that of the original set of p variables. Thus, to omit the next least useful variable, the discrimination of the variable for the new set has to be recalculated. Such a comprehensive analysis requires considerable additional computation.

With high correlation of the variables, one or other of them can be neglected. But with p variables, the gradual elimination of the least valuable ones does not necessarily lead to the best recognition system. This is because no matter how little value the omitted variable may have by itself, it may be of great significance when examined in conjunction with others. Variables may usually be eliminated as a result of known and significant circumstances. If it proves impossible to eliminate specific variables, then those variables whose influence upon the value of multiple correlation is the lowest should be excluded.

Programmes for discriminant analysis are available in ALGOL and FORTRAN (*15*, *26*).

Discriminant analysis provides an opportunity to tackle a number of problems. Of greatest importance is determination of the reliability of differences between groups in general, provided they have been divided on the basis of discriminant function. In order to evaluate the difference between two groups, D^2 may be estimated by tables for the criterion F, considering that:

$$F' = \frac{n_1 + n_2(n_1 + n_2 - p - 1)}{p(n_1 + n_2)(n_1 + n_2 - 2)} D^2$$

where $p =$ the number of variables; and n_1 and n_2 the number of observations in the groups under study. In tables for the criterion F, the indicator F' should be found using the degrees of freedom $v_1 = p$; $v_2 = n_1 + n_2 - p - 1$. When differences among a greater number of groups have to be found, the generalized D^2 may be estimated as χ^2 (chi-squared statistic) if normality of distribution is assumed, with $p(q-1)$ degrees of freedom for verifying the hypothesis on the equality of average value in all q groups for p variables.

Given below is an example of the application of discriminant analysis. Using the data on working conditions in operations in which various types of agricultural machinery were used, together with data on the state of health of the operators, a discriminant analysis was made to study the effects of environmental factors on the development of cardiovascular dysfunction and disease (*27*). The results of a thorough medical examination and the variables of the working conditions were coded for further computer processing (Table 6). The data included levels of noise and vibration, summer temperature in the driver's cab, dust and exhaust

Table 6. Examples of quantitative coding of parameters of the severity of pathological and functional changes and levels of exposure to various environmental factors for farm equipment operators[a]

| Subjects | Severity of pathological and functional changes | | X_{13} Age (years) | X_{14} Length of service (years) | Intensity of exposure to various environmental factors (conventional units) | | Product of intensity and length of exposure (years of service) | |
	X_1 Nervous system	X_{12} ... Modulus of elasticity			X_{15} Temperature	X_{19} ... Exhaust gases	X_{20} Temperature	X_{21} Exhaust gases
A	1	900	30	10	3	3	30	30
B	3	560	36	7	2	3	14	21
C	4	1400	34	16	3	3	48	48
D	1	1250	28	5	2	2	10	10

[a] X = factors under study; the dots show the parameters omitted for the rank of $X_1 - X_{21}$.

Table 7. Discriminant analysis of farm equipment operators: effect of working conditions on cardiovascular pathology[a]

Severity of findings	Age (years)	Length of service (years)	Working conditions—as a product of length of service (years) and intensity of exposure[b]				
			Temperature	Noise	Vibration	Dust	Exhaust gases
None	36.4	11.1	27.4	36.5	37.4	42.2	33.3
Initial functional changes	43.1	14.9	35.3	48.3	50.1	59.5	48.3
Pronounced pathological changes	41.1	17.8	42.1	56.7	58.3	51.6	43.6

[a] After Navakatikyan (27).
[b] Generalized Mahalanobis distance $D^2 = 31.1$, $P < 0.01$.

gas concentration, and the severity of nervous, cardiovascular, and respiratory signs and symptoms, and each was assessed using a four-point scale. The products of the intensity of each environmental factor and the length of exposure (years of service) were also computed. The functional state of the cardiovascular and respiratory systems was determined with the aid of seven variables, each expressed in absolute units.

The group of workers was empirically divided into three classes: subjects with no signs or symptoms, subjects with initial functional changes, and subjects with pronounced pathological changes. Table 7 shows the relationship between cardiovascular disease in operators of agricultural machinery and the effects of occupational factors. The average product of the intensity of exposure and the length of service of the subjects was higher for those with either initial or pronounced evidence of cardiovascular disease than for healthy operators. Subjects with pronounced pathological changes had been exposed more to high temperature, noise and vibration than those with only functional changes. Both groups were of the same age. On the whole, the difference in working conditions between the three groups was statistically significant with a Mahalanobis generalized distance $D^2 = 31.1$ ($P < 0.01$).

Thus, discriminant analysis helped to determine the relationship between the severity of cardiovascular disease, as assessed by functional (prepathological) and pathological changes, and the product of intensity and duration of noise, vibration, and high ambient temperature, all of which contributed to the stress of work.

References

1. WHO Technical Report Series, No. 662 1981, (*Health effects of combined exposures in the work environment*. Report of a WHO Expert Committee).
2. **Barabash, Yu. L.** *Problems of statistical recognition theory*. Moscow, Sovetskoye Radio, 1967.
3. **Bushtuyeva, K. A. & Sluchanko, I. S.** *Methods and criteria for estimation of the state of public health in connection with environmental pollution*. Moscow, Medicina, 1979.
4. **Schwarz, H.** *Stichprobenverfahren* [*Sampling experiments*]. Berlin, Die Wirtschaft Verlag, 1975.
5. **Cochran, W. G.** *Sampling techniques*. New York, Wiley, 1963
6. **Tartakovskaya, L. Ya.** [Combined effects on mercury borers who work with fulminate detonators and are exposed to vibration and mercury]. In: [*Combined effects of physical and chemical factors in the industrial environment*]. Moscow, Institute of Hygiene, 1977, pp. 3–12.
7. **Breslow, N. E. & Day, N. E.** *Statistical methods in cancer research. Vol. 1. The analysis of case-control studies*. Lyon, International Agency for Research on Cancer, 1980 (IARC Scientific Publications No. 32).
8. **Hammond, E. C. et al.** Asbestos exposure, cigarette smoking and death rates. *Annals of the New York Academy of Sciences*, **330**: 473–490 (1979).
9. **Sepetliyev, D.** *Statistical methods in medical research*. Moscow, Medicina, 1968.
10. **Kendall, M. G. & Stuart, A.** *The advanced theory of statistics, Vol. 2: Inference and relationship*. London, Griffin, 1961.
11. **Vainu, Ya. Ya. -F.** *Correlation of dynamics series*. Moscow, Statistika, 1977.
12. **Andreyev, V. A.** *Biometric calculation using "Mir-2" computer*. Nauka, Moscow, 1979.
13. **Bogach, P. G. et al.** *Programming and operation of "Promin" and "Mir" computers*. Kiev. Vyshcha Shkola, 1977.

14. **Bolch, B. W. & Huang, C. J.** *Multivariate statistical methods for business and economics.* Englewood Cliffs, NJ, Prentice-Hall, 1974.
15. *Programmer's manual.* New York, IBM Technical Publication Department, 1970 (*IBM System/360 Scientific Subroutine Package (360A-CM-03X). Version III*).
16. **Lakin, G. F.** *Biometrics.* Moscow, Vysshaya Shkola, 1980.
17. **Osipov, G. V. & Andreyev, E. P.** [*Measuring techniques in sociology*]. Moscow, Nauka, 1977.
18. **Politova, I. D.** *Variance and correlation analyses in agricultural economy.* Moscow, Kolos, 1978.
19. **Navakatikyan, A. O. et al.** [Stress and physical strain among manual workers at modern thermal power stations]. *The UkrSSR Academy of Sciences journal of physiology,* **21**(2): 215–221 (1975).
20. **Navakatikyan, A. O. & Nagornaya, A. M.** [Application of multiple correlation analysis to the determination of the effect of working conditions on morbidity with temporary incapacity for work]. *Vrachebnoye delo,* **12**: 111–113 (1973).
21. **Izmerov, N. F.** [Methodical approaches to the study of free air-pollution effects on population morbidity using methods of mathematical statistics and computers]. *Sovetskoye zdravookhraneniye,* No. 7, 1973.
22. **Tiunov, L. A. & Kustov, V. V.** [Combined biological effects of chemical compounds and physical environmental factors]. *Journal of the Mendeleyev All-Union Chemical Society.* 1974, **19** (2): 164–169.
23. **Anderson, T. W.** *An introduction to multivariate statistical analysis.* New York, Wiley, 1958.
24. **Navakatikyan, A. O. & Karpenko, A. V.** [Informational capabilities of analysis of heart rhythm periodic structure in working man]. *Fiziologia cheloveka,* **7**(2): 214–220 (1981).
25. **Urbakh, V. Yu.** Discriminant analysis: basic ideas and appendices. In: *Statistical classification methods,* Vol. 1. Moscow, MGU Printing Office, 1969, pp. 79–173.
26. **Malinovsky, L. G.** *Classification of objects by means of discriminant analysis.* Moscow, Nauka, 1979.
27. **Navakatikyan, A. O.** Nervous and emotional stress in mental activities and its effect on the cardiovascular system. In: *Occupational hygiene.* Vol. 14, Kiev, 1978, 30–51.
28. **Uberla, K.** *Faktorenanalyse* [*Factorial analysis*]. Berlin, Heidelberg, New York, Springer, 1977.

12

Assessment of occupational stress

R. Kalimo[a]

Increasing awareness of the role played by psychosocial factors in the etiology of health disorders has brought with it a justified demand for intensified epidemiological research on the connections between psychosocial factors related to work and workers' health. The basic assumption for research in this field is that psychosocial factors at work can precipitate or counteract health impairment and affect the results of preventive and curative health measures.

The relation between psychosocial factors and health impairment is most commonly studied on the basis of a stress-theoretical framework. Stress at work is a consequence of a combined exposure to a multitude of factors in the work environment and employment conditions. However, stress and its possible chronic morbid effects at the psychological, physiological and social levels are modified by numerous factors related to individual susceptibility and resistance. It is self-evident that the general social environment also plays a role. The whole process is time-dependent, reactions varying from acute to transitional and long-term outcomes. Research on this complex of factors relies strongly on the behavioural and social sciences, and on the fact that a rational approach to many of the problems is possible only through multidisciplinary efforts.

Epidemiological approach

An epidemiological approach to the study of psychosocial factors at work and their effects on health employs mainly three basic types of study design (1).

1. Study of the relationships between health problems and work-related psychosocial risk factors using cross-sectional data. Many of

---

[a] Institute of Occupational Health, Helsinki, Finland.

these studies are typical examples of descriptive epidemiology aimed at problem identification and often rely on survey techniques and morbidity data. This has been the most typical approach especially in studies at the factory level, but also at the national and community level. On a well formulated theoretical basis, some of these studies have reached the stage of a complex analysis of the interrelations between environmental factors, health outcomes, and other relevant factors. The initial aims of these studies are:

(a) to identify the type and extent of the problems present, e.g., prevalence of mental and psychosomatic disorders, absenteeism, alcohol abuse, turnover of labour, dissatisfaction, and social unrest; and

(b) to identify the psychosocial and physical correlates of the problems in the work setting, e.g., repetitiveness, role conflicts, and noise.

Studies demonstrating correlations between job dissatisfaction, stress reactions, and inappropriate work load, or other problems in the occupational setting, have produced a body of data that can be used on the existing theoretical basis for further hypothesis formulation and testing.

2. Longitudinal, multidisciplinary, intensive field studies of high-risk situations and high-risk groups as compared with controls.

(a) *Cohort studies* of workers exposed to certain psychosocial stressors, e.g., machine pacing, as compared with workers engaged in similar tasks but self-pacing, are carried out increasingly often. However, finding a proper control group is a major problem in the field, in view of the necessity to control a large number of factors other than the exposure in question. Longitudinal studies, which should be given priority on scientific grounds, suffer interference from uncontrollable and often unpredictable social and other changes in peoples' lives, which affect in different ways the groups and individuals under study. Although all longitudinal studies in the health field have this problem, it is particularly disturbing when studying psychosocial factors. While it is not possible to control all interacting variables, the important task is to collect data on them and to take them into account when drawing conclusions on the basis of the main study results.

(b) *Case-control studies* aimed at hypothesis testing have rarely been carried out in the real work environment (2), although a corresponding approach is relatively common in laboratory settings. The main reason for this is that a certain outcome supposedly due to psychosocial factors (e.g., a large number of perceived symptoms indicating frustration, anxiety, and worriedness; an elevated level of catecholamine excretion; or a high turnover of labour) may be so multifactorial in origin that although hypotheses could be formulated, it is often impossible to test them in the field.

3. Controlled interventions, including laboratory experiments as well as evaluation of therapeutic and/or preventive interventions in real-life settings (*evaluation of health action*), making use of results from the two types of study described above.

232

Evaluation of interventions is limited by interference from a number of uncontrollable variables in the same way as prospective cohort studies aimed at clarifying cause-and-effect relations between psychosocial factors and health. Many studies are carried out without suitably controlled comparisons. The working populations subjected to the interventions or whose work environment has been reorganized are often used as their own controls—a procedure known to introduce bias. Although sufficient knowledge of occupational hazards is now available to warrant field interventions at the workplace and at an individual level, community leaders, management, labour unions, occupational safety and health workers, and authorities should be invited to consider, together with the workers, what kinds of intervention are feasible and acceptable to all concerned.

When contemplating interventions in this relatively new area it is recommended that the work environment should at first be changed on a small and experimental scale in order to evaluate all possible benefits and side effects; on the basis of these findings it is then possible to decide what kind of change can be implemented on a wider scale. The same holds true when worker-oriented programmes, e.g., counselling and stress-relaxation techniques, are introduced.

The general guidelines given in other chapters of this volume on carrying out various types of epidemiological studies can in principle also be applied to the study of psychosocial factors and health. In this chapter, therefore, the main emphasis is put on criteria for the selection of indicators and the monitoring of selected parameters as an epidemiological tool, which may pose specific problems in this particular research area. Epidemiological investigations can be facilitated by the development of systematic monitoring for making measurements on health and psychosocial environmental indices and other interrelated factors.

Selection of indicators

The main requirement that must be satisfied by an indicator of psychosocial factors at work and of workers' health is that it gives sufficiently reliable relevant information on the phenomenon under consideration to provide a basis for decision-making and action. This principle is important not only in regard to primarily pragmatic targets but also when the primary goal is scientific.

Many other general criteria exist for the selection of indicators, and they can be applied also to the monitoring of the psychosocial work environment and health; they include validity, conceptual significance, discriminative power, comparability over areas and time, and availability of data.

In discussing criteria for the selection of indicators, reference is sometimes made to the objectivity and subjectivity of the indicators. For many of the psychosocial factors that operate at work and for related health variables there are no indicators that can be measured by objective methods, and this is not even needed. Subjective indicators of working conditions and health are indispensable for identifying the relevant problems and can often help to decide what action should be taken involving workers' participation. Both subjective and objective indicators are valuable when used appropriately and they should be collected simultaneously in epidemiological studies (see, for example, Ref. 3).

Indicators must also be developed to describe the organizational set-up and, to a certain extent, also the general social and economic environment in which the policies or programmes initiated on the basis of the information collected will be implemented.

Both technical and practical considerations must be taken into account in assessing the value of the indicators. An indicator may be "good" as such, i.e., relevant, conceptually valid, sensitive and specific, but at the same time it may be costly, impractical, or impossible to measure. In selecting indicators, the feasibility of methods for gathering data on the indicator must always be evaluated.

In order to overcome the problems arising from the complexity of the subject matter and the risk of bias in the choice of indicators, it has often been suggested that the number of indicators should be increased. However, to facilitate planning and implementation and to make for greater economy in the research effort, as few indicators as possible should be used, but they should reveal as much as possible about the relevant problems. Although sophisticated statistical methods can be of great help in refining the data collected in epidemiological studies, information that is biased from the outset leads to biased conclusions, no matter how skilfully the data are manipulated later.

Indicators for measuring psychosocial factors at work

Effect of physical and chemical factors
It is well known that physical and chemical hazards have effects on workers' physical health. They can also disturb psychological performance and diminish wellbeing. The anticipation of hazard can be a source of fear and anxiety.

The presence of recognized physical and chemical parameters should therefore be determined from the hygienic and health points of view as a first step in psychosocial hazard monitoring. It is also essential to ascertain whether workers perceive the existence of physical and chemical hazards and to what extent the workers are concerned about their effects.

In monitoring indirect and direct effects of physical and chemical hazards on the central nervous system, both unspecific subjective symptoms and psychological dysfunctions are useful indicators. Monitoring of dysfunctions of the central nervous system is difficult, however. Baseline data from the pre-exposure period make intra-individual comparisons possible and increase the validity of the conclusions drawn. An intraindividual comparison is often necessary because interindividual variations in psychological functions are quite considerable. Subjective symptom questionnaires have been developed for monitoring dysfunctions of the central nervous system in workers exposed to neurotoxic agents (4, 5).

Psychological test batteries also are useful as monitoring tools. Standardized test batteries are available for clinical examination of workers with long-term exposure. These test batteries are, however, time-consuming and difficult to interpret; in general, therefore, they are suitable for research purposes only (6).

Ergonomic analysis of the work environment and monitoring of tasks and work organization

Studies on stress reactions and their causes in a wide range of occupational groups have made it possible to identify a number of factors in the work environment that may be hazardous to health (see, for example, Ref. 7). These psychosocial hazards are summarized in Table 1. The methods used for monitoring work conditions fall into two basic types:

(a) methods based on sources other than the workers themselves, such as job description and task-analysis techniques, quantitative measurements and statistics (8–10);

(b) methods that use the worker himself as the source of information, e.g., by the use of questionnaires and interview techniques (11, 12).

The advantages and disadvantages of these two approaches to acquiring data on the work environment have been a matter of some controversy. Both approaches have, however, their own purpose and proper applications.

In order to explain the mechanism by which psychosocial work conditions may have an impact on health, it is essential to gather information about the way workers perceive the work conditions. Once the relationship between psychosocial factors and "typical" or "average" reaction patterns has been established, it is sufficient to gather data by means of objective methods. Such information is especially important when improvements are made in the work environment, while subjective appraisals are necessary for individual-oriented interventions.

Whenever the available resources are sufficient for the assessment of whether a particular parameter of the environment, tasks, or work organization constitutes a psychosocial hazard, the most effective procedure seems to be to begin by measuring the parameter objectively and then to compare the objective measurements with the worker's own criteria of what these values should be.

The resulting description of the objective/subjective discrepancies then makes it possible to draw inferences regarding the "person-environment fit" (13). Ideally, the weighted sum of all such discrepancies would be the index of the "person-environment fit". Research is still necessary to define the appropriate weighting coefficients for each

235

Table 1. Psychosocial hazards in the work environment

Known indicators	Appraisal of current methodology
Stressor related to:	
1. Task, e.g., — repetitive, fragmentary work content, i.e., qualitative underload — excessive demands, i.e., qualitative overload — too much to do and time pressure, i.e., quantitative overload — too little to do, i.e., qualitative underload — lack of control over the work situation, i.e., machine pacing — high responsibility, especially for people	A. Job description and job analysis. Several methods are in use in various countries, many of which go into great detail. Many of the methods can be used only by specialists. Currently available methods are suitable almost solely for analysis of actual work situations.
2. Organization and management, e.g., — lack of possibility for decision making — bureaucracy and authoritarian leadership. — "inadequate" management.	B. Questionnaires, checklists. A large number and variety have been developed. They are often designed specifically for individual studies. There are wide variations in theoretical sophistication and technical outlook. Some methods are theoretically well based and include reliable and valid scales.
3. Work role — role ambiguity and role conflict.	Many of the questionnaires are excessively long.
4. Career problems, e.g., — lack of advancement	
5. Inconvenient working hours, e.g., — rotating work shifts.	
6. Poor person-to-person relationships.	
7. Lack of job tenure.	
8. Physical and chemical hazards.	
9. Problems at the work/home interface.	

measured discrepancy. Existing ergonomic and task-analytical procedures must be simplified and standardized to permit their application by nonspecialists. These procedures should then be used to make objective assessments of working conditions. A corresponding method for determining subjective criteria for each objectively measured parameter should also be developed, as well as a procedure for calculating a single index of the "person-environment fit". Existing questionnaires and checklists should serve as models. These should be validated and standardized.

Indicators for measuring the health effects of psychosocial factors

The presence of an occupational psychosocial hazard, or of many acting in combination, can be recognized by simultaneous monitoring of individual and group reactions as these progress through the following stages.

(1) Persistent change in emotional state resulting from such experiences as frustration, threat, and loss (e.g., increased anxiety, worries, irritability, and tendency to depression), with lowered performance.

(2) A functional change in one or more bodily organ systems, indicating elevated or depressed activity relative to homoeostatic maintenance levels (e.g., elevated heart rate).

(3) Gross behavioural changes indicating failure to cope, desire to escape reality, or an unusually long recovery time after work (e.g., an increased rate of absence from work for ill-defined reasons).

(4) Biochemical and/or physiological changes of a more slowly reversible, nonspecific, and possibly pathogenic nature (e.g., elevated catecholamine and glucocorticoid production).

(5) Frankly aberrant and persistent behavioural reactions that are detrimental to health, interpersonal relationships, and the achievement of occupational goals (e.g., excessive alcohol consumption).

(6) Morphological changes indicative of physical disease (e.g., ulcerative colitis).

(7) Manifest psychiatric illness (e.g., depression).

For all the above stages of reactions, efforts should be made to define the relative specificity and sensitivity of each indication with regard to psychosocial hazards.

Workers' complaints of "distress"
Workers' complaints of distress on the job should always be elicited by monitoring, and examined for indications of the actual or potential presence of some hazard. Less reliable is the worker's subjective identification of the source of his distress. Identified sources may or may not be valid or complete. Workers may be insensitive to certain hazards that produce objectively measurable effects, or may unconsciously deny those effects. Thus, distress monitoring should be considered necessary but insufficient for identifying psychosocial hazards.

A wide variety of self-report stress questionnaires exist and the best provide a highly reliable and valid means of discriminating severely distressed workers from the more normal. However, an improved device is desirable for the purposes of occupational psychosocial hazard monitoring. The ideal questionnaire should ascertain the occurrence of symptoms that are related to psychosocial stress at work and also to the consequences of stress upon work-related motivation and behaviour. It should be of minimum length and unambiguous to workers possessing the full range of literacy skill. It should provide a single index of distress that can be placed upon ordinal or interval scale.

Cognitive functions and work performance

Both inadequate workload and stress due to various causes have been shown to affect cognitive functioning and, consequently, work performance (*14*). While short-term stress may often help an individual to mobilize his own resources and increase his effectiveness, long-term overload, underload, and stress of whatever cause have the opposite effect (*15*).

The use of performance testing to evaluate cognitive functions (e.g., scope of perception, attention, memory, decision-making, creative thinking) holds some promise as a monitoring tool (Table 2). However, there is no general agreement regarding the most sensitive and function-specific tests, and real doubt exists regarding the predictive validity of *any* test that obviously involves an artificial interruption in the worker's occupational routine.

Table 2. Psychological responses to stress at work

Known indicators	Appraisal of current methodology
Cognitive functions: — restriction of scope of perception — lowered ability to concentrate — disturbed memory functions — hesitation in decision making — change in content of thinking — lowered creativity	Although many tests of cognitive functions have been developed, they have not been applied in stress research. They are mostly used to determine the effects of acute stress, especially in laboratory conditions. Applied questionnaires have only a limited coverage.
Emotional reactions: — feelings of deprivation, boredom, guilt, pressure, anxiety, tension, irritation, worry, sadness, pessimistic or hopeless view of future events — apathy	*Short-term effects* Self-rated mood scales are in use for determining the effects of acute stress. The data obtained are usually combined with physiological measurements. *Long-term effects* A large number of questionnaires are in use. Some are theoretically well based. There is great variation in technical outlook. Many versions are developed for individual studies. Comparison of data is difficult.
Self-image: — lowered self-confidence, increased discrepancy between ideal and perceived self-image.	Questionnaires and psychological tests are available.

In a related sense, it is theoretically possible to superimpose a cognitive function test as a secondary task upon the primary operational task and thereby assess some of the worker's functions during actual work. It may even be possible to isolate and measure some operational capability. Both the secondary task approach and that of assessing cognitive functions from operational performance have widely recognized limitations (e.g., a high degree of task-specificity). None the less, a few convincing demonstrations from previous research encourage further applications of these approaches.

Emotional reactions

Emotional reactions to inappropriate workload, role conflicts, inter-personal problems, and other hazards in the work environment are perhaps the best known and commonly recognized indicators of unfavourable conditions at work. Consequently, a large number and variety of methods for monitoring acute and long-term emotional reactions are available. The World Health Organization's modification of the General Health Questionnaire appears to be a reliable and valid instrument for its purpose (16). It is satisfyingly insensitive to cultural differences in the expression of emotional disturbances. Its validity for the measurement of work-related problems and its discriminative power for differentiating work-related emotional reactions from those due to other causes must be further tested.

The questionnaires at present available for assessing the emotional state of workers concentrate upon negative feelings engendered by "bad" occupational conditions. There is a need to expand existing questionnaires or develop new ones for assessing positive feelings produced by "good" conditions. The ideal working environment is not merely one devoid of hazards, but one where factors of the opposite kind engender a high degree of satisfaction based upon the full utilization of existing skills and the development of new ones. At present, methods for monitoring positive factors are scarce, but techniques have recently been developed (17). They should be subjected to field testing.

Self-image

The worker's self-image (e.g., feelings of self-esteem related to employment) is an indication of his state of mental health (see review in Ref. 18). A methodology exists for measuring self-image within certain cultures. However, there is good reason to suspect that the same culture norms do not apply universally. An effort should be made to develop self-image questionnaires and to establish normal and deviant values for different cultures. Cross-cultural studies of the effects of the same psychosocial hazards on workers' self-image should then be undertaken to determine the utility and generality of these indicators.

Physiological functions

Changes in physiological functions are known to be as characteristic of exposure to psychosocial hazards as its psychological effects. The sympathico-adrenomedullary system is the neurochemical pathway responsible for many of the typical physical signs of the stress reaction. The other main mediating pathway of physical symptoms and signs of environmental psychosocial exposures is the pituitary adrenocortical hormonal chain (see, for example, Ref. 19 and 20).

Promising measures of disturbed physiological functions are the urinary excretion rates of the catecholamine hormones epinephrine and norepinephrine. Sample collection and chemical assay procedures are now well standardized and are applied with well defined controls. Catecholamine analysis provides an objective approach for indicating adverse reactions to psychosocial hazards at work. Numerous field studies of workers in a variety of occupations provide reference data and highlight a whole range of reactions possible under adverse conditions (for a review, see Ref. *21*). Catecholamine excretion rates are, however, relatively nonspecific and cannot be used *per se* for determining factors causing stress, or for predicting what its pathological consequences might be. Moreover, while the urine sampling procedure is unobtrusive and readily accepted by workers, the analytical procedures require great skill and expensive equipment. The assessment of occupational stress using catecholamine analysis is well suited to research purposes, but widespread catecholamine monitoring in practical settings would cause great difficulties. This situation would change dramatically if an inexpensive, simple and reliable method could be developed for assaying catecholamine concentrations in urine. For the present, priority should be given to simplifying the available procedures as much as possible. Beyond this, the largely unknown relationship between elevated catecholamine excretion rates and disease should be studied in longitudinal epidemiological investigations.

Similar recommendations can be made with respect to cortisol, the other major "stress hormone" in man, and to plasma concentrations of metabolites and electrolytes controlled by either catecholamines or cortisol or both. The latter indices may be expected to be less sensitive to psychosocial hazards, but more closely related to pathological consequences of hazards. Chronically elevated stress hormone levels may be good early indicators of psychosocial stress hazards, but levels of plasma lipids might prove ultimately to be the best physiological indicators of incipient pathology (coronary heart disease). A study of the full pattern of stress hormones, metabolites, and electrolytes in a working population exposed to psychosocial hazards still remains to be undertaken.

A number of other physiological functions controlled by the autonomic nervous system, as well as central nervous system functions, have been investigated as indicators of psychosocial exposures (Table 3). New possibilities are being continually opened up with advancing technology. By the generally applied criteria, most psychophysiological methods have an acceptable reliability, and many of them are very sensitive. They share, however, one great problem: the difficulty of interpretation. While sensitive to psychosocial exposures at work, physiological functions are also sensitive to many other environmental effects and internal stimuli. Physiological data should, therefore, always be interpreted in relation to psychological parameters and data based on analysis of the work situation (*22*). Whenever possible, a *combination of physiological monitoring and subjective state monitoring* should be

Table 3. Physiological responses to stress at work

Known indicators	Appraisal of current methodology
Biochemical changes in: — neuroendocrinological function (excretion of hormones) — immunological mechanisms — blood lipids and carbohydrates — excretion of essential acids. **Altered activity in:** — brain: electrical activity (electroencephalogram) — muscles (electromyogram) — skin: electrodermal functions (galvanic skin response) — gastrointestinal tract (electrogastrogram) — cardiovascular system (heart rate and other electrocardiographic indices, vasomotor activity, blood pressure) — reproductive system — pupil of the eye (pupillometrics)	Determination of hormones and other biological components in blood and urine samples and measurement of electrophysiological functions have been well tested and widely used. Some indicators are highly sensitive. All indicators are unspecific and thus difficult to interpret. For this reason they must often be combined with psychological measurements and analysis of work stressors. Collection of samples may itself bring artefacts. Methodology is complicated, but technology is continually being simplified. Continuous monitoring of some indicators is already done with a combination of measurements, both in laboratory and real life settings.

employed. However, where facilities and/or experience are limited, subjective state monitoring must, of necessity, be used alone.

Behavioural changes

A change in a worker's behaviour is often the first observable sign of problems in the work environment. Gross changes in workers' behaviour, on or off the job, should be monitored in epidemiological studies. Sleep disturbance is one of the most dependable single indicators of behavioural changes; others are disturbed interpersonal relationships with fellow workers and family members; increased use of alcohol and psychoactive drugs (self-administered or prescribed); sudden, transient changes in productivity at work (increase or decrease); increased absence rates and turnover of labour. Further examples of indicators of behavioural changes are shown in Table 4 (see also Ref. 23 and 24).

Data on some of these indicators are often collected for various purposes other than research and can be made available for epidemiological studies. The existing records and statistics at the national, community, and workplace levels may not, however, always be ideal for research, but should be critically examined to avoid unnecessary effort. There is also a real need to develop a behavioural checklist for providing an index of behavioural disturbances.

Table 4. Behavioural responses to stress at work

Known indicators	Appraisal of current methodology
General:	
— excessive use of coffee, nicotine, alcohol and medicines (especially tranquillizers and stimulants)	Partially and unsystematically covered in questionnaires.
— changes in eating habits	Statistics are sometimes used, but comparison is difficult because of differences in criteria.
— disturbed sleep	
— neglect of physical exercise	Indicators are highly unspecific.
— decreased social participation and activity	
— pretending to be sick, increased use of or neglect of health services	
— change in general lifestyle	
— acting out, antisocial behaviour	
— disruption of interpersonal ties and sexual relationships	
— suicide	
Work:	
— absence from work and postponement of duties	Statistics on absences and accidents are used, but they are unsystematic.
— lowered work performance, qualitatively and quantitatively	Performance is measured mainly in laboratory conditions. In field work, performance is measured mainly with regard to well defined, simple tasks. Sampling is a problem.
— increased number of accidents	
— interpersonal conflicts	
— risk-taking behaviour	

Disease and perceived symptoms

Certain morbidity patterns (e.g., high rates of essential hypertension, coronary heart disease, gastrointestinal disorders, and diseases of muscles, tendons, and joints) may indicate a long-standing presence of severe psychosocial hazards (25). It is necessary, however, to identify the precursors—functional psychological and physiological changes and perceived symptoms—so that these can be monitored to predict and prevent the occurrence of disease (Table 5). These signs and symptoms must fulfil, as completely as possible, the general criteria applying to indicators in regard to validity, specificity, comparability, etc. Many of the perceived symptoms are, however, unspecific and their diverse etiology is common knowledge. This means that perceived symptoms must always be evaluated against the background of working conditions and personal history before any conclusions can be drawn regarding their relation to psychosocial hazards at work.

Data on these indicators for use in epidemiological studies are sometimes available in clinical records. However, a special device is usually needed for research purposes. The problem here is the large number and diversity of questionnaires. Standardized techniques are available and should be used in epidemiological studies.

When the monitoring of occupational psychosocial hazards and their health effects has provided the necessary data base, all "work-related

Table 5. Perceived symptoms and health disorders

Known indicators	Appraisal of current methodology
Physical functional disorders: — muscular tension and pain, e.g., headache and low-back pain — vertigo, dizziness — the gastrointestinal tract. — cardiovascular symptoms, e.g., palpitation, chest pain — respiratory symptoms, difficulty in breathing and "getting air"	Questionnaires, with great variety of background and of form, are used. Clinical examination.
Psychological borderline states: — inclination to depression and to other reactive neuroses — mental and somatic disease	Clinical examination Questionnaires Statistics

diseases" should be studied with respect to the contribution of psychosocial hazards to their etiology. Psychosocial factors potentially play a role in a wide range of noncommunicable, physical diseases, or their disabling sequelae. A few cardiovascular and gastrointestinal disorders appear to involve definable psychosocial hazards in their etiology, but even these diseases may not be the most important or frequent consequences of stress. Work-related diseases should therefore be classified with respect to identified psychosocial hazards, their relative frequencies, and prevention at the individual and group level. Retrospective epidemiological analyses should then be made to uncover both cross-cultural and cross-occupational common factors and differences. Finally, further prospective epidemiological studies should be carried out to confirm the earlier conclusions.

Individual susceptibility and resistance

Individual differences in coping with psychosocial hazards and in susceptibility to health problems are widely recognized (26–28). Yet, research has provided the occupational health specialist with little guidance in attempting to deal with the problems of individual workers. Clinical case-studies will partially reduce this ignorance, but only if they are conducted by adequately trained practitioners. The same instruments (e.g., questionnaires) should be used by practitioners conducting case-studies as by scientists engaged in population surveys; the former can provide new information about the causes of aberrant reactions, and the latter, new normative data. Both should concern themselves first with easily observed or measured factors that might influence individual reactions (e.g., age, sex, membership of cultural majorities or minorities, and the existence of physical or mental handicaps). In addition, possible personality correlates of stress tolerance/susceptibility (Table 6) should be measured from the standpoint of establishing their relevance, validity, reliability and relationship to adverse reactions. It is hoped that the eventual results of such fundamental research will be a definition of the

243

Table 6. Individual susceptibility and resistance to stress

Known indicators	Appraisal of current methodology
Psychological: — resourcefulness and creativity — problem-solving ability — ego strength — flexibility — social skills — level of self-esteem — introversion/extroversion — positive/negative attitude towards future events — ego defences — internal/external control	A large number of test batteries and questionnaires are available with known psychometric qualities for various population groups. Their application in stress research has not proved very successful, however.
Interpersonal: — primary relationships — network of social supports	Questionnaires with great variation in individual items.
Biological: — age and sex — physical strength	Questionnaires Sometimes physical examination
Behavioural: — behavioural habits (e.g., type A/B behavioural pattern) — lifestyle (e.g., eating, exercise, cultural activities)	Questionnaires Observation Statistics

minimum number of relevant personality factors for practical monitoring purposes and standardization of the methods necessary for their measurement.

Socioeconomic and cultural environment

The impact of psychosocial factors at work on workers' health and wellbeing depend to a high degree on the socioeconomic and cultural environment (29). Environmental factors like climate, geographical situation, and technology may also be of decisive importance.

Economic factors greatly modify individual and group reactions, partly because adequate economic resources make it possible to prevent or compensate for many potentially harmful effects. Similarly, cultural factors strongly condition attitudes towards management, fellow workers, subordinates and every other aspect of the work environment and work itself. They also influence female and male roles and relationships at work and outside it, age of entry into working life, passive acceptance of things as they are or active pursuit of improvements in environmental conditions, etc.

Sources of such information vary considerably from region to region. The most economic approach is to survey data collected for various purposes other than research and to collect new data on social indicators only when necessary information is otherwise not available. The value of industrial undertakings and workers as informants, when appropriate, should not be overlooked.

Case study

The following is a description of a study on the work-related psychosocial health risks in an occupational sector (*30*).

Objectives
The overall objective of the study was to carry out an investigation of the relationships between work, job satisfaction, stress, and health in closed institutions (in this instance, prisons) and to discuss the findings within the framework of a psychological theory of the effects of stress at work. For this purpose the following questions were studied and discussed: (*a*) What are the determinants of stress reactions at work? (*b*) What is the relation of stress-induced health problems to work and the occupational background? (*c*) What is the role of job dissatisfaction in the development of stress at work? (*d*) In what ways does the interaction between personality, other individual factors and social support affect the relationships between work, job satisfaction, stress, and health?

Theoretical framework
The framework of the study was a model based on a review and synthesis of various theories of stress at work. The concept of stress was used to denote an imbalanced relation between a person and his environment. The qualities of the environment, divided into the objective and subjective work environment, were regarded as stress factors or stressors, and the response of the individual was defined as a stress reaction covering a wide group of psychological, physiological, and behavioural reactions. The end states of prolonged and/or severe stress were considered as the psychological, physiological, and social aspects of health. The development of stress is partly mediated by a generalized attitude towards work, i.e., job satisfaction. The whole process is modified by individual background and characteristics, social support, and environmental factors other than the original source of the stressors.

Methods
Operational definition of concepts
The variables used in describing the prisons were regarded as characteristics of the objective work environment (size, location, overcrowding, and security). The objective work conditions of the workers were characterized also by their tasks, defined by the occupation, and their position in the organization, as well as length of service in the institution.

The following characteristics of the perceived environment were studied: contents of work (variety, challenge, etc.), mental and physical work load, role in the organization (goal conflict, motivation conflict, role ambiguity and conflict), organizational climate (participation in decision-making, availability of information, feedback, etc.), social relations in the workplace (relations with fellow workers, employers, and prisoners), and characteristics of the physical work environment.

The stress and health indicators were mainly past and present illnesses and perceived psychological and physical symptoms.

Subjective health status, perceived continual fatigue, and physical fitness in comparison with that of fellow workers of the same age were regarded as additional indicators of health. Positive indicators of health covered perceived resources, competence, and social activities, and they were treated as end-state indicators.

Personality, i.e., self-concept, attitudes (towards care of prisoners), and demographic variables (age, sex, education, etc.) were considered to be moderating factors. The buffering function of the interpersonal form of support was also studied.

Sample
The study population comprised all employees in all prisons of Finland in 1975, a total of 2221 employees. As one aim of the health examination was to determine possible risk groups, the study sample was chosen partly by this criterion. The length of work experience in prisons and the workers' ages were considered. Age was finally chosen as the criterion for selection of the study sample. As age was considered a risk factor, most of those selected were in the older age groups, i.e., all those born in 1930 or earlier. However, so that a comparison could be made between age groups, employees born in 1935 and 1940 were also included. Of those selected, 736 persons (94 %) participated. They represented all occupations in prisons, i.e., administrators, guards, foremen, clerks, and teaching and health personnel.

Data and data analysis
The data on work, stress, health, and professional history were based mainly on questionnaires introduced and distributed by the health personnel in each prison. The results of a health examination, including laboratory measurements made by the mobile field clinic of the Institute of Occupational Health, Helsinki, were used. Personality characteristics were ascertained by means of a self-concept test.

Summaries of the descriptive results were reported. The main results were based on multivariate analyses of the dependence of job satisfaction, stress, and health status on work stressors and individual characteristics. The selection of indicators, of which many were on multi-item scales, was based mainly on the factor analyses of each group of indicators. Correlation technique, automatic interaction detector (AID) analysis, and general linear model (GLM) analysis were used in three successive stages of the explanatory analyses.

Main findings
In the explanatory analyses carried out with AID, job satisfaction and global indicators of the occupational background and demographic data were used as parallel independent variables in the analysis of psychic and somatic symptoms, blood pressure, perceived health status, and an indicator of the positive aspects of health.

246

Occupation was found to be one of the most effective predictors, the highest health risk being related to close, continuous contact with the prisoners and responsibility for punishment. Work involving more distant contact, with rehabilitative or administrative functions, seemed to carry a lower risk. Length of service in prison and job satisfaction were other important predictors. Personality factors had a strong influence.

Objective work conditions showed a weaker relationship to wellbeing, but there were certain expected tendencies. (A large size, overcrowding, and a high security level of a prison increased the risk to health).

Further explanatory analyses were done with an emphasis on specific work stressors selected on a theoretical basis and on the basis of the AID analyses.

Perceived lack of challenge in work and conflicts in occupational motives and roles were among the most important predictors of stress reactions and health disturbances. Organizational climate, insecurity, and quantitative workload were also significantly related to some, especially perceived, indicators of stress and health. Only a small proportion of the variation in diastolic blood pressure and none of the variation in serum cholesterol level could be explained by the specific work stressors.

Personality characteristics and perceived social support were found to be important moderators.

Evaluation

The study is a cross-sectional investigation of relationships between health and occupational factors, with some clarification of the individual modifying characteristics. As the study was carried out on an explicit theoretical basis and great emphasis was put on the refined analysis of the data, the study goes beyond the most typical descriptive epidemiological investigations carried out in the field. Selection bias through choice of occupation and reasons for job mobility remain uncontrolled, and exposure and outcome variables are partly based on self-reports, which may involve a risk of the independent and dependent variables being affected by one another. No cause-and-effect conclusions can be drawn on the basis of the study but it provides a basis for explicit hypothesis formulation and testing.

Apparently, epidemiological study of psychosocial factors at work and their relationship to workers' health is at present at a turning point. Available knowledge based on the type of studies described above is ample background for research.

References

1. **Levi, L. et al.** Work stress related to social structures and processes. In: Elliott, G. R. & Eisdorfer, C. *Stress and human health.* New York, Springer, 1982.

247

2. **Kagan, A. R.** A community research strategy applicable to psychosocial factors and health. In: Levi, L., ed. *Society, stress and disease – working life.* Oxford, Oxford University Press, 1981, pp. 339–342.
3. **Kasl, S.** Epidemiological contribution to the study of work stress. In: Cooper, C. L. & Payne, R., ed. *Stress at work.* New York, Wiley, 1979.
4. **Hogstedt, C. et al.** Diagnostic and health care aspects of workers exposed to solvents. In: Zenz, C., ed. *Developments in occupational medicine.* Chicago, Year Book Medical Publishers, 1980, pp. 249–258.
5. **Schneider, H. & Seeber, A.** Psychodiagnostik bei der Erfassung neurotoxischer Wirkungen chemischer Schadstoffe. *Zeitschrift für Psychologie,* **187**: 178–205 (1979).
6. **Lindström, K. & Mäntysalo, S.** Physical and chemical factors as stressors at work. In: *Psychosocial factors at work and their relations to health.* Geneva, World Health Organization (in press).
7. **Cooper, C. L. & Payne, R.** *Stress at work.* New York, Wiley, 1979.
8. **McCormick, E. J. et al.** *Technical manual for the position analysis questionnaire (PAQ).* West Lafayette, University Book Store, 1977.
9. **Rohmert, W. & Landau, K.** *Das Arbeitswissenschaftliche Erhebungsverfahren zur Tätigkeitsanalyse (AET). Handbuch und Merkmalheft.* Bern, Huber, 1979.
10. **Elo, A. L. & Vehviläinen, M. R.** [*Method for health care personnel for evaluation of psychic stress factors at work – a study of reliability and validity of the method*]. Helsinki, Institute of Occupational Health, 1983 (Report No. 196) (Finnish with English summary).
11. **James, L. R. & Jones, A. P.** Perceived job characteristics and job satisfaction: an examination of reciprocal causation. *Personnel psychology,* **33**: 97–135 (1980).
12. **Roberts, K. H. & Glick, W.** The job characteristics approach to task design: a critical review. *Journal of applied psychology,* **66**: 193–217 (1981).
13. **French, J. R. P. Jr. et al.** Adjustment as person-environment fit. In: Coelho, G. V. et al., ed. *Coping and adaptation.* New York, Basic Books, 1974, pp. 316–333.
14. **Hamilton, V. & Warburton, M., ed.** *Human stress and cognition.* Chichester, Wiley, 1979.
15. **Alluisi, E. A. & Fleishman, E. A., ed.** *Human performance and productivity.* Vol. 3: *Stress and performance effectiveness.* Hillsdale, Lawrence Elbaum, 1982.
16. **Goldberg, D.** *The detection of psychiatric illness by questionnaire.* London, Oxford University Press, 1972.
17. **Antonovsky, A.** *Health, stress and coping.* San Francisco, Jossey-Bass, 1979.
18. **Tharenou, P.** Employee self-esteem: a review of the literature. *Journal of vocational behavior,* **15**: 316–346 (1979).
19. **Levi, L., ed.** Stress and distress in response to psychosocial stimuli. *Acta medica scandinavica,* **191**: Suppl. 528 (1972).
20. **Elliot, G. R. & Eisdorfer, C., ed.** *Stress and human health. Analysis of implications of research. A study by the Institute of Medicine/National Academy of Sciences.* New York, Springer, 1982.
21. **Daleva, M.** Metabolic and hormonal reactions to occupational stress. In: *Psychosocial factors at work and their relations to health.* Geneva, World Health Organization (in press).
22. **Wilkins, W. L.** Psychophysiological correlates of stress and human performance. In: Alluisi, E. A. & Fleishman, E. A., ed. *Human performance and productivity.* Vol 3: *Stress and performance effectiveness.* Hillsdale, Lawrence Elbaum, 1982.
23. **Caplan, R. D. et al.** *Job demands and worker health.* Washington, DC, US Department of Health, Education, and Welfare, 1975.
24. **Plant, M. A.** *Drinking careers. Occupations, drinking habits and drinking problems.* London, Tavistock Publications, 1979.
25. **Levi, L., ed.** *Society, stress and disease – working life.* Oxford, Oxford University Press, 1981.
26. **Folkman, S. & Lazarus, R. S.** An analysis of coping in a middle-aged community sample. *Journal of health and social behavior,* **21**: 219–239 (1980).
27. **Jenkins, C. D.** Psychosocial modifiers of response to stress. *Journal of human stress,* **5**: 3–15 (1979).

28. **Kobasa, S. C.** Stressful life events, personality, and health: An inquiry into hardiness. *Journal of personality and social psychology,* **37**: 1–11 (1979).
29. **Dodge, D. L. & Martin, W. T.** *Social stress and chronic illness.* London, University of Notre Dame Press, 1970.
30. **Kalimo, R.** Stress in work. Conceptual analysis and a study on prison personnel. *Scandinavian journal of work, environment and health,* **6** (Suppl. 3): 1–148 (1980).

13

Statistical analysis of epidemiological data: an overview of some basic considerations

M. Nurminen[a]

According to generally agreed definitions, epidemiology is concerned with health-related problems affecting primarily populations rather than individuals; this consequently gives rise to statistical problems. Conceptually, many of these are similar to the statistical problems that arise in other branches of medical science, as well as in other (biological) sciences (*1*).

Medical statistics is thus intertwined with epidemiology but is not equivalent to it. Statistical methods are also applied in branches of medicine other than epidemiology. As examples may be mentioned randomized therapeutic (clinical) trials and the study of the kinetics of substances, both in man and in experimental animals. Similarly, there are branches of epidemiology, such as the development of exposure indicators or simple clinical tests for field surveys, that are not statistical in nature. Nevertheless, epidemiology—whether interpreted as the study of epidemics of infectious diseases, or more broadly as the study of endemic, noncommunicable diseases, or even as covering the general aspects of the organization of health services—cannot proceed far without making demands on statistical methodology, some strategies and techniques of which it shares with medicine and biostatistics.

The statistical approach to epidemiology may be seen as an endeavour to recognize and interpret the empirical distribution of characteristics in a certain population. The first aim is to describe the occurrence of an illness at issue in relation to risk indicators. The next step is the separation between causal and noncausal risk indicators. A

[a] Institute of Occupational Health, Helsinki, Finland.

causal association, as distinct from a chance association or an apparent (pseudo) association, is such that a change in the determinant produces a change in the occurrence of the illness. In the design of a nonexperimental study, the range of the alleged causal factors is selected in such a way as to ensure large variability. In the statistical analysis, the epidemiologist then exploits appropriate methods to explain the variation in the outcome parameter (risk), in particular to estimate the portion that is due to the variation in the exposure variables (risk factors).

Routinely recorded statistics of mortality and morbidity provide relevant information for many epidemiological inquiries. Although of statistical interest, their proper use and awareness of their limitations are bound up with nonstatistical considerations (2) and deserve only passing mention here.

In the text that follows, fairly brief consideration will be given to certain aspects of statistical methods especially developed for the evaluation of associations in etiological surveys.

The epidemiological study as a method of measurement

An epidemiological study can be looked upon as a quantitative method for assessing the demographic aspects pertaining to the frequency, causation, natural course, and prevention of disease (3). In general, therefore, the objective is the quantification of an occurrence, duration and outcome, or effect measure in relation to its determinants (exposures, traits, or intervention) (see Ref. 4). This viewpoint was adopted to facilitate the conceptualization of epidemiological study designs. In fact, however, the terminology and theory of measurements can not only be applied to the characterization of the research plans but can also be put to use to judge the overall performance of the epidemiological method of measurement.

In the discussion that follows, emphasis will be placed on bringing out distinctions between the application of terms and concepts to evaluate the optimization of a study design, on the one hand, and to appraise the propriety of statistical estimators, on the other.

Simple measurement model

The first step is to write a rudimentary mathematical measurement model (5):

$$D = P + E,$$

where D is the measurement data (a single measurement or multiple measurements) of the unknown parameter of interest P, and E is the measurement error (vector). The parameter P may be broadly interpreted as what has been called an occurrence relationship (6). In this model the error term E has to be understood in totality as containing (directly or indirectly) identifiable sources of variation, such

252

as errors of measurement, variations in the modifiers of the relationship, effects of omitted confounders (see below) and so on. Altogether these factors constitute the total error E. In the following, a distinction will be made between two component error variables: systematic error or *bias* (B), and random error or *imprecision* (e). It is the author's opinion that statisticians have given too much attention to reducing imprecision and have not fully appreciated the importance of unbiasedness.

To arrive at an estimation of P, it is necessary to move from mathematical formulation to statistical inference and to make some assumption concerning the random error distribution, for example, that the error variable e has a known *probability density function* $f(e)$, independent of P, with a zero expectation and constant variance V. It is then possible to reverse the above model equation and regard P as being obtainable from the known data D and the measurement error E by means of the "structural equation" (7):

$$P = D - E = D - B - e;$$

and the error distribution:

$$f(e), \quad E(e) = 0, \quad V(e) = V.$$

General goals in study design

To understand the principles of optimizing the study design, it is helpful to describe one of the ways in which component dimensions of desirability may be related. In this classification of criteria, the concept of an erroneous measurement—a combination of bias and imprecision—is replaced by its opposite, namely, *accuracy*. Accuracy refers to the size of deviations from the true value, whereas precision refers to the size of deviations from the mean value obtained by repeated application of the measuring method.

In terms of general attributes of utility, a particular study may be (8, 9):
  (1) feasible (practicable)
  (2) accurate (informative):
      (*a*) unbiased (valid):
          (i) comparable
          (ii) generalizable
      (*b*) precise:
          (i) efficient
          (ii) sizeable.

Feasibility

The *feasibility* of a study plan usually refers to the many practical hindrances in its execution. A suspected nil information yield from the study is likely to be a severe scientific impediment, too, which often results in a wisely calculated sample size of zero.

253

For a statistic or measure to be estimated from the study data, *practicability* could correspond to desirable computing properties of the estimator, such as being obtainable noniteratively. In statistical analysis, this term could also be taken to mean *estimability* of a certain parameter in a given study design. For example, exposure category-specific disease rates can be calculated from case–referent (case–control) studies without making any rare-disease assumption using Bayes' theorem and information on the overall disease rate (*10*).

Accuracy

The term *accuracy*, when used with reference to a study plan, generally means the closeness of agreement between the magnitude of the quantified occurrence or effect, as assessed from the specified occurrence function, and the actual magnitude of the effect.

Unbiasedness

A *valid* or *unbiased* study is so designed that if its empirical base were expanded to cover all possible relevant experiences, it would allow for a truthful answer to the posed question.

This definition of unbiasedness, which is useful for evaluating a study, has its conceptual counterpart in what in statistics is referred to as *consistency* (see below).

Confounding is the mainspring of bias *B* in nonexperimental studies. Other possible causes are the "regression towards the mean" effect (*11*), specification error in the functional form of the relationship between the outcome and covariate (*12*), and measurement error in the exposure ("independent") variable (*13*).

The validity of a study design involves two major extensions (*9*): *comparability* and *generalizability*, or internal and external validity in Campbell's (*14*) terminology. (The validity issues are discussed in detail in Chapter 14.)

Internal validity deals with the correctness of the inference about the object of study. It involves the following distinct aspects: (*a*) unbiased definition of the study base, (*b*) representative sampling of the base population and experience, and (*c*) comparable (possibly inaccurate) information on base representatives (*7*). For demonstration of the practical consequences of special types of bias, the reader is referred to the textbook by Kleinbaum et al. (*15*).

External validity has to do with mental "extrapolation" of the scientific propositions of the study to general abstractions. This conceptual progression is sharply distinguished from the statistical sample-to-population inference, albeit the term "superpopulation" may be employed in both types of generalization.

Precision

The *precision* of a study may be enhanced either by increasing the size of the sample or by improving its efficiency.

254

With regard to the ultimate *size* of the sample, the customary sample size formulae used in epidemiology (*16, 17*) may well give an initial idea of the amount of information that would need to be collected to obtain a desired degree of precision with predetermined probability. However, informal arguments are often more cogent than formal statistical ones in dictating what is a manageable and proper size of the sample. Moreover, in a multivariate situation, there is room for research in the development of techniques for power calculations.

An *efficient* study has a high ratio of information to the number of enrolled subjects, or a high ratio of information to money spent. Both investments may be incorporated into a cost function, which may be specified to permit allocation of subjects to index and referent series (*5, 18*).

In statistics, efficiency has meaning only for unbiased estimators, whereas the amount of information in the sample is defined as the inverse of the variance of the most precise estimator among all unbiased estimators. Thus, the above definition of precision differs slightly from normal statistical usage, which precludes the possible presence of bias in the information (*9*).

It is perhaps pertinent in this context to stress that the concept of random variation (sampling error, residual variation) takes on a completely different interpretation in nonexperimental work from what it has in experimental statistics. Instead of "that component of variation that can be considered random" it becomes "that component of variation that contains the effect of as yet unknown variables on the response being estimated" (*19*). In measuring a physical quantity, imprecision may be rectified by replication. In quantifying an occurrence relation, the corresponding countermeasure is to increase the size of sample (as already discussed).

Unadjusted disturbing variables (not necessarily confounders) may also cause imprecision through increasing the variance of the measured variable D reflecting the parameter P (*8*).

Quantification of information

The concept of *information* is an old one in statistics (*20*). It measures the import of sample data D or some statistic $S(D)$ for drawing inferences about the parameter P, represented by the likelihood $L(P; D)$, thought of as a function of P for observed D.

Intuitively, we would expect that there would be no more information in the statistic $S(D)$ than there is in the total data D, with equality obtaining only if $S(D)$ is sufficient for P. Under certain conditions of regularity (*21, 22*) the lower bound to the variance of the unbiased estimator of P is $1/I(P; D)$). In this sense, the information $I(P; D)$ means formally the accuracy of an unbiased estimator with maximum attainable precision in the extent of the data at hand.

255

When the conditions are not satisfied, it is comforting to know that in large samples the maximum likelihood estimator is consistent (or asymptotically unbiased) and has the minimum variance, or in other words is fully efficient.

Unbiasedness in statistics denotes the simple property that the mean of the sampling distribution of an estimate equals the parameter it is intended to estimate, or $E(S(D)) = P$. It may be said that an estimator $S(D)$ is consistent if it approaches closer to the parameter P (in a probabilistic sense) as the amount of data increases.

Theoretically these two properties (unbiasedness and consistency) tend to be unrelated. Examples are found in statistics of estimators that are unbiased and inconsistent as well as those that are biased and consistent. However, as Miettenen (23) has suggested, in the evaluation of epidemiological studies it seems unlikely that there would be serious practical implications were one to use the statistical notion of consistency in place of the scientific concept of unbiasedness as an essential property of a reasonable estimator. This is because the bulk of the bias in epidemiological studies tends to stem from lack of validity in internal comparability, and such bias is largely independent of the size of the study. (This is the reason why significance testing is less crucial in nonexperimental research than in randomized studies.)

Unfortunately, the maximum likelihood (ML) estimators are usually biased when dealing with small samples, a fairly common situation in epidemiological practice. However, developments in data analysis have rendered this kind of statistical, sample-size-dependent bias immaterial.

In the preceding discussion, the precision of an unbiased estimator of an outcome or effect parameter was measured by the variance of the estimator. The variance of a biased estimator measures only the disturbing effect of random error on the estimator, but gives no hint of the amount of systematic error present in the data, for bias is—by assumption—a constant (i.e., its values do not change in hypothetical repetitions of the study).

On the other hand, the expected (stochastic mean) value of the total error (over an infinite number of repetitions) $E(E) = B$ (with probability one), so that random errors are not reflected in it. However, as pointed out by Miettinen (23), in epidemiology bias is not naturally defined in terms of average over repetitions. Thus, it is necessary either to think of "bias" not as "mean bias" but as "median bias", or to express validity using the criterion of consistency.

An appropriate criterion for the performance of a study design subject to both systematic and random error could be (23):

$$I = 1/(V + B^2).$$

This measure is the *informativeness* of a study. The classical measure of mean square error (MSE), or $E(E^2) = B^2 + V$, as commonly used would be misguided in this context, e.g., rate ratio estimates generally have infinite MSEs, without this carrying any implication of impropriety.

256

Elements of data analysis

The processing of epidemiological research data involves four distinct elements: summarization of essential data or data reduction; invocation of a statistical model; actual data analysis; and inference.

The final aim of the entire data analysis is to summarize the evidence contained in the data with respect to the object studied. The edited data are generally presented in tabular form, in epidemiology commonly. as a frequency table. The assumption of a probabilistic model is essential for the study of occurrence-exposure rates. Such a model must express the relationship between the occurrence or effect rate and its determinants, confounders, and modifiers. The key issues in actual data analysis are usually either hypothesis testing in the qualitative phase or effect estimation in the quantitative phase. These elements of data analysis serve as a basis for inference, which encompasses both the evaluation of the results of the statistical analysis of the data at hand and broader considerations about the consistency of the study findings with all other relevant knowledge.

In general statistics, "data reduction" and "statistical inference" have earlier been regarded by some as separate issues (24). From this perspective, the term "data analysis" carries a somewhat redirected emphasis in statistical methodology; its prime concern is with a computer-oriented approach to describe features of large sets of data (25). At this stage, the inferential and formal modelling aspects of statistics are played down; techniques are drawn from areas of application new to traditional statistical analysis, such as study of multivariate dependencies, clustering (taxonomy), and study of an unstructured multiresponse sample (26).

The ultimate and sole aim of statistics is to provide the wherewithal for making valid conclusions about the vast complexity of factors and relationships in real-life problems. Summarization of evidence in epidemiology, therefore, implies coping adequately with the problems of confounding and imprecision. The prime concern in the treatment of these problems should be the validity of the model of interest in the detailed application of statistical methods. In practice, it frequently turns out that the model either ignores some subtlety in the empirical setting or is otherwise so unsophisticated (in order to make the structure manageable) that it becomes irrelevant or impracticable.

In addition to the well established role of summarization, the meaningful exercise of data analysis presupposes a readiness not only to detect anticipated results but also to reveal unexpected evidence in the data.

Preoccupation with statistical aspects has been characteristic of the overall approach to the principles and methods of evaluating cause-effect relationships in epidemiology. For example, scientific inference has been thought of as a decision-making process and even as a matter of statistical significance testing. Assuming that the control of confounding has been taken care of, statistical significance testing does have a limited role in indicating the probability that an apparent exposure effect could possibly have been dependent on chance alone. In general terms, inference is conjectured as a kind of mental feedback from the experience (or data) at hand (supplemented by all relevant other information) to the scientific problem at issue.

In the following, brief consideration will be given to some specific aspects of each of the broad areas of data analysis listed above without attempting full coverage of any of them.

Data reduction using statistical models

After detection and correction of errors in the "raw" data, or "data editing" (an elegant term to be preferred in a petition for a grant), the observations on the different variables in a group of individuals are arranged in the form of a data matrix in a computer file. A data matrix is, in its simplest form, a two-dimensional array of m variables in n individuals. A voluminous study may result through either m or n becoming large. In a clinical trial there are typically numerous pathological items (m), while in a nonexperimental survey it is the number of subjects (n) that tends to be large. Basically, depending on which dimension is the greater, an investigator may take different courses of action to summarize the evidence in the data.

If the statistical manipulation of the data is approached from a univariate standpoint, that is, treating each variate independently, then the notion of sufficient statistics becomes central. This means that some concise set of results—say the mean and variance, or the percentage of abnormal findings—is considered as a "distillation" of all the information that the data contain about the parameter(s) of interest. The difficulty lies in the fact that the sufficiency of the statistics depends on the details of the probability model adopted.

Consider, for instance, a group of individuals (indexed $j = 1, 2, \ldots$) for whom the probability of the occurrence of illness—treated as an all-or-none random outcome with indicator variables I_j—depends on an exposure variable E_j. For the risk modelled on a logit scale, i.e., as a logarithm of the odds of the risk of illness,

$$\log(\Pr(I_j = 1)/(1 - \Pr(I_j = 1))) =$$
$$\log(R_1/(1 - R_1)) = A + B E_j$$

where A and B are parameters of the relationship, and the minimum sufficient statistic is, for example, the average exposure dosage among those persons who contracted the illness. This result follows from the fact that the contribution to the likelihood for the jth individual can be expressed concisely as

$$\exp(A I_j + B E_j I_j)(1 + \exp(A + B E_j))$$

Hence, on multiplication of the individual likelihoods, it follows that the total likelihood involves data only via (Sum I_j, Sum $E_j I_j$) (see Ref. 27). In the more general case, the model ought to be conditioned for all the relevant extraneous determinants. It is the researcher's difficult task to reduce the list of potential confounders to the smallest number that, in practice, are jointly sufficient.

In the comparison of two sets of data, so-called logit difference parametrization or

$$\log(R_1/(1 - R_1)) - \log(R_0/(1 - R_0)) = \log(OR)$$

is commonly employed; the quantity OR is the odds ratio. The use of such a scale for the parameter is appealing because of the simplicity of the assumed model structure for the binomial distribution (written conveniently in the exponential form), but it would be less so for some other distributions. For instance, the normal distribution function would be a natural choice in probit analysis (28), but the logistic is more tractable mathematically and better suited to computer operations (29). Yet, a consideration overriding technical properties of models and estimation of their "natural" parameters is the directness of biological meaning and range of applicability (see Ref. 30 for illustrations). It is difficult to give generally applicable principles of model building as they depend both on the type of question of interest and on the extent to which it is possible to quantify the problem formally.

An awareness of the problem of sufficiency aids in recognizing the perils of possibly unfounded assumptions underlying the models used to summarize the data. Consequently, statistical analysis is often a trial and error (iterative) process to find the information in the study experience. To give other investigators the possibility of checking whether the (explicit or implicit) assumptions are tenable, the data should ideally be published in sufficient detail to allow repetition of the statistical analyses. (Even if the investigators are willing to comply with this requirement, they are often faced with the formidable difficulty that editorial policy is to omit all detailed data.) Useful forms of presentation for this purpose are, for example, frequency tables, correlation matrices, various graphical presentations, and probability plots.

When data reduction aims to specify the occurrence relation in its full complexity, there is a compelling need to consider many variables simultaneously. Methods to reduce the (multi) dimensionality of the data, such as principal components analysis, factor analysis, and multidimensional scaling techniques, can be resorted to prior to the actual study of multivariate dependences. For a more detailed account of these methods, see Chapter 11. For an application of discriminant function analysis in the field of occupational psychology, see Ref. 31.

For data of the continuous type, such models as the customary (multiple) regression analysis and analysis of variance readily incorporate random components representing various sources of variation (32). Statistical modelling techniques may also be used to allow for the imprecision that results from the numerical restrictions of the sampling frame.

Recent developments in the analysis of discrete data (33) have produced corresponding popular models, notably the logistic and log-linear model analysis techniques, which have been widely used, especially in cancer epidemiology (*34*). For frequency data, these techniques may be used to smooth out random fluctuations by replacing the originally observed frequencies, which may suffer from sporadic or haphazard inaccuracies, by values fitted to the model. The fitted estimates can be further used to obtain meaningful summary statistics, e.g., disease rates for compared exposure groups adjusted for differences in the age and sex distributions of the groups.

Screening tests and cluster analysis

Two kinds of statistical activity peculiar to epidemiology deserve to be mentioned under a separate heading: *screening* and *clustering*.

The older application of statistical techniques to evaluate the performance of screening programmes for detecting diseases at the preclinical stage produced simple indices. The overall validity of a diagnostic test is measured as the sum of specificity and sensitivity, or alternatively by the respective converses of the statistical acceptance and rejection error probabilities.

Parallel with the progress in epidemiology as a science, the application of screening tests has moved from the testing of vaccine effectiveness to the study of chronic diseases. Examples of the latter are the estimation of lead time in cancer screening (*35*) and the prediction of the risk of coronary heart disease in pre-employment medical examinations (see Chapter 18). These developments in the discipline of epidemiology rely rather heavily on statistical theory.

A newer application of multivariate statistical methodology that has become well established in epidemiology is clustering, i.e., analysis of the affinity of people, for instance in space and time. These methods rely on various group similarity (or dissimilarity) multidimensional distance measures (*36*).

Models for dynamic medical systems

Mathematical models have been put forward to describe various dynamic biological processes. The earlier mathematical theory of epidemics (*37*) can be ignored, as it was used mainly for the description of communicable diseases. More recently, both deterministic and probabilistic models have been presented that make it possible to examine the pattern of incidence rates and various types of prevalence rates, and to use the relationship between them to deduce other variables of importance in morbidity or mortality, e.g., duration of sickness (*38, 39*). The many rival theories of the biological mechanisms of carcinogenesis have been modelled in stochastic terms and tested (estimated and validated) in animal experiments, but details of their statistical technicalities and the appropriateness of the basic assumptions remain controversial and subject to development (*40, 41*). In therapeutic studies ("clinical trials") the so-called proportional hazard models (*42*) have been extremely valuable for analysing survival data. These considerations are beyond the scope of the present chapter, however. Nor will any attempt be made to bridge the gap between the mathematical formulation of a model and its application to disease control in a community. Instead, the discussion will be focused on the role and utility of mathematical modelling in modern epidemiology and biometry, with particular reference to the investigation of ill-defined or intractable problems.

The purposes of biological modelling are both to offer tentative suggestions about the causal determinants of an illness and to predict such phenomena as the response of a metabolic system or the outcome of disease events. To a great extent these models are nonparametric in character, i.e., they emphasize the understanding of the biological phenomena qualitatively, not accurately according to mathematical formulae.

It is important to keep in mind the distinction between these two functions of model building. To bring the issue into sharper focus, an example will be taken from the field of biological exposure monitoring in an occupational setting (*43*). This is discussed below in some detail.

A mathematical function describing the dynamic processes involved in the uptake, distribution and excretion of inhaled metallic pollutants was developed, taking the kinetic theory as a starting point. A simple one-compartment model was constructed by combining exponential terms representing consecutive transient responses in the urine and plasma levels of nickel and chromium in workers exposed to time-varying concentrations of these metals in air; the uptake of the metal was expressed as a sum of step functions. The three estimated parameters were: (*a*) half-time of the concentration in the body; (*b*) "baseline" concentration, which described principally the hypothetical concentration in a nonexposed person resulting from dietary metal uptake and body burden; and (*c*) a scaling parameter which, in conjunction with a further assumption regarding the individual's minute ventilation, can be used to compute the accumulation and ventilation rate.

An easier but, in the authors' view, less fruitful approach would have been to fit a simple polynomial to describe the concentrations of these metals in a short period, say during a working day. One advantage of the more complex function was that it allowed a comprehensive description of a person's state of exposure as affected by varying uptake rates and urinary excretion. Simultaneous consideration of all the data points over a period of one week had the added benefit of yielding more stable overall estimates of the model parameters. In other words, it provided more accurate values—for example, for predicting the time needed for the nickel concentration in urine to fall to the "baseline" level—than if an exponential function had been fitted to the exposure

261

and off-exposure periods separately and a value for the parameter of interest had been obtained by averaging.

It is well known in mathematics (44) that over a specified domain one function can generally be made to approximate closely to another function that has a quite dissimilar analytical form. In this particular example, an exponential function could be easily transformed into a polynomial function through series expansion. However, the selection of such a model function for the sake of giving an optimum fit or prediction in one situation is not in the spirit of scientific inquiry, for it reveals nothing about the functioning of the dynamic system itself. Even in the very simple model that was developed, the actual complex phenomena in the body were not described as some kind of "black box", but depicted as a single homogeneous unit in which the inhaled material is mixed within a single compartment and excreted by a passive diffusion-type process; the excretion rate was assumed to be proportional to the amount of metal in the body.

Yet another advantage of basing the form of the model function on a fundamental theory is the possibility of building up the mathematical model iteratively. The kinetic theory initially implied a basic exponential form of the model function. Feedback from the model could modify or improve the theory. New sources of variability, such as different perfusion rates and apparent volumes of distribution, however, would then necessitate the introduction of other parameters into the models. The model assuming one-compartment kinetics was a known over-simplification and might have resulted in quite different assessments of the biological system from, say, a three-compartment model, as has been suggested. However, the limited data did not permit the estimation of a more complex model because of the threat of overparametrization. The authors also wanted to adopt a parsimonious approach to the construction of the model.

Statistical features of the model construction were: (a) estimation of the model parameters and their precisions by the method of least squares, involving the numerical optimization algorithms; (b) assessment of the goodness-of-fit of the fitted models using conventional statistical techniques; and (c) liberation of the model from the deterministic course. Under item (c), a model was tried in which the "baseline" concentration parameter was replaced by an estimated optimum one that liberated the fitted concentration function from passing through the first measured point. This added flexibility to the model function and improved the fit considerably. The reason for trying this technique was purely statistical: an exceptionally low Monday morning concentration forced the fitted function to start its track so much below the average observed concentration that the principal deterministic behaviour of the function was not capable of adjusting to the observed course, a situation revealed by the goodness-of-fit criteria.

Although this example of modelling concerned one individual at a time, the same principles and techniques could be applied to groups of individuals as well. This could be accomplished in two ways: either by somehow averaging the data over the individual fitted curves, or by noting that the observed points represent aggregate data (in the sense that there is no way of knowing individuals' personal values).

The establishment of a tentative empirical occurrence relationship may easily lead to a false impression that a true scientific observation has been achieved. For instance, a log-linear analysis could reveal that there is an "interaction" between occupational exposure to carbon disulfide and diastolic blood pressure in the causation of coronary heart disease. This merely shows that the combined effect of these two risk factors differs from that of their independent effects, or formally that there is a product term in the linear function describing the logarithm of the expected number of outcome events under the best fitting model (the latter being based on some statistical criteria or rules). It may not be possible, however, to attach scientific meaning to this finding unless it provides a starting point for a rational surmise about the biological mechanism. A preliminary analysis of problems of this kind, for which no coherent model exists, should preserve a balance between statistical goodness-of-fit and biological plausibility of the hypothesis. The aspiration in advanced analyses could be the exploitation of the data to identify the specific role of effect modifiers in the potential mechanism of the disease under study and to discuss the credibility of the result in the light of prior medical knowledge and other accrued evidence. For further discussion of this topic, the reader is referred to the paper by Nurminen et al. (45).

Inference in epidemiological research

Statistics may be defined as the study of the processing (collecting, describing, and analysing) of stochastic data in order to provide numerical information on the basis of which inferences may be drawn in a scientific inquiry or actions taken in a practical situation. This definition contains the general terms "information" and "a practical situation" that need to be specified with respect to the studied populations and the questions asked. It also implies a distinction between "inference" and "action". A statistical inference concerns the degree of belief that a scientist places in a hypothesis; it utilizes information to arrive at an assessment of an empirical relationship by means of a probability model.

Information is frequently supplied by the sample data alone, but supplementary prior information may at times be highly relevant to the drawing of inferences. As a rule, no consideration of consequences is attached directly to the descriptive function of statistical inference, but this certainly affects the principal decision-making procedures (46). Consider, for example, the setting of a hygienic standard for a potential carcinogen in the light of results from epidemiological studies. Inference uses these data and, possibly, existing theoretical knowledge. Decision-making also considers other information with a bearing on the practical implications, such as the number of people at risk, or the costs of reducing or removing the effects of the exposure. The attitudes to these questions are the distinguishing features of the three main approaches to statistics: (a) sampling-theoretical (or "frequentist") inference, (b) Bayesian (or "subjectivist") inference, and (c) decision (or "utility") theory (see Ref. 47 for a comparative discussion).

Without going into details regarding these different approaches, attention may be drawn to a few considerations that affect their applicability to epidemiological research practice.

In principle, epidemiology is concerned not with testing hypotheses but rather with asking questions. Hypothesis testing may be regarded as an inferential method, in so far as it has to do with the assessment of some values of the parameter. Also, according to a principle of theoretical statistics, confidence intervals may be based on a significance test; for an example see Ref. 48. But when a test is used for the purpose of giving guidance for action in an uncertain practical situation, it is a decision-making procedure, which is not in the spirit of scientific inference (causal research). The latter activity must always contain subjective judgements of all available evidence.

The ability ("power") of a test to detect an effect is an important consideration. On the other hand, Susser (49) has been concerned about "bias toward skepticism in conventional procedures of inference". He says: "Much statistical strategy aims to avoid false positives, inferences that give credence to causality where none exists". He calls statisticians and epidemiologists "properly professional skeptics". Indeed, avoidance of "negative" or "nonpositive" results has been a largely neglected concern (see Ref. 50 for a wider discussion). However, it has to be emphasized that such concern is in the interests of those directly involved, e.g., the workers being exposed to occupational hazards.

Consider, for example, the study of the effect of carbon disulfide exposure on the incidence of coronary heart disease (see Chapter 18). It can be argued that in medium-sized studies, the usual statistical criterion of "significance" may be relaxed, especially when dealing with a serious outcome such as death. Thus, it would seem to be advisable that the urgency of the practical problem should not be ignored at the inferential stage of the study. It is debatable, however, whether the notion of utility (i.e., numerical values assigned to different possible actions) is the correct way to measure the (humanitarian) consequencies of the different outcomes. It may be asked for example, how is it possible to weigh in precise numerical terms the ethical merit of accepting the chance of saving lives against major monetary savings in postponing the improvement of hygienic conditions of a work environment?

The controversial interpretations of the different schools of statistical inference are reflected, for example, in the stand that the epidemiological community takes with regard to the problem of multiple comparisons, or simultaneous testing of many hypotheses. So-called "hypothesis generation", as against ideas born from "a flash of genius", has been aided by the enormous expansion in the capability and accessibility of modern computers. How should one react to unforeseen associations thrown up by assiduous analysis of (often rather untidy) data? A prominent example is found in the field of monitoring the adverse effects of drugs: the first report of an association between the use of reserpine and the risk of breast cancer (*51*). This novel finding emanated from a massive programme and was only one of numerous possible relations. The publication of the result was delayed until it had been "tested" in two independent studies (*52, 53*) in true "frequentist" spirit.

At the other extreme, the Bayesian inference has been criticized mainly on the grounds of intangibility of prior knowledge, uncertainty, or ignorance, as well as for the technical difficulties in the specification of prior probability distribution (form and parameters) and computation of posterior probability (e.g., numerical integration) (see, for instance, Ref. *54* for a comparison of Bayesian analysis with sampling–theoretical methods).

It can be concluded that any inference made on the basis of data analysis relies on the premise that the study design itself has a sound basis.

Epilogue

The preceding discussion has been more in the nature of a general or philosophical exposition than a detailed technical presentation. The main reason for this is that descriptions of the numerous statistical methods for analysis of epidemiological studies have been widely scattered in the biometric, psychometric, sociological, medical and, more recently, epidemiological literature. The excellent books by Breslow & Day (*34*), Andersson et al. (*12*), and Kleinbaum et al. (*15*) have succeeded in bringing together the various methods in a coherent manner. It may, however, be comforting to end by pointing out that, despite the great variety of numerate methods now available for an epidemiologist, the principle of Cochran (*55*) and Mantel & Haenszel (*56*), through its many extensions, specializations and interpretations (*57–65*) stands out as a general statistical procedure that, by itself, can be sufficient for most data-analysis tasks in epidemiology.

References

1. **Armitage, P.** *Statistical methods in medical research.* Oxford, Blackwell Scientific Publications, 1971.
2. **Acheson, E. D.** *Medical record linkage.* London, Oxford University Press, 1967.
3. **MacMahon, B. & Pugh, B. T.** *Epidemiology. Principles and methods.* Boston, Little, Brown and Company, 1970.
4. **Miettinen, O. S.** Proportion of disease caused or prevented by a given exposure, trait or intervention. *American journal of epidemiology*, **99**: 325–332 (1974).
5. **Cochran, W. G.** *Sampling techniques.* New York, Wiley, 1953.
6. **Miettinen, O. S.** Design options in epidemiologic research. An update. *Scandinavian journal of work, environment and health*, **8** (Suppl. 1): 7–14 (1982).
7. **Fraser, D. A. S.** *The structure of inference.* New York, Wiley, 1968.
8. **Cochran, W. G.** The planning of observational studies of human populations. *Journal of the Royal Statistical Society, Series A*, **128**: 234–255 (1965).

9. **Miettinen, O. S.** *Notes for the WHO course in scientific research methods of the epidemiology of industrial intoxications, Helsinki, October 1975.* Copenhagen, WHO Regional Office for Europe, 1975 (unpublished document).
10. **Neutra, R. R. & Drolette, M. E.** Estimating exposure-specific disease rates from case-control studies using Bayes' theorem. *American journal of epidemiology*, **108**: 214–222 (1978).
11. **Thorndike, F. L.** Regression fallacies in the matched group experiment. *Psychometrika*, **7**: 85–102 (1972).
12. **Andersson, S. et al.** *Statistical methods for comparative studies.* New York, Wiley, 1980.
13. **Johnston, J.** *Econometric methods.* New York, McGraw-Hill, 1963.
14. **Campbell, D. T.** Factors relevant to the validity of experiments in social setting. *Psychological bulletin*, **54**: 297–312 (1957).
15. **Kleinbaum, D. G. et al.** *Epidemiologic research: Principles and quantitative methods.* San Francisco, Lifetime Learning, 1982.
16. **Schlesselman, J. J.** Sample size requirements in cohort and case-control studies of disease. *American journal of epidemiology*, **99**: 381–384 (1974).
17. **Walter, S. D.** Determination of significant relative risks and optimal sampling procedures in prospective comparative studies of various sizes. *American journal of epidemiology*, **105**: 387–397 (1977).
18. **Miettinen, O. S.** Individual matching with multiple controls in the case of all-or-none responses. *Biometrics*, **25**: 339–355 (1969).
19. **McKinley, S. M.** The design and analysis of the observational study—A review. *Journal of the American Statistical Association*, **70**: 503–523 (1975).
20. **Fisher, R. A.** *Statistical methods for research workers.* London, Oliver & Boyd, 1925.
21. **Cramer, H.** *Mathematical methods of statistics.* Princeton, Princeton University Press, 1946.
22. **Rao, C. R.** *Advanced statistical methods in biometric research.* New York, Wiley, 1952.
23. **Miettinen, O. S.** *Notes for the International Advanced Course on Epidemiologic Methods in Occupational Health, Helsinki, August 1978.* Helsinki, Institute of Occupational Health, 1978 (unpublished document).
24. **Tukey, J. W.** The future of data analysis. *Annals of mathematical statistics*, **33**: 1–67 (1962).
25. **Afifi, A. A. & Azen, S. P.** *Statistical analysis. A computer oriented approach*, 2nd ed. New York, Academic Press, 1979.
26. **Gnanadeshikan, R.** *Methods for statistical data analysis of multivariate observations.* New York, Wiley, 1977.
27. **Cox, D. R. & Hinkley, D. C.** *Problems and solutions in theoretical statistics.* London, Chapman & Hall, 1978.
28. **Finney, D. J.** *Probit analysis. A statistical treatment of the sigmoid response curve.* Cambridge, Cambridge University Press, 1947.
29. **Walker, S. H. & Duncan, D. B.** Estimation of an event as a function of several independent variables. *Biometrika*, **54**: 167–179 (1967).
30. **Greenland, S.** Limitations of the logistic analysis of epidemiologic data. *American journal of epidemiology*, **110**: 693–698 (1979).
31. **Hänninen, H. et al.** Psychological tests as indicators of excessive exposure to carbon disulfide. *Scandinavian journal of psychology*, **19**: 163–174 (1978).
32. **Dempster, A. P.** *Elements of continuous multivariate analysis.* Reading, MA, Addison-Wesley Publishing Company, 1969.
33. **Bishop, Y. M. M. et al.** *Discrete multivariate analysis. Theory and practice.* Cambridge, MA, MIT Press, 1975.
34. **Breslow, N. E. & Day, N. E.** *Statistical methods in cancer research.* Vol. 1. *The analysis of case-control studies.* Lyon, International Agency for Research on Cancer, 1980 (IARC Scientific Publication No. 32).
35 **Zelen, M. & Feinleib, M.** On the theory of screening for chronic diseases. *Biometrika*, **56**: 601–614 (1969).
36. **Williams, W. T. & Lance, G. N.** Hierarchical classification methods. In: Enslein, K. et al., ed. *Statistical methods for digital computers.* New York, Wiley, 1977, pp. 269–295.

266

37. **Bailey, T. J.** *The mathematical theory of epidemics.* London, Griffin, 1957.
38. **Haberman, S.** Mathematical treatment of the incidence and prevalence of disease. *Social science and medicine,* **12**: 147–152 (1978).
39. **Haberman, S.** Probabilistic treatment of the incidence and prevalence of disease. *Social science and medicine,* **12**: 159–161 (1978).
40. **Mantel, N. & Bryan, W. R.** "Safety" testing of carcinogenic agents. *Journal of the National Cancer Institute,* **27**: 455–470 (1961).
41. **Hartley, H. O. & Sielken, R. L., Jr.** Estimation of "safe doses" in carcinogenic experiments. *Biometrics,* **33**: 1–30 (1977).
42. **Cox, D. R.** Regression models and life tables (with discussion). *Journal of the Royal Statistical Society, Series B,* **34**: 187–220 (1972).
43. **Tossavainen, A. et al.** Application of mathematical modelling for assessing the biological half-times of chromium and nickel in field studies. *British journal of industrial medicine,* **37**: 285–291 (1980).
44. **Lanczos, C.** *Applied analysis.* Englewood Cliffs, NJ, Prentice Hall, 1964.
45. **Nurminen, M. et al.** Quantitated effects of carbon disulfide exposure, elevated blood pressure and aging on coronary mortality. *American journal of epidemiology,* **115**: 51–57 (1982).
46. **Cox, D. R.** Some problems connected with statistical inference. *Annals of mathematical statistics,* **29**: 357–372 (1958).
47. **Barnett, V.** *Comparative statistical inference.* New York, Wiley, 1973.
48. **Miettinen, O. S.** Comment. *Journal of the American Statistical Association,* **69**: 380–382 (1974).
49. **Susser, M.** Judgement and causal inference: Criteria in epidemiologic studies. *American journal of epidemiology,* **105**: 1–15 (1977).
50. **Hernberg, S.** "Negative" results in cohort studies—How to recognize fallacies. *Scandinavian journal of work, environment and health,* **8** (Suppl. 4): 121–126 (1981).
51. **Boston Collaborative Drug Surveillance Program** Reserpine and breast cancer. *Lancet,* **2**: 669–671 (1974).
52. **Armstrong, B. et al.** Retrospective study of the association between use of Rauwolfia derivatives and breast cancer in English women. *Lancet,* **2**: 672–675 (1974).
53. **Heinonen, O. P. et al.** Reserpine use in relation to breast cancer. *Lancet,* **2**: 675–677 (1974).
54. **Altham, P. M. E.** Exact Bayesian analysis of a 2 ×2 contingency table, and Fisher's "exact" significance test. *Journal of the Royal Statistical Society, Series B,* **31**: 261–269 (1969).
55. **Cochran, W. G.** Some methods for strengthening the common χ^2 tests. *Biometrics,* **10**: 417–451 (1954).
56. **Mantel, N. & Haenszel, W.** Statistical aspects of the analysis of data from retrospective studies of disease. *Journal of the National Cancer Institute,* **22**: 719–748 (1958).
57. **Mantel, N.** Chi-square tests with one degree of freedom: Extensions of the Mantel-Haenszel procedure. *Journal of the American Statistical Association,* **58**: 690–700 (1963).
58. **Oleinick, A. & Mantel, N.** Family studies in systemic lupus erethematosis. II. Mortality among siblings and offspring of index cases, with a statistical appendix concerning life table analysis. *Journal of chronic diseases,* **22**: 617–625 (1970).
59. **Mantel, N.** Tests and limits for the common odds ratio of several 2 ×2 contingency tables: Methods in analogy with the Mantel–Haenszel procedure. *Journal of statistical planning and inference,* **1**: 179–189 (1977).
60. **Dyal, H. H.** On the desirability of the Mantel–Haenszel summary measure in case-control studies of multifactor etiology of disease. *American journal of epidemiology,* **108**: 506–511 (1978).
61. **Day, N. E. & Byar, D.** Testing hypothesis in case-control studies—equivalence of Mantel–Haenzel statistics and logit score tests. *Biometrics,* **35**: 623–630 (1979).
62. **Hauck, W. W.** The large sample variance of the Mantel–Haenszel estimator of the common odds ratio. *Biometrics,* **35**: 817–819 (1979).

63. **Mantel, N. & Fleiss, J. L.** Minimum requirements for the Mantel–Haenszel one-degree of freedom chi-square test and related rapid procedure. *American journal of epidemiology*, **112**: 129–134 (1980).
64. **Hakulinen, T.** A Mantel–Haenszel statistic for testing the association between a polychotomous exposure and a rare outcome. *American journal of epidemiology*, **113**: 192–196 (1981).
65. **Nurminen, M.** Asymptotic efficiency of general noniterative estimators of the common relative risk. *Biometrika*, **68**: 525–530 (1981).

14

Validity aspects of epidemiological studies

S. Hernberg[a]

As already pointed out in earlier chapters, epidemiology is predominantly a nonexperimental science. Ascertaining validity in nonexperimental research is no easy task. While an experimenter can actively manipulate experimental conditions by deciding the quality and intensity of exposure, by randomly allocating the subjects to become exposed or nonexposed, and by reading all the results blindly, the epidemiologist can rarely, if ever, apply these methods. Instead, he or she must be content to observe what Nature has accomplished. This is a very fundamental distinction between experimental and nonexperimental research, and therefore ensuring validity in epidemiological research must rely on other procedures. These are usually much less powerful; hence, the judgement of causality is much more a matter of probability than in experiments. A completely perfect epidemiological study has probably never been done; it is hardly an exaggeration to say that all studies published to date contain at least some systematic errors,[b] whose magnitude and direction must influence the interpretation of the results.

There are two dimensions to validity, internal and external (*1*). The former refers to how "true" the results of a study are with respect to the study itself. The latter expresses the generalizability of the results beyond time and place, that is, to other similar situations and, finally, to the sphere of scientific theories. Take, for example, a mortality study that has shown an excess mortality from coronary artery disease among workers exposed to carbon disulfide. The study has internal validity if systematic errors can be ruled out and external validity when it can be postulated that exposure to carbon disulfide *in general* causes coronary artery disease.

---

[a] Institute of Occupational Health, Helsinki, Finland.

[b] A systematic error (bias) distorts the results of a study in such a way that hypothetical replications of it would produce the same results so that a false conclusions is reached.

The intention of this chapter is to analyse the nature of these validity issues, to give some guidance on how to handle them, and finally to help the reader of a scientific report to evaluate its results.

Internal validity

Internal validity can be broken down into the following three components (1):
  (1) validity of selection
  (2) validity of information
  (3) validity of comparison:
    (a) validity of the reference entity
    (b) absence of confounding.

Validity of selection

Validity of selection means that the probability of a subject being nominated for the study must not depend in any systematic way on the property under study (the disease in cross-sectional and follow-up studies and the exposure in case–referent studies). The avoidance of such selection bias is particularly important in cross-sectional and case–referent studies, where it would lead to considerable distortion of the results.

The effect of selection bias in cross-sectional studies can be illustrated by the following example. Let the problem be: Does dusty work cause bronchitis? In a cross-sectional study, the groups are chosen according to current occupation. Suppose the investigator compares the prevalence of bronchitis (using some fixed criteria) among foundry workers with that of, say, car assembly workers. In this design, the selection is invalid for three reasons: (1) because workers who already have bronchitis know that foundry work is dusty and may prefer other types of jobs; (2) because some of those who develop bronchitis during foundry work may have to quit; and (3) because, since foundry work is physically heavier and "dirtier" than car assembly work, the same severity of bronchitis may cause more sickness absence in a foundry than in a car assembly plant at the time of the survey. Consequently, because of selection, sick persons are under-represented among foundry workers and the possible effect of foundry work may totally or partly escape discovery (negative bias).

In a case–referent study, a selection bias may arise in the following way. Suppose that, at the present day, somebody wishes to study the connection between solvent exposure and disability due to nonspecific neuropsychiatric disease. Patients with this disease are then defined as cases and patients with, say, coronary artery disease as referents. If solvent exposure were over-represented among the cases, this could be an overestimate because the connection has been known or suspected for some time (2). Workers exposed to solvents and having neuropsychiatric problems would then be more likely to get compensation or otherwise to be pensioned off because many physicians would suspect solvent exposure as being causative in such a situation. In this example, the reasons for selection bias were (*a*) the fact that the diagnosis of neuropsychiatric disease is ambiguous and (*b*) the connection between solvent exposure and neuropsychiatric syndromes was known or suspected in advance.

Selection bias may arise in follow-up studies whenever self-selection into or out of an occupation or trade occurs. The effects are insignificant for diseases with long, silent latency times, but they are important whenever symptoms of the disease at issue are provoked or worsened by the work in question. For example, persons with heart conditions are not attracted to jobs known to be physically strenuous, and they may be forced to leave them selectively. Pre-employment examinations, if successful, tend to strengthen selection into the job even further. Complete follow-up must aim at controlling for the effects of selection out of a workplace, while selection into it is best controlled by selecting a reference category similar to the exposed one, except for the exposure.

Selection bias must be avoided at the planning stage of a study, since no method exists by which it can be controlled during the data handling phase.

Validity of information

Validity of information means that the accuracy, or rather inaccuracy, of the information gathered from both the cases and the referents, or both the exposed and nonexposed, is similar. *Asymmetrical* inaccuracy weakens validity. However, the sensitivity (the power to detect a causal association, if present) of the study suffers from *symmetrically* inaccurate information. Information bias may affect all types of epidemiological investigation.

In case–referent studies that derive information from questionnaires and interviews rather than from measurements, the greatest problem is how to ensure comparability between the histories obtained from the cases and the referents (or their relatives). In general, it is thought that cases "remember" better than the referents (recall bias). This asymmetry then leads to an exaggeration of the effect under study. However, while the possibility of asymmetrical recall of exposures must always be kept in mind, this source of error must not be inappropriately exaggerated. In fact, the present author is not aware of a single study providing controlled documentation supporting this generally held belief. In the absence of hard data, it would seem conceivable to presume that recall bias depends on the "specific weight" of the event. For example, there must be great differences between the quality of remembering past medicine intake and, say, "exposure" to lumbering work for many years. In the occupational health setting, it is usually possible to double-check the information given by cases and referents by contacting the employer. When interviewers are used in case–referent studies, another possible source of information error is that interviewing of the cases is conducted in a different manner from that of the referents (observer bias). For example, the interviewers may become aware of the hypothesis to be tested, and they may tend to be more careful in taking histories from the cases. The direction of this bias would, again, be positive.

In case–referent studies, a very common method of avoiding information bias is to abstain from the use of healthy persons as referents. Instead, patients having some other disease than the one being studied are selected. The rationale is that such referents are thought to "remember" past events in the same manner as the cases. However, this introduces the possibility of confounding, because the referent condition may be influenced by the exposure(s) under study, although the connection may be unknown for the time being. Hence, the control of one source of bias may cause another.

Retrospective follow-up studies are also subject to information bias because there is a general tendency to observe "exposed" populations more thoroughly than "nonexposed" ones—by periodic health examinations, for example. A particular warning must be given against the practice of checking the death certificates of the exposed cohort (and perhaps "finding" some auxiliary causes of death, such as cancer) and then using the national rates (uncheckable) as the reference. A prospective follow-up study can be more easily designed to avoid information bias if care is taken to secure a similar quality of data from both the exposed and the nonexposed cohorts. In both instances, special care should be taken to ensure a minimum loss of subjects from follow-up, since lost (unknown) information may be asymmetrical and hence affect validity even more than inaccurate information.

Information validity may also be ensured by the "blinding" of readings whenever possible (this is feasible for laboratory tests, X-rays, ECGs, EEGs, etc., but not for clinical examinations, history-taking, psychological tests, etc., which require oral communication). Other methods include interobserver error control, calibration of equipment, and mixed runs (equal proportion of samples from the exposed and unexposed groups) for biochemical analysis.

Validity of comparison
The problem of validity of comparison affects all types of epidemiological study equally. The first consideration is the choice of reference group. An ideal reference group should share all the characteristics of the study group relevant to the problem at issue, except for the properties by which the groups are defined. In cross-sectional and follow-up studies this means that the exposure and its possible effects should be the only relevant properties that differ between the groups. In other words, the reference group should provide an estimate of what would have happened in the exposed group had there been no exposure. For example, if the effects of exposure to chromates on the incidence of lung cancer are being studied, the groups should be similar with respect to all other factors influencing this disease, e.g., smoking, age distribution, and urban air pollution. In case–referent studies, the referents should reflect the general exposure pattern in the population from which the cases are drawn. If another disease is chosen

as a referent condition (to avoid information bias), this condition must be one that is neither caused nor prevented by the exposure(s) in question. An ideal referent category is sometimes extremely difficult to find, not least because of practical circumstances, and if a less-than-ideal group has to be used, the greatest emphasis must be placed on controlling those properties that are likely to possess the strongest distorting effects.

Confounding is the other component of comparison bias. A confounder is an external factor that is intermixed with the scientific problem. Because of this intermixing, the confounder may create a false association where no association, in fact, exists (positive confounding) or it may mask a true association (negative confounding). It may also disturb the quantitative assessment of an effect that is qualitatively detectable. To exert this confounding effect, a factor must (a) be a causal risk factor of the illness in general and (b) be associated with the exposure *in that particular study*. For example, smoking is a confounding factor when studying the connection between exposure to chromates and the incidence of bronchial cancer *only* when the smoking habits of the exposed cohort are different from those of the reference group *in that particular study*. Confounding becomes especially problematical when the relation is rather weak (rate ratios of the order of, say, less than 2 or 3). When the relation is strong, it is not likely that such powerful confounding would pass undetected.

In occupational medicine, confounding is often much less of a problem than is ususally believed, because many occupational factors themselves may produce strong effects. In so far as smoking is concerned, Axelson (3) has calculated the effect of various hypothetical smoking distributions on the rate ratio of lung cancer. As can be seen from Table 1, the asymmetry must be rather extreme to produce serious bias. More specifically, the effects of differences in smoking habits between common categories of workers only slightly affect the rate ratio for lung cancer. In practice, then, the control of smoking in a study becomes particularly important when the rate ratio is less than about 2. The share of confounding caused by smoking becomes less and less significant the more the rate ratio exceeds 2, which often occurs in research on occupational lung cancer.

The control of confounding takes place either at the planning stage of a study (restriction, matching) or in the analytical stages (stratification, modelling), on the assumption, evidently, that the relevant data are available. For example, if alcohol use is to be controlled as a confounder, information on drinking habits is necessary, and the investigator must already foresee this when planning the study.

External validity

External validity has to do with the generalizability of the results beyond the particular study population, meaning that it allows scientific inferences. Assessment of external validity is very much a matter of

Table 1. Estimated crude rate ratios in relation to fraction of smokers in various hypothetical populations[a]

Population fraction (%)[b]			
Nonsmokers	Moderate smokers	Heavy smokers	Rate ratio
100	–	–	0.15
80	20	–	0.43
70	30	–	0.57
60	35	5	0.78
50	40	10	1.00[c]
40	45	15	1.22
30	50	20	1.43
20	22	25	1.65
10	60	30	1.86
–	65	35	2.08
–	25	75	2.69
–	–	100	3.08

[a] After Axelson (3).
[b] Two different risk levels are assumed for smokers, i.e. 10 times and 20 times that of nonsmokers for "moderate" and "heavy" smokers, respectively.
[c] Reference population (similar to the general population in countries such as Sweden). It is rare to find groups whose smoking habits are more extreme than those within the dotted lines.

judgement. An absolute requirement is good internal validity, since otherwise the generalization would lead to a false inference. To be externally valid, the data obtained from the study should be generalizable to an abstract "superpopulation" having the same characteristics (e.g., intensity of exposure, age distribution, type of work). Next, there should be "scientific truth"; for example, if there is found to be an excess occurrence of coronary artery disease among viscose rayon workers, the study must be so designed that one can hypothesize that exposure to carbon disulfide (of sufficient intensity and duration) *in general* causes or accelerates the development of coronary artery disease. If a reasonable amount of evidence is available (epidemiological, experimental, theoretical) a scientific theory can then be formulated and further tested by new studies.

The question of external validity must be considered at the planning stage of a study, and the study population and the reference category must be defined accordingly. Only then can the results be generalized. This kind of generalization differs from the statistical sample-to-population generalization in sampling surveys in the sense that specific mechanisms and cause-effect relationship are concerned, not the extrapolation of the findings to people at large. In other words, if one finds that spray painters show neurological abnormalities, it is scientifically more interesting to formulate the hypothesis that solvent exposure causes neuropathy than to estimate that, say, 10% of all spray painters have this disorder.

The healthy worker effect

This conceptually vague term describes the comparison error arising from contrasting the mortality experience of an exposed cohort with that of the general population (4, 5). The latter is heterogeneous, not completely free from the exposure under study, and rarely, if ever, represents the same social stratum as the exposed cohort. In other words, the general population does not fulfil even the most elementary requirements of comparison validity. Nevertheless, in spite of this, such an invalid reference category is used very often, simply for reasons of practicality and economy. An *ad hoc* reference cohort would double the costs of the investigation, and besides, valid and suitable reference cohorts are hard to find.

The healthy worker effect causes the standardized mortality ratio (SMR) of the exposed cohort to fall well below 100 if no life-shortening occupational hazard exists. In fact, ratios of the order of 60–90 have often been reported. These figures are obvious underestimates of the "true" mortality (6). The main reason for the better-than-expected mortality experience of occupational groups lies in the fact that the general population also includes unemployable and unemployed persons. Among them are those in institutions, those with congenital abnormalities, those handicapped during childhood, those otherwise ill at the time job seeking commences, and those unemployed or with unstable employment. All these groups have a higher mortality rate than the active population. Table 2 shows some published SMRs from occupationally active cohorts. Attention should be drawn to the fact that the interpretation of an SMR between, say, 90 and 110 is extremely difficult. Is it due to the fact that no life-shortening hazard exists, or does the healthy worker effect mask such a hazard? This problem had to be faced in the Finnish foundry worker study (7). No certain conclusions could be drawn, but the investigators intuitively favoured the explanation that the healthy worker effect masked a slightly elevated overall mortality.

The healthy worker effect is not constant but varies depending on a number of circumstances (6). It is strongest in younger age groups, and

Table 2. Some SMRs showing the healthy worker effect (general population = 100)

Occupational group	SMR
Finnish foundry workers (7)	90
Typical foundry occupations in Finland[a]	95
American steel workers (12)	82
American rubber workers (13)	87
Finnish granite workers (14)	83
Finnish dock workers (15)	81
American chemical workers (16)	81

[a] Excluding maintenance, repair, cleaning and similar occupations.

275

it declines with age until it is no longer significant (and sometimes even reversed) in the post-retirement age range. It is stronger for men than for women, stronger in higher social categories than in lower ones, and strongest during the first years after the beginning of employment. Even more important, it is different for different causes of death. In general, diseases with silent early stages and a rapidly fatal course do not cause a healthy worker effect (except during the first one or two years after cohort identification). Cancer is a typical example. By contrast, the likelihood of death from coronary artery disease can be established in advance by early diagnosis of the disease or identification of its risk factors. Furthermore, self-selection occurs because of symptoms or earlier warnings from physicians that demanding jobs should be avoided.

What still weakens the use of the general population as a reference group is that no details of possible confounders are known; for example, there is no information on smoking or drinking habits in national mortality statistics. Consequently, the healthy worker effect poses a serious methodological problem when occupational groups are compared with a general population. The interpretation of such studies is therefore very difficult unless the effect under study is very pronounced. Because systematic errors of unknown strength are involved, tests of significance are difficult to interpret; sometimes they are even meaningless. This is so because they measure *random* variation only, ignoring systematic errors. In fact, there is a danger that they mislead the unwary reader if used uncritically.

Whenever possible, a better reference group than the general population should be used (4, 5). If this is not feasible or otherwise impossible, the rates for the occupationally active population, if available, are to be preferred. Local comparisons are better than national ones, especially if there are geographical differences in the incidence of the disease of interest. If the cohort is large, within-cohort comparisons between different exposure categories may increase validity. If the use of national data (or some other invalid reference category) cannot be avoided, the first five or ten years of follow-up should preferably be disregarded. However, no surrogate can replace a valid, *ad hoc* reference group (5).

False positive and false negative results.

The commonest reasons for false positive results are comparison bias and information bias. Just as the reference group selected may be too "unhealthy", it may also be too "healthy". This leads to false positive results. False positive findings may also arise through confounding. Large studies are the most dangerous in this respect, since even a small difference yields statistical significance. Then even a weak confounding factor may be decisive. By contrast, in smaller studies the difference must be so large that a statistical significance due to confounding is easily revealed, as already discussed. For example, a rate ratio of 1.5 is significant ($P < 0.01$) in a study of 1000 exposed and 1000 referents if the mortality is 10 %, whereas the same level of significance in a study of 50 exposed and 50 referents requires a rate ratio of 11. Hence, in large studies the magnitude of the rate ratio is more important than the statistical significance of the difference, while in small studies the magnitude of the rate ratio is very much influenced by chance and the statistical significance is therefore more important.

The most important reason for false positive results in case–referent studies is probably the so-called "recall bias". It arises whenever cases "remember" better than referents, as discussed earlier.

A very important but neglected problem is the evaluation of "negative" studies. These are at least as important as positive ones in occupational medicine because one is often interested in defining a noneffect level for harmful exposures. However, a clear distinction must be made between truly negative and "nonpositive" studies (8).

A true negative study must fulfil three criteria: (1) it must be large; (2) it must be sensitive; (3) it must have well documented exposure data. Obviously, it must also be well designed in other respects and there must be a clear awareness of whether only a specific condition is to be excluded (say, cancer) or if all types of adverse health effect are to be excluded. Only investigations that comply with these criteria can be considered true negative studies.

A small negative study is uninformative in the sense that it can exclude the presence only of very powerful hazards. Of course the terms "large" and "small" are ill-defined concepts. If "large" is defined as a "sufficient" number of exposed cases, some specification is provided. If a rather small but otherwise well designed study does not reveal any excess of, say, lung cancer in an exposed cohort, it can at most be concluded that this rules out the possibility of very strong carcinogenic effects of the exposure at issue. If a large study shows the same negative result, the possible carcinogenicity of the exposure in question can be rejected with greater probability.

Insensitive studies, i.e., studies with a crude design or crude measuring methods, are also uninformative. They can be compared with insensitive laboratory tests in clinical diagnosis, e.g., the sole use of haemoglobin determination for diagnosing lead poisoning. Moreover, negative results can be referred only to the actual or lower levels of exposure; hence, the availability and prudent use of accurate exposure data is crucial. It is known that this condition is seldom met. The requirement that the study design should be good applies, of course, to all epidemiological studies. However, more errors lead to false negativity than to false positivity; furthermore, negative bias is often more difficult to detect. In addition, investigations designed to study one or a few conditions cannot be interpreted as demonstrating "complete safety."

Some common errors leading to false negativity are reviewed below (see also Ref. 8). Inappropriate design often results in an inefficient study that fails to reveal an existing effect. A small sample size (and/or reference group) is the commonest design weakness (but it is often unavoidable for reasons of availability). In general, such a weakness is easy to identify, but uncritical authors or unwary readers of a report still often misinterpret the failure to obtain "statistical significance" as being evidence of negativity. In such a setting, only effects leading to very high rate ratios (of the magnitude of, say, 10–30) can be excluded. A "negative" result may thus emerge when the rate ratio is moderate only because statistical significance has not been established. Hence, the results should never be automatically classed as "positive" or "negative" on the basis of whether or not statistical significance has been reached. Especially if the P-value is based on rather small numbers, one case more or less can decide whether or not the study is "positive" or "negative". A few wrong diagnoses, for example, can then severely distort one's judgement. Consider, for example, a 20-year follow-up study of 500 men, aged 45–59 years. Based on figures for the Finnish general male population, the expected number of pancreatic cancers is 1.5. Table 3 shows the rate ratios and the P-values for different numbers of observed cases. By conventional reasoning, four observed cases would yield a "negative" result, while five cases would give a "positive" one. It is quite obvious that artificial cut-offs for statistical significance should not be abused. Instead, one should consider the number of exposed cases, the magnitude of the rate ratio, the confidence limits, and above all, how consistent the data are with previous studies and if they are biologically plausible.

Table 3. Effect of different numbers of observed cases of pancreatic cancer on the statistical significance

Observed No. of cases	Expected No. of cases	Rate ratio	P-value
3	1.5	2.0	0.191
4	1.5	2.7	0.062
5	1.5	3.3	0.019
6	1.5	4.0	0.004

False negative results can also arise if the study centres on incorrect categories of exposed workers. For example, if retired workers are left out of occupational cancer studies, crucial information will usually be lost and the "truth," which may actually be positive, may appear negative. Likewise, if a study is restricted to retired workers only, it may fail to detect pathognomic manifestations of the exposure in question in younger age groups (see, for example, Ref. 9).

All too often workers with too short an exposure time and too low an exposure intensity, or sometimes even nonexposed, misclassified workers, are included in the exposed cohort. Usually, the motive is to increase the cohort size. However, this may be inappropriate because the

278

study becomes insensitive, i.e., an existing effect becomes diluted as long as the cohort is analysed as a homogenous population. The only justification for such a procedure is the study of exposure–response relationships, and this, of course, is important information provided the qualitative aspects of causation are already known. However, basing a qualitative negative conclusion only on the outcome in subjects with low exposure intensity and/or short exposure time is unjustifiable.

Whenever diseases with long latency periods (e.g., occupational cancer) are studied, a false negative result emerges if the follow-up time is too short. Of course, no occupationally induced excess can be found before the latency time has passed—this is biologically impossible. It should be noted that this error has nothing to do with lack of statistical power; hence it cannot be overcome by increasing the sample size. Unfortunately, however, most latency periods are unknown. It is believed that many cancers have latency periods of several decades. An arbitrary rule would be not to start computing person-years before half the supposed mean latency period of the cancer in question has passed. Alternatively, several different latency times can be tried.

In contrast to systematic errors, whose effects may be either positive or negative, random errors always mask existing effects. Such errors arise, for example, from poorly standardized measuring methods. Poor analytical precision of the independent (exposure) variable always decreases existing group differences and levels off the regression slope in exposure–response studies.

A choice of insensitive indicators of illness also results in falsely negative results when the purpose of the study is to demonstrate complete safety (e.g., for the documentation of hygienic standards). For example, mortality is too crude an indicator of health risks associated with exposure to many conditions in modern industry. Failure to find increased mortality has unfortunately been misinterpreted as total absence of health hazards. This misinterpretation becomes especially dangerous when the conclusion is based on a comparison with the general population.

Errors of comparison are common in cohort studies. A very problematic comparison error arises when the general population is used as a reference category, as discussed earlier. In case–referent studies, where the referent group should be one whose condition is neither caused nor prevented by the exposure(s) at issue, insufficient knowledge of the etiology of that condition can cause a comparison bias. The larger the number of exposures studied in the same study, the greater is the possibility that the reference condition will not fulfil this requirement for one of the exposures.

Finally, inappropriate interpretation of the results of a study can account for falsely negative (and of course also for falsely positive) conclusions. For example, failure to appreciate that a negative finding is negative only in relation to actual or lower exposure levels is a common error in "negative" epidemiological studies. Wrong thinking is also involved when a general negative statement of "safety" is made on the basis of too few, too insensitive, or even inappropriate measures of effect. For example, if no changes in the blood picture or of the hepatic and renal functions are seen at a certain intensity of lead exposure, it does not follow that such levels are safe in all respects (for a good example of wrong interpretation, see Ref. *10*).

Judging the causality of an association

It should once more be emphasized that definite cause–effect inferences cannot be derived from any non-experimental research, no matter how large the accumulated evidence is. Causality, therefore, can be viewed in terms of probability only. Bradford Hill (*11*) has elaborated some rules for judging the causality of an association. The use of this list self-evidently presupposes reasonable certainty that the study is valid.

1. The stronger the association (the higher the rate ratio) the greater the probability that it is a causal one. However, "small" series must be interpreted with great caution because their rate ratios are subject to substantial random variation.

2. The consistency of the observed association speaks in favour of causal relationship. If the observation has been made repeatedly by different persons, in different places, and under different circumstances and times, these are all strong arguments in favour of causality.

3. The specificity of the association is important. If the association is limited to specific workers and to particular sites and types of disease, and there is no association between the work and other diseases, this supports causation.

4. The temporal relationship of the observed association is also important: it is straightforward when the effect is immediate but becomes subtle when there is a latency period. In other words, exposure must have commenced early enough to be a "cause" (for example, earlier than half the mean latency period, if known).

5. The presence of an exposure–response relation is also listed by Bradford Hill as supporting a causal association. However, confounding would invariably cause, or contribute to, an exposure–response relation. Therefore, this argument loses some of its power.

6. Biological plausibility is helpful for the evaluation, although lack of knowledge may complicate the use of this argument. For example, no biological explanation was known to support Pott's observation in 1775 of the excess of scrotal cancer in chimney sweeps. On the other hand, computer technology has made it possible to compute correlation coefficients for a multitude of variables, and this inevitably results in plenty of "significant" coefficients. Such coefficients should be looked upon with suspicion if there are no biological explanations.

7. Coherence of the evidence from experimental work, for example, strongly supports the causality of an epidemiological association.

8. If a change in the exposure brings about a change in morbidity, which can be observed from interventive epidemiological studies, this must be judged as strong evidence in favour of causality.

9. In some circumstances, reasoning by analogy helps in judging the causality of an association. For example, a compound with structural similarities to a known carcinogen, when found to be associated with excess cancer, is more easily accepted as the cause of this excess than one without structural analogy.

References

1. **Campbell, D.** Factors relevant to the validity of experiments in the social setting. *Psychological bulletin*, **54**: 297–312 (1957).
2. **Axelson, O. et al.** A case-referent study on neuropsychiatric disorders among workers exposed to solvents. *Scandinavian journal of work, environment and health*, **2**: 14–20 (1976).
3. **Axelson, O.** Aspects on confounding in occupational health epidemiology. *Scandinavian journal of work, environment and health*, **4**: 98 (1978).
4. **Hernberg, S.** Evaluation of epidemiologic studies in assessing the long-term effects of occupational noxious agents. *Scandinavian journal of work, environment and health*, **6**: 163–169 (1980).
5. **Wang, J. D. & Miettinen, O. S.** Occupational mortality studies: Principles of validity. *Scandinavian journal of work, environment and health*, **8**: 153–158 (1982).
6. **McMichael, A. J.** Standardized mortality ratios and the "healthy worker effect". Scratching beneath the surface. *Journal of occupational medicine*, **18**: 165 (1976).
7. **Koskela, R. S. et al.** A mortality study of foundry workers. *Scandinavian journal of work, environment and health*, **2** (Suppl. 1): 73–89 (1976).
8. **Hernberg, S.** "Negative" results in cohort studies.—How to recognize fallacies. *Scandinavian journal of work, environment and health*, **7** (Suppl. 4): 121–126 (1981).
9. **Enterline, P. et al.** Mortality in relation to occupational exposure in the asbestos industry. *Journal of occupational medicine*, **14**: 897–903 (1972).
10. **Ramirez-Cervantes, B. et al.** Health assessment of employees with different body burdens of lead. *Journal of occupational medicine*, **20**: 610–617 (1978).
11. **Hill, A. B.** *Principles of medical statistics*, 8th ed. London, The Lancet, 1967, Chapter 24.
12. **Lloyd, J. W. et al.** Long-term mortality study of steelworkers: IV. Mortality by work area. *Journal of occupational medicine*, **12**: 151–157 (1970).
13. **McMichael, A. J. et al.** An epidemiologic study of mortality within a cohort of rubber workers, 1964–72. *Journal of occupational medicine*, **16**: 458–464 (1974).
14. **Ahlman, K. et al.** [Working conditions and health of granite workers.] *Kansaneläkelaitoksen julkaisuja*, **4**: 128 (1975) (Finnish with English summary).
15. **Ahlman, K. et al.** [The health and working conditions of dock workers.] *Työterveyslaitoksen tutkimuksia*, **105**: 64 (1975) (Finnish with English summary).
16. **Ott, M. G.** Determinants of mortality in an industrial population. *Journal of occupational medicine*, **18**: 171–177 (1976).

15

Experimental epidemiology

P. Lazar[a]

Epidemiology is primarily an observational science. However, the methods that have been described in the earlier chapters of this manual are mostly aimed at a better understanding of the possible relationships between risk factors and the incidence of a disease or of a group of diseases, i.e., they are concerned with the establishment of causal inferences. It must be recognized, however, that there is a certain fundamental contradiction between this goal and the constraints inherent in the available methods. It is only by careful study design (choice of the controls and of the confounders, stratification, matching, etc.) or by sophisticated statistical analysis of the data (e.g., multivariate analysis, checking for consistency between the hypotheses, with all their consequences, and the collected observations) that this major difficulty can be partly overcome. Consequently, experimental data are highly desirable, if they can be obtained. But, as it is obviously impossible to introduce new risk factors deliberately, it is immediately apparent that actual opportunities for adopting a true experimental approach occur only in the case of a controlled *decrease* in exposure.

The problem of experimental epidemiology appears then to be closely related to the primary prevention of the incidence of diseases. Prevention can indeed be regarded as a problem in itself, without any direct reference to epidemiological research, since in many cases the setting up of a prevention programme is more a question of applying present knowledge than of obtaining new information. However, it can be argued that, since so few opportunities do exist for obtaining a genuine proof of the etiological (direct or interactive) role of a factor on the incidence of a disease, it would be a pity not to use the opportunity offered by the implementation of a preventive programme to achieve a more conclusive demonstration of a cause-and-effect relationship. Moreover, and this is still more convincing, it might be quite rewarding not only to try to reduce occupational hazards, but also to assess the actual impact, efficiency, and other implications of the proposed

---

[a] Institut national de la Santé et de la Recherche médicale, Paris, France.

prevention programme in itself. The best way to achieve such a goal is obviously to design the operation as an experimental study. To what extent is this possible? Which are the specific constraints related to occupational health? How should such a design be conceived and managed? These are some of the basic questions that will be treated in this chapter.

Biological experiments

It is not necessary to discuss in great detail the various kinds of relevant biological experiments, the different conceptual models, and the results that they yield. Indeed, it does not seem desirable to devote too much space to such a field of investigation under the heading of "experimental epidemiology", despite the word "experimental", since epidemiological investigations concern human diseases specifically and not all the sources of information that can be used to improve our knowledge of them.

In relation to occupational health, biological experiments can be for three purposes: (a) to detect possible adverse effects of exposure factors (chemicals, dusts, radiations, etc.); (b) to measure dose–effect relationships (effects of high and low doses, existence of a threshold, role of the rate at which the dose is received); and (c) to simulate some more complex situations (joint action, synergism, or antagonism, etc.). Thus, they can provide information on potential hazards before any epidemiological evidence of an adverse effect, or they can confirm suspicions coming from epidemiological data, or they can help in defining "safer" levels of exposure.

At this point, the question that is usually raised is the relevance of animal or cell culture experiments to the problem of human sensitivity to adverse effects of environmental hazards, in other words, the question of the validity of any extrapolation from *in vivo* or *in vitro* experiments to man. In some respects, however, this question is not as important as it might appear, since it is quite clear that almost all human decisions, at any level, are made in a state of partial and controversial knowledge only. The more consistent the data from epidemiological and experimental studies, the higher the probability that some relevant decision may be taken in order to reduce the risks. In the field of cancer research, for instance, it is of interest to note that the adverse properties of some specific human carcinogens were first discovered from animal experiments (1).

Biological experiments also provide basic models and technical guidelines for the setting up of experiments in humans, the main difference obviously lying in the ethical aspects. The most decisive contribution was undoubtedly the introduction of the concept and techniques of *randomization*, a basic tool for experimental design, first introduced by Sir Ronald Fisher in agricultural studies (2) and then extended to all situations where interindividual variability of the "experimental units" is the rule. It is known that when the effect of a "treatment", T, is to be compared with that of a control intervention, C—or when two treatments A and B are to be compared—in a situation where there is intrinsic variability between the responses of the available experimental units (plots in agriculture, animals in biological experiments, patients in therapeutic research) the best solution is to allocate the treatments "at random" between the units. Such a procedure gives the assurance of providing, on the average, comparable groups for *any* factor susceptible to influence the observed results. It is often possible to take into account in the design or in the analysis of the experiment, factors that are already known to play a role. However, the main value of randomization is to ensure equalization, not only for known factors but also for all those that are still unknown, and then to permit causal inferences even when most parameters of the problem remain obscure.[a] Randomization is the only method that can be considered to remove bias from the comparison of two groups when the determinants of the response parameter are only partially known and/or measured.

Randomization is nowadays a crucial concept. There is no doubt that, when it is possible to achieve randomization, it is the best way of approaching the primary goal, that of establishing a causal relation. When it is not possible, for any reason, the more "randomlike" the design the better it is, but it can never reach the same level of confidence as a randomized one.

Design of the study

Explanatory versus pragmatic approaches

From the preceding remarks, it follows that the most usual epidemiological situation in which an experimental approach is conceivable is within a prevention programme. Two conditions need to be fulfilled: (a) the existence of at least some knowledge about the adverse effects of certain occupational hazards, and (b) the existence of a strategy to decrease or to suppress such effects, either by reducing the hazards themselves or by protecting against their action. The main goal is obviously to increase the efficiency of means of reducing damage to health, in a more or less specific situation; at the same time, it is also of interest to obtain more information about the disease process, the precise role attributable to the suspected causes, and the effect of a controlled reduction of exposure on the evolution of the incidence of the disease in the observed population or on the progress of the disease among already affected individuals, if they go on working after their health has been impaired. In brief, two major objectives can be distinguished: to reduce some specified adverse effects, and to increase general knowledge.

Such a situation is quite similar to the one encountered in the design of controlled trials of preventive measures in medical research, and the possible solutions can take their inspiration from those used in this field. Most of the steps in the elaboration of the study can follow the usual

---

[a] Other important features of experimental design can easily be found in any basic textbook, such as those mentioned in the references.

rules, even if major departures from these rules are necessary at some stages, owing to the specific constraints of occupational health.

A major contribution to the design of controlled clinical trials has been made by Schwartz & Lellouch (3, 4) and their arguments are quite relevant to the problem under discussion. These authors established a clear distinction between two kinds of controlled trial, those they call *explanatory* and those they call *pragmatic*. In the explanatory kind, the main aim is to increase therapeutic knowledge by means of experimental data derived from observations in man, even if the patients and conditions in which this knowledge is established are highly selected, thus reducing the immediate practical implications of the results. In pragmatic trials, the main goal is, conversely, to formulate practical rules suitable for use in the near future, even if the value of this is offset by some deficiencies in the biological interpretation of the data. Schwartz & Lellouch show that the choice of either of these two opposite extremes introduces important differences in the design, conduct, and analysis of the trial.

For instance, in an explanatory trial, in order to avoid any misinterpretation of the data, the highest priority is to be sure that subjects given treatment T (or C) do, in fact, take it and do not withdraw before the end of the trial. This has the disadvantage, however, that it creates a certain selection bias. For example, in those countries where an informed consent form has to be signed by the patient before receiving a treatment under experimentation, this form has necessarily to be presented at the beginning of the trial, before randomization can be effected. In a pragmatic trial, however, acceptance of the treatment can be considered as part of the question being asked and can be "confounded" with the main question of the existence of any difference between T and C. Similarly, in an explanatory trial, it is recommended that all conditions—except for treatment T or C—should be kept the same in the two groups. In a pragmatic trial, this is by no means imperative, since the aim is to compare two approaches for possible application in the future, which does not necessarily imply such a constraint. As far as the final analysis of the trial is concerned, in an explanatory design one is usually trying to answer some specific question about the disease and its evolution under treatment, whereas in a pragmatic trial, it is more a question of establishing an overall balance between the advantages and disadvantages of the two procedures being compared, in order to be able to make a choice between them. It can be shown that this distinction can have important implications for the statistical analysis itself (4).

From this brief comparison of the two alternative approaches, it can be deduced that the pragmatic trial is probably the most commonly applicable to occupational health problems, since the practical constraints do not leave much room for a more academic approach. Careful consideration should be given, however, to the advantages and disadvantages of each experimental design in relation to a given problem before any final choice is made.

Randomization process

The main difficulty in setting up the study seems to be the constitution of comparable groups using standard randomization processes. In the related field of medical research this is never an easy task, since it raises multiple problems: technical, practical and, obviously, ethical. However, hundreds of randomized trials have been carried out, without any major obstacle. In the field of occupational health, the situation still seems to be quite different (5) and it may be asked why this should be so.

If one tries to imagine the practical difficulties, the main one is likely to be that it is almost impossible to constitute two different groups within the same workplace. This was the case, for instance, in a study by Gosztonyi (6) of the efficiency of devices for protecting hearing when it was not possible, for obvious ethical reasons, to divide a group of workers exposed to hazardous levels of noise into a protected and an unprotected control group. In some cases, such ethical problems do not exist and the two "treatments" to be compared appear quite similar ("blind" trial). This would be the case, for instance, if one wished to compare two apparently similar hand-protection creams. Even in such a case, however, is it likely that people working in the same place would easily accept being treated differently? This is not impossible, but requires a high level of social confidence. Where the "treatments" are dissimilar, it might be quite difficult to maintain a real difference between the two groups if some "contamination" is possible, that is if workers belonging to one group could spontaneously adopt, at least partly, the "treatment" of the other group. This would especially be a problem if the aim of the study was to introduce, as part of the protective process, an attempt to improve the information of workers on the risks and how to avoid them. A more fundamental problem arises if, as is quite conceivable, the target of protective action is located not at the individual level but at the level of a group (workroom or factory, for instance), either because it requires some technical device that cannot be introduced at the individual level—as in the intervention study on the steaming of cotton to prevent byssinosis by Merchant et al. (7)—or because it has no real meaning at this level (modification of work organization, for instance).

In brief, for various reasons, some of which have been explained above, there is a high *a priori* probability that an experimental study in occupational health must necessarily located at a *group* level. This constraint introduces various technical difficulties.

In theory, it is quite possible–and easy–to randomize at the level of the group rather than the individual. In practice, one of the main difficulties is to find enough "units"—teams, workrooms, factories, etc.—susceptible of being compared, since the statistical comparison ought to be done at the same level as the randomization. Particular emphasis needs to be placed on this remark since the constraints it refers to are very strong. Another difficulty, at the level of the statistical analysis of the data, lies in the discrepancies that might exist between the sizes of the various units, a usual situation in the field of occupational health.

A technically similar but less complex problem has been extensively studied in animal experiments where, for various reasons, the animals have to be kept not in individual cages but in collective ones, in which each subject receives the same treatment. It is known that, if a cage factor does exist, the statistical analysis of the results of different treatments must take account of the intercage variations. Obviously, the larger the number of animals per cage, the better the accuracy of each cage estimate of the response to treatment. But the main source of variance, the one from which the intertreatment comparison should be made, now includes the (confounding) cage factor. This means that the actual number of degrees of freedom of the analysis is determined by the number of cages, not by the number of mice or rats, and this can imply a considerable reduction of power (9).

The basic difficulties mentioned above suffice to explain why so few genuine experimental designs have ever been set up to demonstrate the efficiency of a preventive action and why "quasi-experimental" designs, consisting for instance in "measuring" the effect of an intervention by comparing the situation before and after it, have usually been preferred until now (5).

Randomized versus quasi-experimental designs

The principles and techniques of quasi-experimentation have been extensively discussed by Cook & Campbell (9). They define quasi-experiments as "experiments that have treatments, outcome measures, and experimental units, but do not use random assignment to create comparisons from which treatment-caused change is inferred". It is clear, from the principles established by Fisher, that such a design is defective with respect to giving a correct causal interpretation of the data. However, it would be unwise to reject such "experiments" since most present knowledge about the efficiency of interventions is based on them.

In order to be in better position to judge what is lost in the absence of randomization, it is helpful to examine how randomization itself, even when it has apparently been performed, can fail to overcome some of the basic difficulties of the problem. Two critical questions have to be answered: what has been demonstrated by the experiment, and to what extent can these results be directly applied to other "similar" situations?

Randomization, if it is performed using correct techniques for allocation of the treatments and if it is not rendered invalid by some major bias introduced later, either during surveillance of the experimental units or examination of the results, ensures that the differences that appear between the treated groups are causally related to the actual differences between the treatments themselves (subject to the unavoidable statistical risk, usually fixed by the level of significance proper to the main comparison of the study). The exact meaning of this assessment remains somewhat questionable, since in practice it is very hard to be quite sure that no parasitic factor has been unintentionally associated with only one of the treatments, which could by itself explain the final difference. Incidentally, this is one more reason to doubt that, except in very specific situations, it is possible to achieve genuine *explanatory* trials in occupational epidemiology using groups of people as experimental units: control of all the possible confounders seems to be usually unattainable. But this restrictive statement does not mean, in any sense, that it is not possible to compare, using a much more *pragmatic* approach, two sets of preventive measures distributed at random between two sets of experimental units. Formulating the problem in this way reduces the importance of the question concerning the interpretation of the difference between treatments, since no specific explanation is now looked for, but it leaves unsolved and even reinforces the question of the wider application of the results.

Theoretically, indeed, such an application would require a high degree of similarity between the conditions of the experiment and the conditions under which the results will be applied. However, it is clearly begging the question to suppose that, even if some precautions are taken to avoid too high a selection of the units sampled for the controlled experiment, the units to which the results will be applied will be strictly comparable with the initial sample. The only general statement that can be made is that the more diversified the units used for the experiment— that is, the less selective for the various confounders that can interfere with the results—the higher the probability that the conclusions will be applicable to a variety of situations, whether these arise later in the same factories or in other factories.

In practice, it is very likely that intervention studies will often be based on before–after comparisons. In order to avoid the possibility of a systematic concurrence of the intervention with some other temporal variation in the occupational or general environment, the time of the intervention can be randomized among units; when feasible, such a procedure could be considered as a way of diversifying the units studied in order to increase the possibility of extending the results to other circumstances. An example of the necessity of repeating before–after comparisons in order to draw a sound causal inference is provided by a study by Imbus & Suh (10). These authors compared pulmonary function tests on workers exposed to cotton dust before and after steaming was interrupted owing to the installation of new equipment in the plant concerned. They observed a larger drop in forced expiratory volume (FEV) over the six hours of the work shift during the unsteamed period as compared with the steamed period among the same workers. They noted, however, that the baseline FEV also varied between the two periods. They finally concluded that other studies of this type should be made to rule out the possible interference of some concurrent variations in risk factors unrelated to the steaming of cotton.

Finally, whether or not randomization is performed, common sense should not be forgotten at each step in the preparation, the conduct and the analysis of an experiment. A randomization process, if correctly applied in spite of the difficulties, ensures to some extent that the observed results do correspond to the controlled treatments, and is therefore highly to be recommended, even if it does not constitute an insurance against all risks. Conversely, a careful quasi-experiment, even if fundamentally less convincing *a priori*, is not to be rejected systematically, especially if it is the only realistic way in which an experimental design can be achieved; indeed, most epidemiological knowledge does result from purely observational surveys.

Standardization of methods

As previously described, the most common "natural" unit on which observations can be made in experimental epidemiology applied to occupational health is a group of people, working together in the same area. This does not mean, obviously, that individual measurements are not made, but that the intervention and measure of the intervention can be better conceived and applied at the level of the group. This raises problems of measurement methods and, moreover, of standardization of methods, in order to increase the comparability between periods for the same group or between groups at a given time.

It is well known that in any medical problem, major inter-individual sources of variation exist. However, even if appreciably different from one subject to another, an exposure factor (e.g., smoking) or a specified disease (e.g., lung cancer) can often be described with the required level of accuracy and standardization: the mode of questioning about the number of cigarettes smoked, or the manner of smoking, and the clinical and histopathological classifications of the tumors can be specified in a way that ensures a high degree of uniformity throughout the study. Obviously, such precautions can also be taken when a set of people are to be examined, if the exposure and the effects measured for the group can be considered as a sum of the individual ones. But this is sometimes questionable.

For instance, the "level of exposure of the group" to a suspected occupational hazard might be more logically described with reference to the workroom than at the individual level. This opens the way to "objective" measurements of environmental exposures at work, if technically possible, but immediately raises the question of accuracy, relevance, and reproducibility of such information. As far as "effects" are concerned, complex problems can result from the possible interindividual interactions within each group. This is especially true, for example, when some subjective factors can interfere with the target health parameters or when education is a part of the prevention process. All this means that the problem of choice of methods of measurement should be raised directly at the level they will be used (in this case, usually group level) and that the problem of possible confounders should be treated at a higher level of complexity: the community level.

290

Special references should also be made to one final problem. This, again, it is not specific to occupational experimental epidemiology, but it takes on a crucial importance in this field: it is the question of stability of the populations exposed and of exposures with time, since it is well known that many causes of instability do exisit for both these variables. At the population level, selection factors can intervene, without any guarantee that no differential behaviours will affect the different groups compared. Similarly, the question of retirement—and particularly of early retirement during the period of the trial—should be looked at carefully, and health parameters of retired people measured, since, especially with chronic diseases, in a significant number of cases it is to be expected that the onset of the disease will not occur until after the cessation of activity. At the exposure level, technical and economic conditions can exert a strong modifying influence on some work parameters during the course of the study itself, with a high probability that such modifications will interfere with the problem being examined.

Finally the acuteness of the difficulties evoked here raises the question of the expected length of a study; it may be wondered to what extent it would be realistic to start experimental studies that are supposed to last one or two decades. No doubt a better way is to try to organize shorter, multicentre studies, which would also increase the full implications of the results and decrease the probability that the speed of development of new techniques would deprive the studies of any significance even before their results are known. This problem is also closely related to the question of the statistical power of the study, which will be discussed in the paragraph concerning the evaluation of the results.

Implementation of the study

As is the current practice in controlled therapeutic trials, it is wise to start an experimental programme in occupational health by writing a detailed protocol of the planned study. This should include the following ten headings:

(1) Present knowledge concerning the question(s) raised.

(2) Specific questions to be answered by priority; main orientation of the study (explanatory versus pragmatic).

(3) Choice of populations and units (individuals or groups) for inclusion in the study.

(4) Precise definition of the occupational factors and cofactors to be measured or estimated: specific work loads or hazards; environmental measurements, etc.; level of measurement (individual or collective).

(5) Precise definition of the health parameters to be taken into account; methods of collecting this information during and after the period of activity.

(6) Strategy concerning the follow-up of work conditions and the people under observation.

(7) (Crucial) choice of the type of study, i.e., randomized or quasi-experimental.

(8) Expected duration of the study; expected power, possibilities for multicentre studies.

(9) Mode of analysis of the results.

(10) Expected type of distribution and extrapolation of the results.

Beyond these technical aspects, it is obvious that an experimental epidemiological study raises multiple problems, including the basic ones of ethics and social consent, which will be discussed later (see p. 294). Furthermore, it seems unrealistic to imagine that such a project could be carried out successfully unless supervised by a special team, responsible for ensuring that the protocol is respected at all stages of the study and able to introduce, without major disturbances, any unavoidable modifications imposed by unforeseen difficulties.

Analysis of the results

Statistical significance of the differences

At the end of the study—or at an intermediate stage—two (or less often several) groups have to be compared in order to determine the extent to which the observed results confirm or refute the hypothesis being tested. General rules of statistical analysis provide the necessary tools and it is not the intention to review them in detail here. The main aim of this section is to give warnings of some pitfalls in order to avoid major errors.

Obviously, statistical analysis is conditioned by the way the study has been structured. In the present case, the main constraint will often be that the "experimental unit", as already noted, is not an individual but a group: team, workroom, factory, etc. Since it is a general rule that statistical comparisons should be made at the level of experimental units (8), this clearly means that the number of degrees of freedom on which the main comparisons are to be made is determined by the number of groups and not by the total number of individuals within the groups. This requirement undoubtedly creates a severe problem with regard to the "power" of the study, i.e., its ability to detect differences, and leads to a consideration of possible ways of increasing it.

First, it is obviously desirable to increase, as far as possible, the number of experimental units for a given total number of individuals entering the study. This means that the smaller the size of the unit that can be subjected to the various treatments to be compared, the better. From this point of view, randomization at the level of each individual, when possible, is the optimum solution (but, as has been seen, this is not possible as a rule). Thus, an effort should be made, when preparing the study, to optimize the size of the working unit taking all the constraints into account, including the one referred to here.

The second way of increasing the power of an experiment for a given number of experimental units is to describe the results in a manner as appropriate as possible to the aim of the study, i.e., to show a significant difference attributable to the difference in treatments. This means that an effort should be made to improve the analysis by a choice of relevant parameters to express such a difference and by taking prognostic variables into account. The former goal can often be achieved by measuring variations in state rather than final states, that is by taking measurements before and after the intervention has taken place. As far as prognostic variables are concerned, the only general rule that can be stated is that it is worth taking them into account if, and only if, they are *strongly* related to the expected results (8).

Using measures relating the final state to the initial one (either differences or ratios) is a particular application of the rule concerning prognostic variables. This means that is it worth doing it if final values do depend closely on initial ones. Another point to consider is that before-after differences can be misleading if interpreted too hastily: it is important to be on one's guard against taking the well known statistical phenomenon of "regression towards the mean" as a proof of the efficacity of the intervention. The basic argument of this common statistical trap is the following. If one selects in a distribution the subjects with the highest values and if one later takes a second measurement on the same subjects on the assumption that there has been no change in the overall distribution, these second measurements will be, on the average, lower than the first ones, unless there is complete correlation between the two measurements. Symmetrically, the subjects who at first had the lowest values would later give higher ones, on the average. Obviously such a phenomenon, even if quantitatively not negligible, cannot introduce any bias if the experiment includes comparable groups. But it does draw attention to the risk of assuming, as is sometimes done, that before-after variations can be taken as an absolute criterion of efficacity. More generally speaking, it is clear that many factors can change between the beginning and the end of a period of intervention, and that a change in the measurements cannot be directly interpreted as resulting from the intervention itself.

Causal interpretation of the results
Taking prognostic variables into account is a classical way to improve the power of an experiment, since this means that the differences between the various "treatments" compared are looked for within more homogeneous subgroups (8). There is no way, however, in which such a process can completely "equalize" groups that would not have strong *a priori* reasons to be comparable since, by definition, it can take account only of known prognostic variables, not of all those that generate the "unexplained" part of the variance of the results. Hence, it is again evident that it is quite illusory to imagine that it is possible to replace randomization by "perfect" stratification.

Evaluation of the results
Basically, the randomization process (or, in its absence, the alternative design that can be set up) provides comparable (or "quasi-comparable")

groups. But it is clear that the interpretation of the results cannot be reduced to the demonstration of a formal statistical "significance" of the differences measured between the various groups compared. Generally speaking, a comparative experiment, particularly if the adopted design is of the pragmatic kind, requires taking a set of criteria as a whole into account. This can be achieved in various ways, the simplest one being, when all the relevant parameters are available, to combine them into global score (4). More generally, the various measured positive and, if any, negative effects of the preventive process studied are included in a larger set of "softer" data (psychological and sociological acceptance of the solutions, economic and financial costs, political implications, etc.), thus rendering any attempt to arrive at a definitive conclusion at the end of the experiment somewhat artificial. It is already an outstanding achievement if it is possible to reach clear and extrapolatable conclusions with respect to some major health dimensions of the problem. The role of experimental epidemiology loses its importance at this point, giving way to sociopolitical decision-making.

Ethical problems

These last considerations raise the problem already hinted at of the ethical justification for the study. As is known from many experimental fields involving the participation of human beings, it is not possible to devise universal rules for deciding whether a given experiment is ethical or unethical. The only rule about which there is unanimity is that it is indispensable to consider this problem every time an experiment is planned. The answer obviously depends on how much is known about proposed alternatives, the level of social consent to the idea of comparing various solutions in human subjects, and the scientific quality of the planned design.

To enlarge the problem a little, ethical considerations have to be extended to the question of confidentiality, which is also extensively treated in relation to therapeutic trials and, more generally, to epidemiological surveys. There is an apparent conflict between the right of any individual to a legitimate preservation of privacy and the absolute necessity of obtaining records susceptible to statistical treatment. This conflict can be solved by the establishment of precise conventions of access to the data. In the case of experimental epidemiology applied in the occupational field, it is clear that such rules are even more necessary, since it is imperative that there should never be any risk of data recorded in an experiment being used against any worker. In a somewhat analogous way, one should also try to establish conventions concerning the "protection" of those firms that accept to participate in a cooperative project: this seems to be the only way of convincing them to collaborate without reservation in such experiments.

The best way of solving all these problems is certainly to try and bring together all the "social partners" in the team responsible for the conception and implementation of the project.

294

Examples of epidemiological experiments

The difficulties mentioned above are such that it is not surprising that the number of published epidemiological experiments in occupational health is not very high. In some fields, however, it is quite conceivable that such trials could be undertaken, particularly if their results could be known quickly. It is certainly rather unrealistic to start a controlled experiment on a disease that takes one or several decades to become manifest – except, naturally, if the treatment considered could also influence the speed for clinical onset of the disease (e.g., development of a cancer). Conversely, short-term effects are particularly well suited to an experimental approach, for at least two reasons: first, the answer can be known reasonably fast and, second, the expected frequency of withdrawals from the study is obviously smaller when the time of observation is relatively short.

A project for reducing the rate of preterm deliveries among pregnant working women

For all the preceding reasons, pregnancy (a condition lasting less than one year) can be considered a particularly favourable medical entity for investigation. As far as women's work during pregnancy is concerned, it is known from several studies that if working women have slightly fewer preterm deliveries than women staying at home (*11*), the probability of a premature baby is directly related to the number of hours of work (per day or per week) and to the strenuousness of the work. It is now possible, from accumulated epidemiological knowledge, to describe the risks quantitatively for various occupational conditions: they can range from less than 3 % to more than 20 %. Nevertheless, there is no easy solution to the problem, since any modification of women's work has psychological, sociological, and economic consequences. For instance, it is clear that pregnant women do not usually like to change their occupation for a lighter one, since such a change is not "free" (sometimes in financial terms, but also in terms of adaptation to work or of perceived risk of future unemployment). At the sociological level, any specific protection for women is seen as disrimination and thus as a backward step in women's emancipation. At the economic level, it can be argued that all protective measures increase production costs.

However, all these negative arguments do not necessarily imply that nothing can be done, here and now, to improve the conditions of pregnancy, especially among the less favoured women, and this idea is gaining more and more acceptance among both women and employers. The situation now seems ripe, therefore, for organizing a multicentre trial, the structure of which might be, in each centre:

(1) to measure the present rate of preterm deliveries (from the medical records or by means of a retrospective survey) according to the type of occupation;

(2) to define a specific strategy for preventing preterm deliveries among pregnant women, depending on the technical possiblities of the

firm: changes in occupation, progressive reduction of work duration during pregnancy, earlier prenatal cessation of work, etc.;

(3) to measure the rate of preterm deliveries, and more generally, obstetrical risks among those women pregnant during the period of the trial;

(4) to try to evaluate the other factors liable to influence the prematurity problem (at the psychosociological or economic level)

The main difficulty of such a study is, obviously, as mentioned earlier, the problem of controls: which groups should be used as controls, how should they be identified, and should randomization be used? The answer depends on national conditions. Two possibilities should be explored. One would be to refrain from any attempt to make an absolute judgement and to concentrate on the comparison of various solutions for reducing preterm delivery rates. The second one would be to accept the need for a short delay in some (randomized ?) firms in introducing preventive measures, the argument in favour of this solution being that it might be worth getting some practical experience of what can be done in some places before extending it to the whole set of participating centres. If neither of these two solutions is possible or acceptable, it is still necessary to ask whether a preventive programme should not, in any case, be organized, since a reduction in preterm deliveries can be considered as a socially and ethically imperative duty. However, the study would then become a simple before-after comparison without any control, probably one of the less satisfying types of quasi-experiment.

A multifactorial preventive programme in a working population

As has been pointed out by Wilhelmsen (*12*): "when living habits should be changed, it is well appreciated that the individual habits are strongly influenced by the surrounding community. Individual allocation to actively treated and control groups may be difficult to use in such circumstances. This problem has been elegantly solved in the European Collaborative Trial by randomizing factories instead of individuals." Indeed, occupational situations do offer attractive opportunities for organizing multifactorial prevention campaigns since they make it possible to reach healthy, relatively stable, and willing-to-participate populations. Even if such campaigns cannot be considered as pure examples of experimental epidemiology in the occupational field, since occupation is here more a convenient tool than an object of study in itself, they show many conceptual similarities with occupation-oriented experiments. Thus, in the multifactorial trial for the prevention of coronary heart disease conducted by a WHO European Collaborative Group (*13*), 88 factories located in five countries were randomized into 44 "intervention" units and 44 "control" ones, by matched pairs. The usual precautions of a randomized study were applied: examination of the workers of each factory pair "by the same team, within a short period of time", with methods "carefully standardized according to

WHO recommendations". Data collected on 63 732 men aged 40–59 years showed that the method of random allocation was "successful in producing two well balanced groups" despite the major differences existing between the factories (differences that, as noted above, are a favourable factor for a further application of the results of the trial).

The main difficulties of such a trial – beyond the problems of organizing an international collaboration on this scale – probably lie in the statistical analysis of the results, since the main (conservative) comparison has to be done on 88 units of unequal size, that is, with differences in the precision of each average observation. However, a simple non-parametric analysis (a sign test for comparing results in the matched pairs) gave promising information in a preliminary examination of the data. No doubt a successful final result would be a great encouragement for carrying out other occupational health trials in the future.

References

1. **Tomatis, L.** Contribution de la recherche cancérologique expérimentale à la prévention du cancer humain. In: Lazar, P., ed. *Pathologie industrielle: Approche epidémiologique.* Paris, Flammarion Médecine, 1979.
2. **Fisher, R. A.** *The design of experiments.* Edinburgh, Oliver & Boyd, 1935.
3. **Schwartz, D. & Lellouch, J.** Explanatory and pragmatic attitudes in clinical trials. *Journal of chronic diseases,* **20**: 637–48 (1967).
4. **Schwartz, D. et al.** *L'essai thérapeutique chez l'homme.* Paris, Flammarion Médecine, 1973. (English translation: *Clinical trials.* New York, Academic Press, 1980).
5. **Liddell, F. D. K. & McDonald, J. C.** Survey design and analysis. In: McDonald, J. C., ed. *Recent advances in occupational health.* Edinburgh & London, Churchill Livingstone, 1981, pp. 95–105.
6. **Gosztonyi, R. E.** The effectiveness of hearing protective devices. *Journal of occupational medicine,* **17**: 569–580 (1975).
7. **Merchant J. A. et al.** Intervention studies of cotton steaming to reduce biological effects of cotton dust. *British journal of industrial medicine,* **31**: 261–274 (1974).
8. **Lellouch, J. & Lazar, P.** *Méthodes statistiques en expérimentation biologique.* Paris, Flammarion Médecine, 1974.
9. **Cook, T. D. & Campbell, D. T.** *Quasi-experimentation.* Chicago, Rand McNally College Publishing Company, 1979.
10. **Imbus, H. R. & Suh, M. W.** Steaming of cotton to prevent byssinosis – a plant study. *British journal of industrial medicine,* **31**: 209–219 (1974).
11. **Rumeau-Rouquette, O.** *Naître en France.* Paris, Editions INSERM, 1979.
12. **Wilhelmsen, L.** Commentary on the European collaborative trial on the prevention of coronary heart disease. *European heart journal,* **1**: 80 (1980).
13. **World Health Organization European Collaborative Group.** Multifactorial trial in the prevention of coronary heart disease: 1. Recruitment and initial findings. *European heart journal,* **1**: 73–79 (1980).

16

Accident epidemiology

J. Saari[a]

Occupational accidents still constitute a major health problem for the working population. It has been estimated that over 15 million occupational accidents occur the world over every day (*1*). In view of this statistical background, it is essential that all available information be utilized for accident prevention. Our knowledge about the etiological factors responsible for accidents and about how accidents can be avoided in practice must be greatly increased despite the vastness of this problem.

The present state of knowledge of occupational accidents is the result of a complex process of development. The next major step in this process is the perfecting of methods for the estimation and analysis of accident risk in advance (*2–4*). However, for a long time to come, the main source of improved knowledge will continue to be those accidents that have already occurred.

The role of epidemiology in accident prevention

Epidemiological studies present a natural and promising approach to the further elucidation of the accident phenomenon. Accidents resemble infectious diseases. The individual is diseased (= injured) soon after exposure to causal factors. When occupational accidents are compared with modern occupational diseases, the difference appears to be greater than it really is because the latency period between exposure and occupational disease is long, whereas injuries may result immediately from accidents. However, some determinants of accidents may also exist for long periods before an accident occurs. The problems of accidents and diseases are, in many respects, so similar that the same methods should yield useful results.

Epidemiology has been used in accident research since the Second World War. It is natural that it was persons with a medical education

---

[a] Institute of Occupational Health, Helsinki, Finland.

who introduced epidemiology into the field of accident prevention (5–8).

The application of epidemiology to accident prevention has three aims (9):

(1) description of the distribution and rate of accidents in human populations;

(2) identification of the etiological factors responsible for accidents;

(3) provision of the data essential for the planning, implementation, and evaluation of services for the avoidance of occupational accidents and for the correct assignment of priorities among those services.

These aims are similar to those of epidemiology in general (10).

The study of accidents is a multidisciplinary task (see Fig. 1). The information derived from epidemiological studies must be supplemented by data from other disciplines, such as technology, psychology, medicine, and sociology.

Epidemiology is primarily a methodological tool in accident research. For the most part, it is used for designing studies, analysing data, and combining data to give such characteristics as incidence and prevalence rates. The variables of the studies, which are generally determined by other disciplines, differ from case to case and depend on the objectives of study.

Criteria of safety performance

Criteria and their requirements

Generally speaking, the objective of accident epidemiology is to measure the safety of a system. As direct measurements are not possible, indirect indicators (i.e., accidents that correspond to diseases in epidemiology in general) must be used as criteria for the safety of the object of study. For reasons described later, criteria other than accidents are also used; examples are near-accidents, critical incidents, and errors. Such criteria can be considered as "attenuated" forms of the actual "disease-accident".

Epidemiology usually studies the diseases of human beings. In accident epidemiology the "diseased" objects are not necessarily human or even animate. Machines, specific areas of a workplace ("black spots") or entire companies may also be "diseased". The inclusion of inanimate "diseased" objects does not alter the methodological performance of a study, but better efficiency may result from the process of data collection.

Whatever criterion is used, it must fulfil the usual requirements of reliability and validity. Table 1 lists the most essential requirements for a criterion in accident epidemiology. The same requirements apply to the determinants (exposures) under study. The minimum requirement is that the validity and reliability of the criterion used are known, either from previous experience or from their determination with special techniques.

300

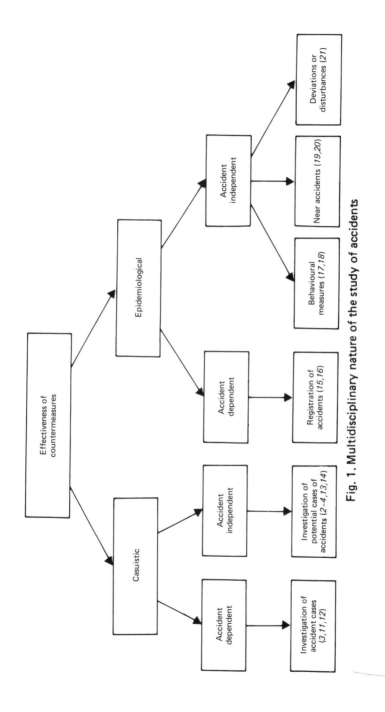

Fig. 1. Multidisciplinary nature of the study of accidents

Table 1. Characteristics of effective measures of safety performance

Validity
— produces information representative of safety
— validating a measure requires the use of an outside criterion

Reliability
— capable of duplication with the same results
— the changes observed should reflect actual changes in the criterion variable and not internal fluctuations in the measurement technique itself

Error-free results
— variable error (which tends to cancel out when readings are made an infinite number of times)
— constant error (not eliminated by replication); sources of constant error:
 (a) the investigator's failure to identify all the casual factors
 (b) the analyst's failure to break down the causal factors being evaluated to the point where they are relatively homogeneous
 (c) failure to consider all behaviour and conditions with the potential to produce future injury or loss
 (d) failure to report all accidents

Constant units of measurement throughout the range of evaluation
— measures preferably at least on an interval scale

Quantifiable measurement criteria capable of statistical analysis
— qualitative evaluation limits statistical inference and makes individual interpretation possible

Stability
— involves the maintenance of a given range of values when repeated measurements are made of workers' behaviour and environmental conditions

Sensitivity
— by detecting changes in process and performance, sensitivity serves as a criterion for evaluation

Administrative feasibility
— resources available
— practical answers in a very short period of time

Efficient and understandable
— the cost is in proportion to the benefit
— understood by persons responsible for approval and use

[a] After Tarrants (20).

Registered accidents

Registered accidents are normally a readily available source of information. They are usually defined as being registered injuries. Registration varies in different countries and companies.

Registered accidents are criteria that can be used for many purposes. The main advantage is that they are obtainable as a result of daily routines. Information on the number of the cases is more reliable than information on the descriptions of events. Therefore, when studies requiring information on accident types are being planned, it is desirable to consider establishing a special registration system for the particular purposes.

Another advantage of registration is that recorded accidents are real cases that, without any assumptions, have led to harmful and unwanted consequences.

For some studies, registered accidents are an inadequate measure because such reporting is related to injuries rather than to safety in a more general sense. For example, a falling object may cause an injury if it hits someone. But if the object does not hit anyone, the event is not reported, even if the interesting phenomenon itself—the fall of an object—has occurred in the same way in both cases and is undoubtedly associated with safety.

The predisposition to injuries depends on the individual and the part of the body in question. The same amount of energy may or may not cause an injury depending, for example, on the individual's age. Different tissues and parts of the body are vulnerable to different amounts of energy (22–24). If this is not taken into account, the measurement is neither valid nor reliable.

Registered accidents also have drawbacks as criteria. The extent of harm (usually measured as the duration of absence from work) is a more or less subjective decision made by the examining physician, the injured worker, and his dependants. Their decisions are affected by:

— prevailing opinions concerning the acceptability of accidents and the length of absence;
— the hindrance caused by the injury on the job and the possibilities of temporary replacement;
— the compensation schemes applied in each case.

Misreporting and lack of reporting are other drawbacks that sometimes cause the information to be completely invalid (25). According to Powell et al. (26) only 5–7 % of all accidents are reported at some workplaces. Some investigators feel that registered accidents measure the tendency of reporting rather than the degree of safety of the system being studied (27, 28). The following are some of the reasons for misreporting and lack of reporting:

— feelings of guilt and a sense of responsibility;
— medical services are not available;
— negative attitudes and prevailing opinions towards medical personnel;
— inefficiency of the reporting system (voluntary reporting is probably less efficient than compulsory reporting);
— an accident is a rapid and mentally shocking event, the recall of which may be distorted and partial or even a rationalization of what really happened;
— accidents that should be reported are poorly defined and described; thus the persons concerned do not realize the need for reporting.

Because there are so many interfering factors, the term "accident" has to be carefully defined for each study. This is even more important in studies using such criteria as near-accidents, critical incidents, disturbances, or deviations.

One way of overcoming some of the difficulties is to restrict the study to mortality rather than morbidity. However, some investigators think that conclusions drawn from severe accidents cannot be extended to minor accidents (27, 28, 30). But many research workers emphasize the importance of studying all accidents regardless of the duration of the disability (31).

An accident registration system should fulfil the following requirements (32).

1. The information contained in the reports of incidents should be:

(a) based only on genuine in-plant occurrences;

(b) based on events that are significant in terms of potential severity or frequency of occurrence;

(c) as free as possible from reporter bias;

(d) capable of easy extraction from reports;

(e) in a form that permits categorization without prior reinterpretation.

2. The information analysed and summarized from such reports should be in a form (or directly translatable into a form) useful for:

(a) direct accident prevention activities—specific or general;

(b) cost-benefit analysis of preventive strategies;

(c) meeting production, administrative, and legal requirements for specific information;

(d) providing planners, engineers, job designers, and consultants with data about situations, features of design, or activities with high accident potential;

(e) providing a factual basis for the planning of individual section, shift, department, or plant-wide safety programmes.

3. When the design of a system or any of its components is undertaken, it should be recognized that the system already in use, however deficient, has evolved to fit the dynamic functioning of the organization of the rest of the plant, and therefore may have much to offer. As stated earlier, informal methods of communication often develop at the local level to counterbalance any deficiencies in the formal systems.

4. Recognizing that the quality of information depends to a significant extent both on its value as perceived by those who produce it and on the value they themselves obtain from it, the system should incorporate useful feedback.

Specially reported accidents

Instead of routine registration systems, a special system of reporting can be established. Visits to hospitals, medical centres or physicians, police records, and the records of insurance companies have been used as data sources. A special system of reporting is the only alternative if people not covered by compulsory compensation systems are being studied. In many countries, examples of such people are to be found among the self-employed, e.g., shop owners, undertakers, and farmers.

The disadvantage of special systems of reporting is that the population in question is seldom known. Nor is it known how representatively the cases are registered with respect to number, severity, or type. Therefore, it is not often possible to calculate accurate incidence rates, let alone trends. However, a qualitative view of the problem can be obtained for planning more thorough studies.

One advantage of these systems is that reports can be specially planned for a particular study, thus avoiding irrelevant information. Another advantage is that the quality of the reports is often better than in routine systems of registration, provided that sufficient time is allocated for people to fill in the reports.

Special systems of reporting have been used most often for studying leisure-time and traffic accidents (*33, 34*), but the method is also worth considering for occupational accidents—especially if information about a large number of severe accidents has to be collected quickly.

Accident-like events

For many research purposes, accidents leading to injury are rare. One has to wait for months or years before a sufficient number of injuries has been collected if the study is restricted to one or a few medium-sized companies and if the retrospective material available is not reliable enough. This is one reason for using accident-like events as criteria for safety performance. Recently, it has become more popular to use them instead of accidents in studies. The main reasons are:

— material collection is quicker in terms of the total length of the study;
— because nothing harmful really happened, people report such events more frankly;
— persons involved may remain anonymous if necessary;
— persons involved stay at work and can be interviewed;
— data collection can be directed specifically to factors relevant to the study, and the amount of irrelevant data is reduced.

These measures suffer from severe drawbacks as well:

— the predictive value of the results of studies of accident-like events is controversial;
— a special data collection system always has to be established;
— the events may remain unreported if they are not recognized by all persons involved; an exact definition is compulsory;
— it may be difficult to motivate people involved in the study;
— long-term studies are not generally feasible;
— statistical data about the population base are seldom available.

The key to success lies in the definition. If the events under study cannot be described well to the participating population, the study will fail. Unfortunately, not all events can be described well, but as techniques continue to develop the future looks promising.

The most commonly used accident-like events are:

— near accidents (*19*)
— critical incidents (*20*)
— disturbances, including:
    (*a*) deviations from ordinary work routines (*21*, *35*);
    (*b*) unsafe conditions (*36*);
    (*c*) specific types of behaviour (*17*);
    (*d*) errors (*37*);
    (*e*) unsafe acts (*17*, *18*, *36*).

Data collection can be organized in several ways. The workers can be interviewed. The interviews should take place several times a week, preferably daily. This technique involves reporter bias, especially if the event studied has not been properly defined. Questionnaires have also been used instead of interviews.

Observation is another method (*17*, *18*, *26*) but it is rather laborious and costly for the investigator. The reliability of observations, however, is better than the reliability of interviews. Some studies have used safety officers or safety representatives as observers (*18*).

Etiological factors

A model of accident causation should be kept in mind during the selection of the accident determinants to be studied. Accidents are multicausal, with many interactions between causal factors. Confounding factors and effect-modifying factors must be analysed carefully. A model of accident causation (borrowed from previous publications or created specifically for the study) helps to avoid confusing results.

The determinants cannot always be measured directly. Variables indirectly related to safety may be a more natural or a more practical choice. For example, a worker's experience is fairly easy to determine; for this reason it has often been the subject of study. Experience is related to other determinants, such as knowledge of the work environment and the ability to perceive incipient hazards. As these variables are much more difficult to measure, experience may be a practical substitute for them. Without a proper model of accident causation, it is not possible to evaluate the results if such substitution has been used.

A number of accident models have been presented in the literature (*3*, *16*, *27*, *31*, *38–43*). As no single model can explain all accidents, a model must be selected on the basis of the objectives of the study.

Accident model

The structure of an accident

An accident is a process of parallel and consecutive events leading to a harmful consequence. The time sequence of this process can be divided into four parts (Fig. 2):

(1) the normal phase: the work process functions as planned and is under control;

(2) the preceding phase: control is lost during this phase;

(3) the contact phase: injuring factor (= harmful energy) begins affecting man;

(4) the injury phase: injury or other harm occurs; this phase includes the lessening or intensification of harm.

In an accident situation, injury is always inflicted by energy—physical or chemical. The injury comes about when the energy exceeds man's tolerance limit. Thus, the severity of injuries depends on:

(1) the amount of energy released and its spatial distribution;

(2) the tolerance limit of the person involved, which depends, among other things, on:

(a) the part of the human body afflicted by the energy; and

(b) individual characteristics.

A contact event precedes the injury. During this event, man and the injuring factor (= energy) come into physical contact, after which injuries are inflicted either immediately or some time later. Contact events are those typically expressed in statistics as the type of accident, e.g., struck by, caught in or between, stepping on, striking against, contact with extreme temperatures.

A number of antecedents precede the contact event. Most of the preceding events are abnormal, unwanted, or unplanned. Frequently their nature cannot be determined until afterwards. Fig. 3 is a schematic representation of an accident, illustrating the character of the preceding events.

Determinants (causal factors)

It is important to remember that the determinants causing events leading to injury are always present, whether or not an accident occurs. For example, a simple falling accident may contain the events depicted in Fig. 4. In such a case, the determinants (in other words, the causes) of stepping on a slippery spot on the floor might be:

— the person's poor eyesight;

— other simultaneous activities, e.g., thinking about private affairs or the next task;

— a false mental model of the environment because oil has never been spilled on the floor before;

— the oil cannot be differentiated from the rest of the field of vision owing to the colour or the surface structure of the floor;

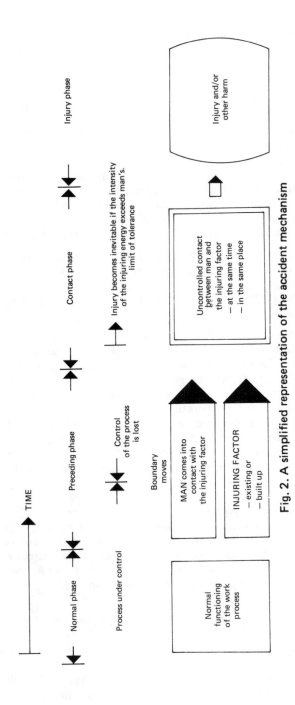

Fig. 2. A simplified representation of the accident mechanism

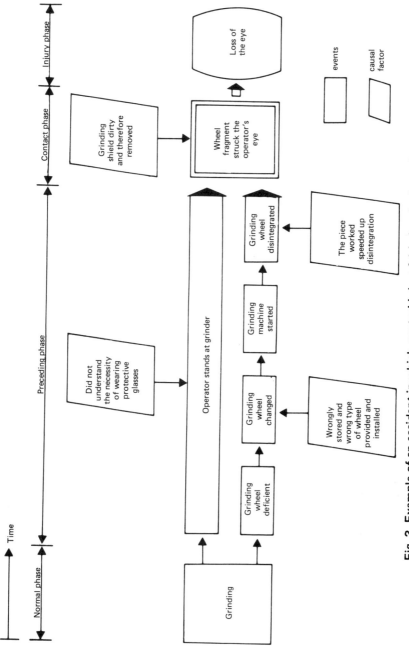

Fig. 3. Example of an accident in which control is lost fairly late in the accident process

309

310

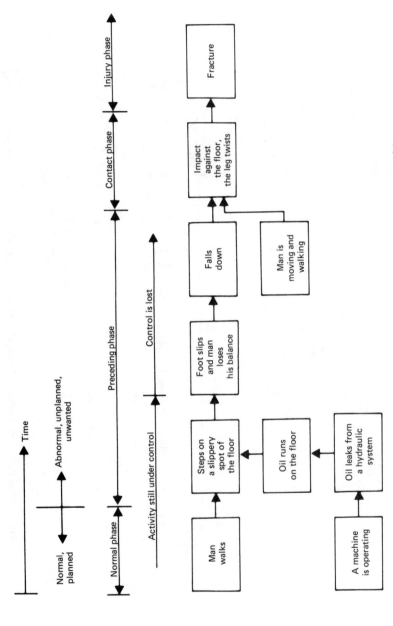

Fig. 4. Example of the events leading to a simple falling accident

— the oil cannot be seen owing to poor illumination;
— the person's perception is affected by alcohol, drugs, disease, etc.
— the person is in a hurry because of some urgent tasks;
— the person steps on the spot deliberately, e.g., he wants to test the frictional characteristics of his new safety shoes.

There are many such possibilities; only a few have been listed here. The example is an attempt to illustrate the multifactorial and multicausal nature of accidents. Accident epidemiology must carefully examine the effect modifiers and confounders. For example, the relationship (occurrence function) between walking and accidents involving falling is influenced in a complicated manner by many factors, including the surface characteristics of floors, the types of task, individual differences in perception and behaviour, the instructions given, the technical apparatus used, and environmental factors.

In cases like this, it would also be preferable to seek an occurrence function relating falling to its determinants (causes) rather than one relating injuries to their determinants, as the severity of injuries is affected by the person's age and tolerance to impact forces. One study concerning fatal accidents due to falling reported that all the people who died were more than 40 years old. Yet one cannot assume that younger people do not fall at all; in fact, they fall almost as frequently as older people. The difference is that younger people are not fatally injured. However, from the point of view of data collection, it is more practical to study injuries (i.e., the "disease") and not merely falling itself, with no concern for its consequences. Which of the two is actually selected depends on the objectives of the study.

In the past it was often thought that accidents have fairly stable determinants (38). Accidents were thought to have a common universal set of determinants. It must be emphasized, however, that accidents are not a uniform set of occurrences. They vary widely with respect to internal structure and especially with regard to causal factors.

Current thinking is that causal factors are not stable in the sense that the strength of the association between a determinant and an accident remains constant if the characteristics of the determinant are unchanged. A classic example in accident research is the concept of accident proneness.

It was believed that each person has an individual stable tendency to become involved in accidents. This tendency might perhaps be dependent upon an inherited set of characteristics that cannot be substantially altered during the person's lifetime. A tempting objective for a study would then be to find an occurrence function relating accident proneness to accidents. The amount of harm to be expected could then be estimated by measuring the individual's accident proneness.

Despite intensive studies, no occurrence function could be determined. Even if accident proneness existed, an issue that was hotly debated in the literature (*27, 29, 41, 43–48*), the function could be determined only for specified environments and tasks.

According to current thinking, man, technical apparatus, and the environment form a system. As Surry (*43*) expresses it: "man and his environment are inseparable. His behavior is a result of past and present interaction with his surroundings. It is for this reason that the study of 'human factors' in accidents cannot divorce itself from these interactions and must constantly refer to them".

Thus, if a person is accident prone the association between proneness and accidents depends on other characteristics of the work system. A strongly accident prone individual might be saved from accidents because other parts of the system adjust themselves, whereas an individual with low proneness might sustain several accidents because other parts adjust themselves differently.

In addition to man, machines, and environment, a fourth influencing factor is the way the system is operated. An example borrowed from computer technology will serve as an illustration. A computer, its operator, and the surroundings (the environment) form a system. This system can be put into operation in many different ways, depending simply on the program fed into the computer. Calculations may be done faster or more slowly, the risk of errors may be higher or lower, the operator may or may not be given tasks; all of this depends on the way the system is put to work, in other words, on the program.

The same applies to all man-machine-environment systems. The tasks assigned to the system (rules, instructions, and other management factors) finally determine the extent to which various properties of the system are utilized and stressed. The previous example of accident proneness is seen in a new light when these factors are taken into consideration. For example, some action of the management might prevent accident-prone employees from being exposed to conditions in which accidents could occur. In such cases, no association between proneness and accidents would be revealed, although accident proneness did, in fact, exist. This demonstrates that serious attention must be paid to possible confounding or effect-modifying factors.

Thinking in terms of systems has another essential implication for the formulation of the determinants to be studied: determinants should be formulated in relation to other parts of the system. A study is usually condemned to fail if the determinants are measured in absolute terms. For example, associations between man's visual perception and accidents might be an object of study. If visual acuity were measured independently on an absolute scale, failure might be expected. Acuity should be related to something else, such as the individual's visual acuity required on the job, in relation to the size of the smallest object handled on the job, and so on.

The lively debate concerning accident proneness in earlier decades was due partly to insufficient understanding of the interrelation between systems and accidents. Later, a new concept, accident liability, replaced accident proneness. Accident liability was defined as an individual's tendency to sustain accidents when environmental factors were taken into consideration. Thus, an absolute measurement was replaced by a relative measurement. This new concept was also related to the time factor, and more importance was given to the individual's past history.

Study designs

Although analytical epidemiological methods have already been described in Chapters 9 and 10, it is necessary to consider here a few special aspects related to accident epidemiology.

Case–referent studies

Case–referent studies provide a rather efficient design for accident epidemiology from the point of view of cost benefit. The basic problem is the selection of referents. (The selection of cases, which can also be a problem, was discussed previously.) In accident epidemiology, it is difficult to determine whether or not referents are "accident-free". The difference between injured persons and referents might be stochastic in nature.

For example, if the exposure to be studied is the slipperiness of walking surfaces, it is difficult to define clearly the event that constitutes the "accident". A loss of balance may occur to different degrees without an actual fall. At least three different events could be considered as cases: a loss of balance without falling; a loss of balance leading to falling but not injury; a loss of balance leading to both falling and injury. Cases of the last type are easy to record, but the referents may also have experienced losses of balance that were not recorded. Thus, it is possibly more likely that the results were due to determinants other than slipperiness.

Two procedures have been used to select referents: (1) selection of referents from groups of people who have not had an injury leading to an accident for a specified period of time (e.g., six months or one year); (2) selection of referents from groups of people who are at work at the time of study. These referents may have experienced an accident in the near past, which disturbs the analysis, but the practical simplicity of the procedure is an advantage.

Matching is a method by which confounding factors can be controlled. It has been used in accident epidemiology. A common difficulty is finding a sufficient number of matched referents. In practice, the maximum number of matched variables is five or six, and severe difficulties arise after the second variable.

A common problem with case–referent studies is the difference between series in information collection. Information about cases is typically obtained by interviewing, whereas referents can also be studied by observation. Thus, data are collected by different methods, resulting in artificial differences between the cases and the referents. In principle, the same methods should be used to obtain information about both series. Although this appears to be a straightforward way of solving the

313

problem, it may mean the loss of information that could be gathered because more reliable methods are available for the referents. A solution seldom adopted is to use other accidents as referents. For example, when accidents that involve man's behaviour are being studied, the referents could be accidents caused by purely technical factors. However, it may be difficult to find a sufficient number of cases.

Another possible solution is to redefine the concept of cases. In other words, cases could be a sample of workers or worksites from companies where the accident rate is high. Referents are then taken from companies with a low accident rate (49). The cases are not accidents *per se*, but they are more likely accident candidates than the referents. This design makes it possible to obtain information from both series by observations and measurements, rather than relying on the less reliable recall of past events.

Cohort studies
When cohort studies are undertaken in accident epidemiology, special consideration must be given to the dynamic character of exposure. The intensity of determinants varies widely. The same applies to confounders and modifiers. The exposure, however, is not sufficiently standardized by such factors as industrial branch, occupation, or job.

Another typical feature of accident epidemiology is that, normally, the population is also dynamic. A fatal or permanently disabling accident definitely removes the person from the population. On the other hand, work conditions, technical arrangements, and work procedures tend to change as a consequence of accidents. These changes have to be taken into consideration when studies are planned and when their results are analysed.

The dynamic nature of both the exposure and the population restricts the availability of retrospective studies. Prospective studies are preferable from the point of view of data collection. If recollections of past accidents are used as the source of data, the maximum time limit is one year after the accidents occurred. Even over this period, a person's recall tends to become defective to the extent that the reliability of the information is questionable (7, 50).

References

1. **International Labour Office.** *Accident prevention, a workers' educational manual*, 7th ed. Geneva, 1972, pp. 1–2.
2. **Firenze, R. J.** *The process of hazard control*. Dubuque, IA, Kendall/Hunt, 1980.
3. **Johnson, W. G.** *MORT safety assurance systems*. New York, Marcel Dekker, 1980.
4. **Petersen, D.** *Analyzing safety performance*. New York, Garland STPM Press, 1980.
5. **Gordon, J. E.** The epidemiology of accidents. *American journal of public health*, **9**: 504–515 (1949).
6. **Paul, H. H.** Beiträge der epidemiologischen Forschung zur Unfallverhütung. *Zentralblatt für Arbeitsmedizin*, **22**: 7–13 (1972).
7. **Suchman, E. A. & Munoz, R. A.** Accident occurrence and control among sugar-cane workers. *Journal of occupational medicine*, **9**: 407–414 (1967).

8. **Waller, J. A.** Current issues in the epidemiology of injury. *American journal of epidemiology*, **98**: 72–76 (1973).
9. *The epidemiology of road traffic accidents. Report on a Conference.* Copenhagen, WHO Regional Office for Europe, 1976 (European Series No. 2).
10. **MacMohon, B. & Pugh, T. F.** *Epidemiology.* Boston, Little Brown & Company, 1970, p. 13.
11. **Ferry, T. S.** *Modern accident investigation and analysis.* New York, Wiley, 1981.
12. **Kuhlman, R. L.** *Professional accident investigation.* Loganville, GA, Institute Press, 1977.
13. **Denny, V. E. et al.** Risk assessment methodologies: an application to underground mine systems. *Journal of safety research*, **10**: 24–34 (1978).
14. **Swain, A. D.** *Improving human performance.* London, Industrial and Commercial Techniques, 1973.
15. **National Safety Council.** Work accident records and analysis. *National safety news*, **111**: 81–105 (1975).
16. **Shannon, H. S. & Manning, D. P.** The use of a model to record and store data of industrial accident resulting in injury. *Journal of occupational accidents*, **3**: 57–65 (1980).
17. **Komaki, J. et al.** A behavioral approach to occupational safety: pinpointing and reinforcing safe performance in a food manufacturing plant. *Journal of applied psychology*, **63**: 434–445 (1978).
18. **Laner, S. & Sell, R. G.** An experiment on the effect of specially designed safety posters. *Occupational psychology*, **34**: 153–169 (1960).
19. **Rockwell, T. H. et al.** Development and application of a non-accident measure of flying safety performance. *Journal of safety research*, **2**: 240–250 (1970).
20. **Tarrants, W. E.** *The measurement of safety performance.* New York, Garland STPM Press, 1980.
21. **Kjellen, U.** *The deviation concept in occupational accident control—theory and method.* Stockholm, Kungliga Tekniska Högskolan, 1982 (TRITA – AOG – 0019).
22. **Baker, S. P.** Determinants of injury and opportunities for intervention. *American journal of epidemiology*, **101**: 98–102 (1975).
23. **Cooke, W. N. & Blumenstock, M. W.** The determinants of occupational injury severity: the case of Maine sawmills. *Journal of safety research*, **11**: 115–120 (1979).
24. **Smith, K. & O'Day, J.:** Simpson's paradox, an example of using accident data from the state of Texas. *Accident analysis and prevention*, **14**: 131–132 (1982).
25. **Wilson, W. L.** Estimating changes in accident statistics due to reporting requirement changes. *Journal of safety research*, **12**: 36–42 (1980).
26. **Powell, P. et al.** *2000 accidents.* London, National Institute of Industrial Psychology, 1971.
27. **Hale, A. R. & Hale, M.** *A review of the industrial accident research literature.* London, H. M. Stationery Office, 1972.
28. **McFarland, R. A.** Injury—a major environmental problem. *Archives of environmental health*, **19**: 244–256 (1969).
29. **Arbous, A. G.** Accident statistics and the concept of accident proneness. *Biometrics*, **7**: 340–391 (1951).
30. **Williams, M. J.** Validity of the traffic conflicts technique. *Accident analysis and prevention*, **13**: 133–145 (1981).
31. **Skiba, R.** *Die Gefährentragertheorie.* Dortmund, Bundesanstalt für Arbeitsschutz und Unfallforschung, 1973 (Forschungsbericht No. 106).
32. **Adams, N. L. & Hartwell, N. M.** Accident-reporting systems: a basic problem area in industrial society. *Journal of occupational psychology*, **50**: 285–298 (1977).
33. **Honkanen, R. et al.** A case-control study on alcohol as risk factor in pedestrian accidents, a preliminary report. *Modern problems of pharmacopsychiatry*, **11**: 1–5 (1976).
34. **Shealy, J. E. et al.** Epidemiology of ski injuries: effect of method of skill acquisition and release binding accidents. *Human factors*, **16**: 459–473 (1974).
35. **Monteau, M.** *A practical method of investigating accident factors: principles and experimental application.* Luxembourg, Commission of the European Communities, Directorate General of Social Affairs, 1977.

36. **Rockwell, T. H. & Bhise, V.** Two approaches to a non-accident measure for continuous assessment of safety performance. *Journal of safety research*, **2**: 176–187 (1970).
37. **Corlett, E. N. & Gilbank, G.** A systematic technique for accident analysis. *Journal of occupational accidents*, **2**: 25–38 (1978).
38. **Benner, L., Jr.** Five accident perceptions: their implications for accident investigators. *Hazard prevention*, **16**(11): 16–20 (1980).
39. **Goeller, B. F.** Modeling the traffic-safety system. *Accident analysis and prevention*, **1**: 167–204 (1969).
40. **Heinrich, H. W. et al.** *Industrial accident prevention*, 5th ed. New York, McGraw-Hill, 1980.
41. **Hoyos, C.** *Psychologische Unfall – und Sicherheitsforschung.* Stuttgart, Kohlhammer, 1980.
42. **Näätänen, R. & Summala, H.** *Road-user behavior and traffic accidents.* Amsterdam, North-Holland, 1976.
43. **Surry, J.** *Industrial accident research.* Toronto, University of Toronto, 1968.
44. **Burkardt, F.** Vergleiche zwischen beobachteten und erwarteten Häufigkeitsverteilungen von Bergbau–Unfällen bei gleichem Unfallrisiko. *Psychologie und Praxis*, **3**: 97–100 (1960).
45. *Human factors and safety.* Geneva, International Labour Office, 1967 (Information Sheet No. 15).
46. **Froggart, P. & Smiley, J. A.** The concept of accident proneness: a review. *British journal of industrial medicine*, **21**: 1–2 (1964).
47. **Häkkinen, S.** *Traffic accidents and driver characteristics.* Helsinki, Suomen Teknillinen Korkeakoulu, 1958 (Tieteellisiä Tutkimuksia No. 13).
48. **Häkkinen, S.** Traffic accidents and professional driver characteristics: a follow-up study. *Accident analysis and prevention*, **11**: 7–18 (1979).
49. **Saari, J. & Lahtela, J.** Characteristics of jobs in high and low accident frequency companies in the light metal working industry. *Accident analysis and prevention*, **11**: 51–60 (1979).
50. **Sheehy, N. P.** The interview in accident investigation: methodological pitfalls. *Ergonomics*, **24**: 437–446 (1981).

17

Uses of epidemiology in occupational health

S. Hernberg[a]

Study of community health problems

A workplace can be considered a community with its specific health problems. Some of the problems may be caused by occupational exposures or the work conditions in general, but most health problems in working populations are actually part of "general" morbidity. These problems, of course, also have a bearing on occupational health, since almost any disease can interfere with the working capacity of the employee or interact with his or her ability to cope with occupational exposures and stresses. Hence, a complete picture of the "community health" of a workplace is crucial for any well functioning occupational health service, and therefore "occupational health" must be given a broad meaning in this context.

Survey of the general state of health of employees
A survey of the general health of employees can be designed to identify:

(1) persons who are at special risk of contracting a specific disease (e.g., coronary artery disease) in order that preventive action may be taken;

(2) persons with minor curable or correctable disorders (e.g., iron deficiency, refraction anomalies);

(3) early stages of chronic diseases requiring regular health services (e.g., diabetes, hypertension);

(4) persons whose suitability for certain jobs is restricted or contraindicated (e.g., those with allergy, bronchitis, low-back problems);

(5) persons with unfavourable health behaviour (e.g., heavy smokers, obese and/or physically inactive persons).

---

[a] Institute of Occupational Health, Helsinki, Finland.

Because resources are limited, the purpose of the survey should be made clear before the start. In other words, is the intention one of initiating regular health control, one of moving workers with health problems to safer jobs, or some other? Equally important is the selection of screening methods of sufficient sensitivity and specificity. It is self-evident that insensitive methods do not detect the cases they should detect, and methods with low specificity produce a high proportion of false positive findings. This leads to unnecessary further examinations, which may be costly, even pose some risk, and may cause unnecessary anxiety.

For feasibility reasons, the survey should not employ too many screening tests either. Furthermore, there should be guarantees that the occupational health service or, alternatively, some other health service system, can take care of those in need of further diagnostic examinations, regular surveillance, or therapy. For example, diagnosing refraction anomalies is of no use if ophthalmological services are lacking or overloaded, and for a person to know he has hypertension may be psychologically damaging if adequate and regular treatment cannot be provided. If transfer to less demanding tasks is impossible, there is no use "identifying" (and worrying) handicapped individuals who themselves are usually well aware of their ailment anyway.

If the scope of the survey is prevention, the disease must indeed be preventable, or its course must be capable of being prolonged, otherwise the survey may cause more harm in the form of unnecessary anxiety than benefit in terms of prevention. Here it may be noted that it is usually prudent to await the results of well planned randomized clinical trials before initiating "prevention" at the workplace level, *with the important exception of reducing harmful exposures.*

Finally, since the working population itself and also its health state are dynamic, any action should be planned with continuation in mind. The survey may be repeated at regular intervals or, even better, the activities may be focused on worker categories with a high likelihood of positive findings, e.g., those 40 years and over, smokers, and obese persons. Commonly those aged 40, 45, 50, or more years are selected for examination, thereby creating a cyclic system covering everybody in that age range in a 5-year period. In addition, other risk groups may be examined more often.

The meaningful use of many of the data collected by various health surveillance methods obviously requires them to be evaluated against information on the work conditions. Hence, the occupational health service should always be well aware of the characteristics of each individual's job, not forgetting that the situation may change rapidly in modern industry.

Observation of morbidity in relation to occupation, work area, or specific exposures
Morbidity (here defined as *any* deviation from "health") can be measured by means of an almost endless list of indicators, starting from subjective symptoms at the "soft" end and terminating with the event of

death at the "hard" end. In the "community diagnosis" of workplaces, the harder indicators, such as death and severe disease, are not usually used; however, nationwide occupational mortality statistics may demonstrate occupational differences in mortality and even suggest explanations for them (see below). Occupational health services at workplaces, or alternatively, descriptive *ad hoc* studies of some health problems performed by external research teams, usually employ self-administered questionnaires, interviews, the records of the occupational health services, absenteeism statistics, and a variety of clinical, functional, and biochemical tests. Since the aim of such activities is to describe morbidity in relation to work area, occupation, or some specific exposure, these variables must be well defined. Furthermore, the tests employed must describe the types of health effect that may arise from the exposure in question. Such health effects may be rather specific (e.g., disturbances in protoporphyrin synthesis among lead workers), semi-specific (e.g., tremor among workers exposed to mercury), or quite nonspecific (e.g., chronic bronchitis in foundry workers exposed to dust).

Specific indicators of toxic exposures are often used for regular monitoring purposes. In such a situation the causative agent is already known and regular effect monitoring is a means of preventing the occurrence of serious adverse effects. For example, the measurement of the concentration of delta-aminolaevulinic acid in urine has been recommended for the surveillance of lead workers (*1*). Similar tests may also be used for the detection of uncontrolled exposure in previously unscreened populations of workers (preferably in combination with exposure measurements, as discussed later). For example, the concentration of zinc protoporphyrin in the erythrocytes appears to be an inexpensive, rapid, and reliable test for early lead effect (see, for example, Ref. *2*).

Semispecific and nonspecific tests may reveal the presence of too high concentrations of noxious agents, when groups of workers exposed to such concentrations exhibit more effects than other workers not so exposed. For example, a survey of Finnish foundry workers showed an excess of chronic bronchitis (as defined according to the British Medical Research Council's criteria) among workers exposed to high dust concentrations as compared with other workers (*3*) (see Table 1). Smoking interacted with dust exposure.

Such cross-sectional data are invalidated by selection, with the result that the true prevalence is underestimated, since many of those with outspoken symptoms seek other employment. Hence the "true" occurrence of bronchitis was probably higher than indicated by the data shown in Table 1. A prospective follow-up of newly employed workers would certainly have yielded a better estimate, but would also have meant a shift from descriptive to etiological epidemiology.

Risk identification
Descriptive epidemiological observations usually provide the first clue to hitherto unrecognized occupational health hazards. For example, a sudden increase of dermatoses in a group of workers may indicate that a

Table 1. Age-standardized prevalence (%) of a history of bronchitis in different categories of foundry workers[a]

Category	No. of workers	Chronic bronchitis (%)
Non-smokers		
no dust	67	3.3
dust	83	8.3
Smokers		
no or possible dust	352	12.6
dust	359	24.6

[a] Adapted from Kärävä et al. (3).

new chemical with a harmful action on the skin has been taken into use. The unexpected occurrence of several cases of a rare tumour, e.g., angiosarcoma of the liver, may give the first intimation of the presence of a carcinogen in the work environment. Short influenza-like episodes confined to a certain department may suggest, for example, contamination of the air conditioning system with an agent causing "humidifier fever". Such observations require both alertness on the part of the plant physician and a good system of record-keeping.

Nationwide occupational morbidity and mortality statistics
Official statistics and national or regional morbidity and mortality registers may provide some clues to the epidemiological study of work-related diseases (see Chapter 3). However, direct conclusions are usually difficult to draw from such registers because (a) they frequently misclassify the occupation, (b) occupational titles as such are ill-defined and their meaning (in terms of type of work and of specific exposures) changes over time, and (c) only one occupation (the end occupation) is usually recorded. In addition, diseases tend to be misclassified, and the numerator (occupation-specific mortality) and the denominator (the population at risk) come from different sources and often refer to different time periods. In England and Wales, the Registrar General has reviewed mortality in relation to information provided by decennial censuses on deaths in a period before and after the census for more than 100 years (4). The direct effects of some occupations are readily seen, such as various dust diseases among miners, potters, foundrymen and cotton workers, while the influence of occupations on certain other diseases is less clear because factors outside work, such as social class and smoking habits, may affect the rates. Hence, national registers on occupational morbidity have their main use as starting points for etiological epidemiological studies, because the crude descriptive data they provide are too inexact for specific conclusions and are, moreover, confounded by non-occupational factors.

320

Observation of changes in the occurrence of illness

As already discussed, regular health surveillance programmes can be used to follow changes in the occurrence of illness among employees. Two main applications can be distinguished. The first is aimed at the rapid detection of new health problems, and requires a good, preferably computerized recording system. It also requires well standardized diagnoses and medical alertness. Since new hazards may often affect a rather small number of workers only, the "detection limit" of the recording programme improves if the classification according to working areas or type of work is detailed enough.

The second application focuses on known health hazards. In this setting, the morbidity of similarly exposed workers is regularly followed and changes are recorded. This, of course, requires that the exposure in question causes easily measurable and reversible changes, such as disturbances of the porphyrin synthesis (lead), metal fume fever (e.g., zinc) or bronchitic symptoms (certain dusts). Changes in the occurrence of such symptoms and signs may indicate that preventive measures have been efficient or, alternatively, alert the plant physician to suspect negative developments. However, in principle, there is a conceptual flaw in this second application, since the only recommendable surveillance method of workplace hygiene is *monitoring of the exposure*, not waiting for adverse health effects to occur. Hence, only severely inadequate resources for occupational hygiene justify reliance on health records for revealing a deteriorating health situation.

Survey of exposure data

Occupational epidemiology should relate indicators of illness at least to qualitative measures of exposure, but preferably to quantitative ones. Furthermore, true preventive action at the workplace aims at eliminating or reducing harmful exposures or isolating the workers from them (which is quite different from the often held but wrong opinion that monitoring, health surveillance, and early diagnosis of occupationally induced disorders add up to prevention, without any further action). Quite obviously, in both instances obtaining systematic and accurate information on conditions of exposure is crucial.

Exposure surveys fulfil the following functions:

— identification and location of hazards;
— quantification of hazards;
— providing a basis for preventive action (reduction of exposure);
— assessment of the need for regular health surveillance;
— monitoring of the trend of the hazard; and
— provision of data for cause–effect inferences and exposure–response relationships (see the section "Dimensions of exposure", p. 334).

Exposure surveys may focus on samples from the ambient air (area sampling or personal sampling) or on measuring the toxic agent (or

321

sometimes its specific effects) from biological samples (blood, urine, exhaled breath, hair, etc.). Assessment of the total exposure for individuals or groups and consequently their risk assessment is best achieved using biological samples, while measurements of concentrations in the ambient air are best suited to engineering purposes, to identification of areas with exceptionally high exposure, to regular surveillance of the workplace hygiene, and to the frequent instances where no standardized biological tests exist.

Exposure surveys may be occasional or regular. They may cover one of a few departments, one factory, an entire type of industry, or a geographical district (e.g., a city or even the entire country). The larger the source population (of workplaces or individuals) the more important become the statistical sampling procedures. The following example illustrates how a nationwide survey of lead exposure in Finland was performed (5).

The survey was carried out between 1970 and 1973. Occupational lead exposure was studied in all types of work with even the slightest potential health hazard, as judged on the basis of *a priori* knowledge. Two main sampling methods were employed. First, all workplaces from a particular industry were included whenever lead exposure was estimated to be heavy enough to cause a danger of poisoning. Secondly, different statistical sampling systems were used to select types of work with lead exposure. Two sources of information were used for the sampling, namely, the Enterprise Register of the Central Statistical Office of Finland for the Helsinki region, and the industrial statistics from that office for the rest of the country. From that source, all workplaces that had used lead as a raw material during a certain year could be located. Fig. 1 shows how the sampling proceeded; a detailed description can be found in Tola et al. (5). All workers from workplaces known to pose a high risk were included, and also all those from small workplaces irrespective of prior assessment of the risk. If some of the workers were not exposed, only those doing specific exposed jobs in the plant were included, and if the plant operated shift work, only one shift was chosen. Individual statistical sampling was applied for large workplaces. Altogether 2209 workers representing 30 different types of work and 182 workplaces were studied. Personal exposure to lead was assessed from venous blood samples. Strong emphasis was put on quality control. The results of the survey are shown in Table 2.

Although, with a few anticipated exceptions, the median blood lead levels were fairly low, the upper range exceeded the maximum level recommended at that time ($70 \mu g/100$ ml) in many industries. In other words, although the average hazard was low, some individuals or workplaces were still at high risk. On the basis of the survey, it was estimated that some 15 000 to 20 000 workers in Finland out of a total work force of 2.2 million (about 1 %) would have a median blood lead value exceeding $40 \mu g/100$ ml (the action level for regular monitoring in Finland) at any point in time.

322

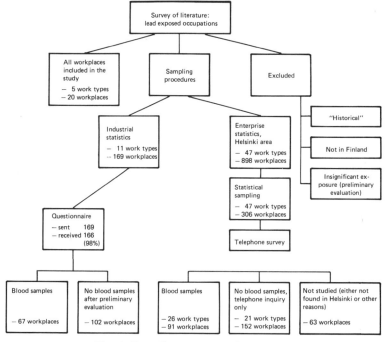

Fig. 1. Sampling strategy of the study

Reproduced from: Tola, S. et al. (5).

Determination of normal values

Most chemical agents used or produced in industry do not occur in the natural environment, so usually there is no problem with the determination of normal values. Nevertheless, a few exceptions exist.

Some synthetic organic compounds do occur in the organs or tissues of the general population, e.g., chlorinated hydrocarbon pesticides and polychlorinated biphenyls, although those compounds should rather be regarded as residues than as "normal" components. Hence, their occurrence in the general population reflects environmental pollution, not "normal" values.

However, many metals occur normally in persons who are not occupationally exposed. Variations in the levels of these metals in the body depend primarily on their geological distribution and the amounts otherwise present in the environment. On the one hand, there are essential trace metals, e.g., zinc, chromium, manganese and cobalt, which are natural components of the organism and whose absence or occurrence in too small quantities can cause deficiency syndromes. For such metals the concept of "normal values" is straightforward. On the other hand, some other metals, e.g., lead, mercury and cadmium, have become widely

323

Table 2. Median values and ranges of the blood lead concentrations in different types of work in Finland*

Type of work	Number of workplaces	Number of workers	Median value (μg/100 ml)	Range (μg/100 ml)
Lead scrap smelting	2	32	79	35–118
Storage battery manufacturing	3	137	66	19–101
Metal founding	7	45	53	6–108
Shipbreaking	2	13	49	26–106
Crystal glass manufacturing	3	49	41	12–82
Car radiator repair	5	56	38	17–83
PVC plastic manufacturing	10	157	37	10–126
Lead glazing	1	26	37	16–58
Cable manufacturing	5	36	36	12–68
Manufacturing and repair of storage tanks	10	48	34	11–99
Storage battery repair	4	18	33	12–70
Scrap metal shop work	10	30	33	15–62
Treating metal surfaces	6	45	29	10–97
Car repair	13	165	27	10–80
Machine shop work (railroad)	5	130	26	6–72
Ammunition manufacturing	3	28	25	8–62
Sheet metal work (construction industry)	7	53	25	10–47
Paint manufacturing	12	88	24	5–85
Shipbuilding	4	255	24	10–65
Sheet metal work (miscellaneous)	8	47	22	9–45
Machine shop work	13	206	21	9–98
Iron and steel founding	8	42	21	11–58
Painting	12	80	20	5–72
Service station work	10	76	20	11–40
Plumbing	8	92	19	5–55
Manufacturing of electric lamps	3	22	19	12–33
Telephone repair and installation	1	67	17	7–41
Manufacturing radio and telephone equipment	5	52	13	7–32
Traffic police work	1	28	13	9–20
Street sweeping	1	86	12	7–30
Totals	182	2209		

* After Tola et al. (5).

dispersed in the environment as a result of man's activities. Present levels found in the blood and organs of the general population in industrialized countries are certainly not "normal", if this term is taken to describe natural conditions, although they may be considered "normal" in the light of today's situation. It has been shown repeatedly that the concentrations of these metals in the blood or urine of different population groups vary according to the level of pollution of the general environment. Hence, a survey of "normal" values of such metals should specify what type of population groups are studied: urban dwellers, rural populations, persons living in the vicinity of heavy polluters, etc. The possibility of present or past occupational or accidental exposure, even if minimal, should not be overlooked either. Also seasonal variations may occur. Since there are so many known circumstances influencing "normal" exposure, it is evident that random sampling from large populations is not very informative, besides being highly impractical. More pertinent information is gained by drawing samples from well defined representative source populations (e.g., those living in busy downtown areas, those living along highways, or rural agricultural populations). A nationwide survey by Nordman (6) of environmental lead exposure in Finland can be used as an example.

The purpose of the study was to compare geographical and urban-rural differences. Occupational lead exposure was scrupulously ruled out by a questionnaire and by utilizing information from the concomitant survey of occupational lead exposure by Tola et al. (5) described above. The population groups were selected from rural (two counties), urban (downtown and suburban dwellers, and two additional target groups, i.e., street sweepers and traffic policemen), and industrial districts. The blood lead concentration was used as the indicator of lead exposure. Extreme care was taken to ensure accuracy and precision of the analytical method (inter- and intralaboratory checks, blind samples, etc.). This is especially important when analysing the low values occurring in populations not occupationally exposed. Altogether 1378 persons were studied. The results showed that the blood lead levels were low from an international standpoint and that no systematic differences existed between urban and rural populations. Only those living in the vicinity of a lead smelter, and to a slight extent also the street sweepers and the policemen, showed elevated mean values. Since the highest individual value (except for the group living around the smelter) was as low as 23 μg/100 ml, it could be concluded that in Finland values in excess of 20 μg/100 ml indicate abnormal (usually occupational) exposure. The experience gained from this survey has helped evaluate what proportion of the blood lead concentration in occupationally exposed workers is due to "natural" exposure and how much to occupational exposure. This information is helpful in judging the feasibility of recommendations for blood lead standards for occupational exposures.

Generation of hypotheses for the study of cause–effect relationships

A scientific hypothesis can emanate from a variety of sources, including analogies, experiments, clinical observations, and descriptive epidemiological data; often it arises from a synthesis of prior knowledge derived from many different disciplines. In the present context only epidemiology will be considered.

Four main ways in which descriptive epidemiology may generate hypotheses can be distinguished. First, occupational health records may arouse suspicion of a causal connection between disease and exposure to new—and sometimes old—hazards, as previously discussed. Second, cross-sectional *ad hoc* surveys comprising differently exposed workers as well as unexposed reference groups may show a varying prevalence of symptoms and signs, which may relate to some specific exposure(s). Because cross-sectional studies suffer from the well known problems of invalidity (e.g., selection bias and information bias connected with the use of "soft"effect indicators), the results of such studies can usually be no more than suggestive of a cause–effect relationship but they may generate a hypothesis for further testing. It must again be emphasized that in cross-sectional studies selection bias usually tends to mask existing relationships, whereas information bias tends to cause false positive results (e.g., through a tendency of "exposed" workers to complain more).

Thirdly, computerized linkage of exposure and morbidity registers may occasionally provide clues for etiological research. However, in general, strong diluting effects, mainly arising from misclassification and/or crudeness of the exposure parameter, tend to mask existing effects (see Chapter 3). Furthermore, "positive" findings are also difficult to interpret. For example, statistical significances (low P values) from populations of tens or hundreds of thousands of individuals are completely without meaning. The reason is that such large numbers inevitably produce "statistical significances" because of the extremely small differences needed.

A recent exercise based on a case–referent study of congenital malformations illustrates well how important the negative bias can be when conventional, crude exposure data, obtainable from large registers, are used. Holmberg (7) showed in a careful case–referent study based on detailed personal interviews and hygienic checks of the work conditions, that out of 120 mothers of children with congenital abnormalities of the central nervous system (hydrocephaly and anencephaly) 14 had been exposed to industrial solvents during the first trimester, against 3 out of 120 referent mothers ($P < 0.01$) (for a more detailed description, see p. 331). The same material was then studied in a "conventional" way (8), using routine information obtainable from registers, namely (1) working or not working outside home, (2) social groups, (3) industrial classification, (4) maternal occupation, and (5) paternal occupation. No consistent or significant differences could be shown using these classifications, although it was *known in advance* that solvent exposure was over-represented among the mothers of the affected children. The insensitivity of conventional register parameters was thus clearly demonstrated.

The fourth way of using descriptive epidemiology for creating hypotheses is the study of national occupational mortality or pension statistics, as discussed earlier. Such statistics may provide clues for etiological research, but they do not usually permit direct cause–effect conclusions because of the many interfering factors, such as misclassification, effects of non-occupational conditions, etc.

In conclusion, descriptive epidemiology may generate useful hypotheses when based on good basic data, but if these are crude, it more often tends to mask existing effects. In particular, the uncritical use of data collected for other purposes is risky. The availability of computers has unfortunately increased the opportunities for malpractice. Meaningful interpretation of descriptive data requires deeper insight into epidemiology and better knowledge of the many sources of error involved in their use than is generally realized.

Search for causal relations: etiological epidemiology

Epidemiological hypotheses

Etiological epidemiology is, as already stated, generally concerned with the testing of some hypothesis. Usually, the problem consists of revealing a causal connection between some occupational exposure and a nonspecific disease (e.g., cancer, lumbar pain). Quite often the problem involves the study of multifactorial etiology, the occupational factor being only one of several causes. When the occupational exposure is the "only" (i.e., clearly dominant) cause, the hypothesis is so trivial (e.g., "lead causes lead poisoning") that etiological research is not needed. The only exception, perhaps, is exposure to newly introduced agents whose action is not completely known. However, when it comes to completing the clinical picture (e.g., does lead cause subclinical neuropathy before the onset of poisoning?) etiological epidemiology is useful, as will be discussed in a later section (see p. 332). Occasionally, etiological research can be initiated without any well defined prior hypothesis. A typical example is a case–referent study with the aim of finding out which different exposures are associated with a particular disease, or a follow-up study to determine which different diseases are caused by a given exposure.

An epidemiological hypothesis may be specific or semispecific. A specific hypothesis states that "exposure A causes disease B", e.g., asbestos exposure causes bronchial cancer. Semispecific hypotheses are of two types. The first has the form "exposure A causes the diseases B_1, B_2. . . . B_i", e.g., asbestos causes bronchial cancer, mesothelioma, stomach cancer, . . . and asbestosis. The second type postulates that "disease B is caused by factors (exposures) A_1, A_2. . . ., A_i", e.g., bronchial cancer is caused by smoking, asbestos, chromates, . . . and foundry dust (note the conceptual distinction from studies of the "fishing expedition" type without any prior hypothesis exemplified in the preceding paragraph). A nonspecific hypothesis can be formulated "exposures A_1, A_2. ....A_i cause the diseases B_1, B_2 .... B_i", e.g., exposure to lead, arsenic, nickel, copper, zinc, cadmium and sulfur dioxide in a smelter causes bronchial cancer, stomach cancer, prostatic cancer, cardiovascular disease, etc. Quite obviously, this is not an etiological but a descriptive problem, related to time and space, and procedurally such a constellation forms a data system for research rather than a study.

Use of epidemiology to test specific hypotheses

The specific hypothesis "A causes B" can, in principle, equally well be tested by means of either a follow-up or a case–referent study. The choice is determined by validity and feasibility aspects. If the disease is rare, a case–referent study is definitely to be preferred, and conversely, if the exposure is rare, a follow-up study is more efficient. However, exposure can sometimes be "enriched" (see Chapter 10). If both the exposure and the disease are common, both designs are feasible and the decision depends on matters such as the availability and/or reliability of retrospective data, possibilities of tracing drop-outs (necessary in follow-up studies, but not in case–referent studies), financial resources and the number of personnel available (case–referent studies are often less expensive), etc. The method of "nesting" a case–referent study in a cohort is being more and more widely used as it is both cost-saving and efficient (9). An example of a study testing a specific hypothesis is presented in Chapter 18.

Use of epidemiology to test semispecific hypotheses

The semispecific hypothesis of the type "exposure A causes diseases B_1, B_2. ....B_i" can only be tested (epidemiologically) by a follow-up design or, as a subcategory of this, an intervention study. The main question then is whether one should choose a prospective or a retrospective follow-up study. Since the basic design is the same, this is actually a matter of *timing*, not one of different designs.

Retrospective follow-up studies offer the advantage that the results become available rather quickly, usually within a few years. In this type of follow-up study, the starting point of which lies in the past, the cases have already occurred. The problem is to trace the cohort members, both the exposed and nonexposed, and to determine their state of health (usually survival) rather than to wait for the development of disease. A disadvantage of retrospective timing is that past exposure data are usually scanty, unreliable, or even nonexistent. The same is usually true of data on soft morbidity indicators, such as blood chemistry and blood pressure. For this reason, a retrospective follow-up study must be confined to hard indicators of disease, usually death.

Prospective studies offer the following advantages: (1) the study can be better designed to meet the investigator's requirements, (2) exposure data can be systematically collected, (3) several indicators of the diseases under study can be measured, (4) the measurements can be repeated over time if needed, (5) the methodology can be standardized and its validity checked, and (6) an intervention can be combined with the follow-up. The main inconvenience of prospective timing is the long period sometimes required before the results become available. In the study of chronic diseases, such as occupational cancer, this period may be 20–40 years. High cost is often said to be a disadvantage of prospective follow-up studies, but this is not strictly true in terms of cost per amount of information. Because a prospective follow-up study, in contrast to a retrospective one, allows the collection of a variety of both exposure and morbidity data, the absolute costs will, of course, be high. But much more information is gained and hence, *in relative terms*, the costs do not necessarily differ greatly from those of a retrospective follow-up.

Prospective follow-up studies testing a semispecific hypothesis are difficult to find in the literature and therefore a retrospective follow-up study of the rubber industry will be used as an example. Here it must be noted that, for didactic purposes, "rubber work" has been considered in this example as one single exposure, although it is well known that it actually involves exposure to a multitude of agents.

Several different studies had indicated that, although no significant excess of total mortality seemed to occur, rubber workers exhibited an excessive mortality from certain malignant neoplasms and some types of cardiovascular disease. This prior knowledge prompted a retrospective mortality study of a tire manufacturing plant in the United States (*10*). The exposed cohort comprised all the hourly paid active and retired male employees, aged 40–84 years on 1 January 1964, within the tire manufacturing plant. Company records of payrolls, pensions, death claims, severance awards, transfers between hourly and salaried status, and employment exits were used as data sources. A cohort of 6678 male rubber workers was identified from those sources. During the full nine-year period 1964–1972, 1783 life insurance death claims were paid on behalf of deaths occurring within the cohort. Death certificates were obtained from the company's insurance carrier for all but 0.5 % of the 1783 deaths. "Expected" mortality figures were derived from the US male death rates of 1968, which was the midpoint of the 1964–1971 period. Analyses of annual national death rates throughout the period showed that the 1968 mortality data were 3 % higher than the period average, thus producing a slight underestimation of SMRs within the study cohort.

The main results of the study are shown in Table 3. Eleven of the cause-specific SMRs showed statistically significant deviations from the expected figure of 100. The highest SMRs were for malignant neoplasms of the stomach and the lymphatic and haematopoietic systems. In each case, the SMR was higher for the active age range 40–64 than for the full age range. Leukaemia showed an approximately threefold excess.

Table 3. Standardized mortality ratios (SMR) of rubber workers for selected causes of death, for the full and the "active" age-range[a]

Cause of death (ICD Code)	Active age range (40–64)			Full age range (40–84)		
	No. of deaths observed	No. of deaths expected	SMR	No. of deaths observed	No. of deaths expected	SMR
All neoplasms (140–239)	110	108.9	101	351	336.9	104
Malignant neoplasms (140–209)	108	107.3	100	344	333.3	103
Cancer of stomach (151)	12	5.5	219[xx]	39	20.9	187[xxx]
Cancer of large intestine (153)	10	8.3	121	39	31.8	123
Cancer of rectum (154)	2	3.3	60	7	11.7	60
Cancer of pancreas (157)	4	6.4	62	17	19.8	86
Cancer of resp. system (160–163)	33	45.8	72	91	109.3	84[x]
Cancer of prostate (185)	6	4.1	147	49	34.4	142[x]
Cancer of bladder (188)	2	2.5	80	9	12.3	73
Cancer of brain, CNS (191–192)	1	3.5	29	4	5.9	68
Lymphosarcoma (200)	6	2.4	251[x]	14	6.2	226[xx]
Leukaemia (204–207)	11	3.5	315[xxx]	16	12.5	128
Diabetes mellitus (250)	13	8.3	157	43	30.0	143[x]
Ischaemic heart disease (410–413)	211	205.9	103	737	741.4	99
Cerebrovascular disease (430–438)	26	30.6	85	182	176.5	103
Arteriosclerosis (440)	4	1.8	223	34	22.1	154[x]
Chronic resp. disease (490–493)	13	13.6	96	53	55.4	96
Liver cirrhosis (571)	14	17.8	78	35	28.6	122
Accidents, poisoning & violence (except suicide) (E800–999, minus E950–959)	20	34.0	59[x]	42	66.9	65[xx]
Suicide (E950–959)	10	9.5	105	17	16.7	102
All other causes	69	94.5	72	289	324.0	89
All causes	489	524.9	93	1783	179.5	99

x = $P < 0.05$, xx = $P < 0.01$, xxx = $P < 0.001$
[a] After McMichael et al. (10).

There was also a moderate excess mortality from diabetes mellitus and the number of deaths from atherosclerosis was considerable higher than expected. In conclusion, the study supported the hypothesis that exposure to "rubber work" causes an excess of several malignant and nonmalignant diseases. (However, there was also a "protective" effect on some cancers, e.g., brain and rectal cancer. This is difficult to explain. Possibly it was due to misclassifications, leading to a shift from one diagnostic category to another.)

The semispecific hypothesis of the type "disease B is caused by exposures A_1, A_2.... A_i," can only be tested (epidemiologically) by means of a case–referent study. An ongoing Finnish study (11) on the association between congenital defects of the central nervous system (CNS) and parental exposures to occupational toxic substances can illustrate this type of hypothesis testing. This study also provides an example of how hypothesis testing and a nonspecific "fishing expedition" type of study can be combined. Prior knowledge provided the hypothesis that CNS defects may be caused by exposures to solvents, heavy metals, and radiation, so several questions were asked concerning these exposures. In addition, fixed and open-ended questions regarding many other occupational exposures were included. The study was based on incident cases of CNS malformations. It started in June 1976, covered the whole of Finland whose current population is about 4.8 million, and used data from the Finnish Register of Congenital Malformations (12). This Register is based on compulsory notification of all malformations detected in live-born infants during the first year of life. The primary notifications and the death certificates are surveyed by a trained pathologist who selects certain marker defects, among them CNS defects. The "case-mothers" (i.e., mothers of affected children) are routinely interviewed by midwives at the time of their first postdelivery visit to their local maternity and child welfare centres. The routine questionnaire consists of 80 items, including information on the family, prepregnancy history, and details of the latest pregnancy. No routine enquiries are made into specific occupational data, however. For each case-mother, the mother whose delivery immediately preceded that of the case-mother in the same maternity welfare district serves as a referent. Because the Malformation Register provides a fixed setting of case and referent mothers, it cannot easily be changed for the purpose of a specific study. Therefore, the detailed interview on occupational history needed for this specific study had to be incorporated into the routine system.

Two trained interviewers visited both the case and referent mothers especially for the purpose of conducting the occupational interview, which was composed of questions on specific exposures (e.g., solvents, metals, pesticides, wood preservatives, paints, dusts, fumes, ionizing and nonionizing radiations both at work and during leisure time, as well as "dummy questions" to check recall bias. The mother was also asked, in the form of an open-ended question, to describe a typical working day. Whenever there were possibilities for exposure at work, an experienced industrial hygienist checked the data reported by the mother by a telephone call or a visit to the workplace, without being aware of whether a case or a referent mother was concerned. This was considered particularly important for the purpose of minimizing asymmetrical inaccuracy of information (observer bias), a well known source of error in case–referent studies of congenital malformations.

The data for the first two years showed that, out of 120 pairs, 14 case mothers and 3 referent mothers had been exposed to industrial solvents during the first trimester of pregnancy ($P < 0.025$). A similar analysis of exposure to metals, pesticides, and wood preservatives revealed no statistically significant difference, and the same applied to occupational exposures to noise, low and/or high temperature of the working environments, "fumes" or nonionizing and/or ionizing radiation. In conclusion, this study supported the hypothesis of a connection between solvent exposure and CNS effects, but did not do so for the other postulated exposures.

Etiological fraction of occupational exposure

As already pointed out above (see p. 327), etiological epidemiology is usually concerned with diseases of multifactorial etiology. Hence, a causal relationship between some exposure and a disease usually means that the exposure in question is one of the causes of that disease, but not necessarily the *only* cause. Under such circumstances it may be interesting to know the relative share of an occupational cause in relation to other causes, i.e., the proportion of cases among the exposed that are actually due to the exposure. This *etiological fraction* (EF) can be calculated from the rate ratio (RR) in the following way (3):

$$EF = (RR - 1)/RR.$$

For example, if 120 cases of bronchial cancer are observed in a cohort of chromate workers against 44 expected, the RR is about 2.8. The $EF = (2.8 - 1)/2.8 = 0.64$, i.e., 64 % of the cases are due to chromate exposure. Quite obviously, the etiological fraction is not generalizable, since its magnitude depends strongly on the intensity and duration of the exposure and the specific population studied. Furthermore, it is difficult to interpret. Being a percentage, it is dependent on the effects of other factors, e.g., smoking (14). A reduction in those factors would lead to an increase in the etiological fraction of interest, even if the absolute number of excess cases remained unchanged. Furthermore, since the concept(s) of multiple etiology are complex (e.g., they presuppose several "sufficient" causes, which can be split up into "component causes", *contributing to* but not necessarily *causing* the disease) the sum of different etiological fractions is often in excess of 100 %. In fact, it has a lower limit of 100 % while its upper limit could be any value (14).

Completing the clinical picture

A common use of epidemiology is to complete the clinical picture. Many occupational diseases are insufficiently well understood, especially as regards their early stages. Conversely, the variety of manifestations that toxic agents can cause are only partially known. This deficiency is by no means typical of occupational medicine; other fields of medicine share this defect because so much of our knowledge is derived from clinical studies. A thorough discussion of this problem in general can be found in Morris's book *Uses of epidemiology* (*15*). In this context, only a few illustrative examples from occupational medicine will be given.

Morris distinguishes between completing the clinical picture in breadth and in depth. Completing the picture *in depth* has to do with learning more about the early stages. Many so-called classical occupational diseases are described in textbooks mainly from the clinical point of view, reflecting the state of the art of clinical medicine and also the old attitude still prevailing among some clinicians that slight manifestations of ill-health are insignificant. Today, subclinical changes caused by occupational factors are beginning to be considered unacceptable by more and more experts. Because early changes are often nonspecific, epidemiological studies are needed to determine their association with a particular exposure.

A typical example is that of carbon disulfide (CS_2) poisoning. The clinical picture is well known, being characterized by psychoses, polyneuropathy, extrapyramidal symptoms, hemiplegia, neurasthenia, and gastrointestinal disturbances. Such severe cases are rarely seen today, at least in developed countries. However, the question arises whether absence of frank disease is, in fact, proof of complete safety. That this is not so has been shown by several epidemiological studies. According to these, subclinical, nonspecific dysfunctions of both the central and the peripheral nervous systems are common among clinically healthy workers exposed to CS_2 in viscose rayon production (*16–18*). These manifestations can be shown epidemiologically as an excess of various symptoms, such as insomnia, nightmares, sleep disturbances, fatigue, headache, numbness of the limbs, and impotence, or as objective signs, e.g., reduction in the conduction velocities of peripheral nerves, impaired performance in psychological tests, and excess of nonspecific ECG changes. Such studies have completed the picture of CS_2 poisoning by describing its early stages, which would have been very difficult or impossible to identify without epidemiological methodology.

Completing the clinical picture *in breadth* corresponds to asking the question "What different types of manifestations are caused by this exposure?" i.e., the typical "cohort question" discussed on pages 328–329. In order to avoid repetition, only a short example will be given. The clinical picture of arsenic poisoning has been known for centuries, not only from occupational experience but also because arsenic trioxide was earlier widely used by physicians as a medicine and also by assassins for quite opposite purposes. The poisoning is characterized by gastrointestinal colic, diarrhoea, hepatic degeneration, polyneuritis, palmar hyperkeratosis, and pigmentation of the skin. More recently, epidemiological studies have shown that arsenic *also* causes skin cancer, bronchial carcinoma, and probably some other tumours (for a review, see Ref. *19*). A recent Swedish study has also demonstrated that smelter workers exposed to arsenic manifested an excess of deaths due to coronary artery disease (*20*). Epidemiological research has thus shown that arsenic causes not only classical arsenic poisoning but also different cancers and coronary artery disease. It would not have been possible to link these nonspecific manifestations to arsenic exposure by means of clinical medical research.

Exposure–effect and exposure–response relationships

Once the qualitative causal association between two phenomena is known, there arises the second-order question of determining the quantitative relationship between different degrees of exposure and the different qualities, intensities, and frequencies of its effects. Besides being scientifically relevant, this knowledge is of the utmost importance in the scientific–administrative procedure of setting standards.

Some definitions and concepts

First, it must be pointed out that the use of the term "exposure" instead of the more familiar pharmacological concept "dose" is motivated by the trivial fact that the dose is not measurable in occupational situations, e.g., skin absorption and the effect of physical activity cannot be estimated accurately. Secondly, "effect" and "response" are different concepts. The Task Group on Metal Toxicity, at its meeting in Tokyo in 1974 (*21*) defined "effect" as a "biological change caused by an exposure". When data are available for both exposure and effect on a graded scale, an exposure–effect relationship can be established. An example may be the relationship between blood lead levels and the concentration of delta-aminolaevulinic acid (ALA) in the urine. The term "response" was defined as the *proportion* of a population that demonstrates a specific effect (say, a concentration of ALA in the urine in excess of 5 mg/litre), and its correlation with estimates of exposure provides the exposure–response relationship. The exposure–effect relationship exhibits an *average* effect in all individuals at the same exposure levels, provoking the false impression that the population is homogeneous. By contrast, an exposure–response relationship takes into account the variation in susceptibility within groups of individuals, since it indicates the proportion of persons affected. For preventive purposes it is extremely important to be able also to define the most sensitive members of a group, and hence the exposure–response relationship is more informative.

From the point of view of preventing adverse effects, the no-effect or no-response limit is of particular interest. However, many conceptual and practical difficulties are inherent in the demonstration of a negative; these aspects are discussed in Chapter 14.

Dimensions of exposure

It is well known that lack of valid exposure data is the most serious obstacle to the study of exposure–effect and exposure–response relationships. Therefore, the need to make every effort to ensure that such data are obtained whenever a quantitative study is planned cannot be overstressed (see also p. 321). A hierarchical order (from the point of view of usefulness in exposure–response research) of the dimensions of the exposure variable can be constructed in the following way.

1. Biological measurements (blood, urine, etc.).
2. Air samples collected from the breathing zone with the aid of portable samplers.
3. Area sampling by means of static samplers.
4. Classification of the exposed subjects by:
— work area
— type of work
— occupational title.
5. Classification into "exposed" and "nonexposed" subjects.

Exposure should ideally be classified not only according to intensity, but also according to duration, fluctuations (peak exposures may be highly relevant) and calendar time (important, for example, when studying occupational cancers with a long latency time). Concurrent other exposures should also be accounted for. The worst problem is usually that retrospective exposure data are lacking or inaccurate. While it is always possible, at least in theory, to measure present levels, there is little remedy for a lack of past information. This lack may, in fact, render the study completely unfeasible. Some information can sometimes be obtained, however, by comparing the current situation with judgements of past conditions provided by former workers, foremen, and safety personnel (e.g., "four times worse in the 1950s"). In addition, the time when a major improvement, such as installation of exhaust ventilation, was made is usually listed in the company's records. The result, of course, will be a rough classification into, for example, heavy, medium, light, and no exposure, which may be sufficient for qualitative studies although quite insufficient for exposure–effect and exposure–response studies. Sometimes it has been possible to simulate past conditions by reconstructing old workplaces, but such opportunities are rare exceptions.

It should be realized that poor precision of exposure data tends to produce a systematic flattening of the regression slope by broadening the horizontal scatter. By contrast, poor quality of the effect broadens the vertical scatter, thereby decreasing the correlation coefficient, but this does not influence the slope of the curve in any systematic way. Poor quality of exposure data is therefore the most common reason for failing to find an exposure–effect or exposure–response relationship where such a relation actually exists.

Measures of effects
Some aspects of effect measurement have already been discussed (see p. 319). When selecting parameters for a study, the investigator should always consider their optimum "hardness". In general, the harder the

parameter, the more reliable the information. As mentioned before, the event of death is subject to no ambiguity, but the cause of death is a less reliable parameter. The softer the parameter, the greater the possibility of error. Questionnaires, for instance, may give highly misleading data, and their use requires careful standardization of the method and critical interpretation of the results. Yet, it does not necessarily follow that hard parameters should always be preferred. Death is such a crude measure that mortality statistics can reveal only the most severe health hazards. In addition, several ailments, such as disorders of the musculoskeletal system, mental disease, and eczema, do not appear in mortality statistics at all, and when milder health effects such as fatigue, dizziness, nausea, neurasthenia or eye irritation are being studied, no better methods than interviews or questionnaires exist. Thus, the ideal hardness of a parameter depends on the problem to be studied.

Some occupational disorders are characterized by *specific* symptoms and signs. For example, lead causes certain well defined disturbances of haem synthesis measurable, for example, as an increased excretion of ALA in the urine. On the other hand, many occupational disorders are characterized by quite *nonspecific* effects. For example, occupational carcinogens cause cancers that usually do not differ at all from cancers having other origins. Both specific and nonspecific effects can be used for studying exposure–effect and exposure–response relationships. It is also important to study in which order various effects appear. In general, protection of workers should aim at the prevention of the earliest effects. However, some effects may not be connected with biologically significant health impairment and cannot therefore be considered adverse. A WHO Study Group (*1*) defined the following types of effect as adverse:

— effects that indicate early stages of clinical disease;
— effects that are not readily reversible, and indicate decrement in the body's ability to maintain homoeostasis;
— effects that enhance the susceptibility of the individual to deleterious effects of other environmental influences;
— effects that cause relevant measurements to be outside the "normal range", if they are considered as an early indication of decreased functional capacity;
— effects that indicate important metabolic and biochemical changes.

Effects not complying with any of these criteria can be defined as nonadverse effects (e.g., inhibition of ALA-dehydratase activity, yellow colouring of smoker's fingers). Obviously, the study of exposure–effect and exposure–response relationships to provide a basis for the setting of standards should focus on early *adverse* effects.

Studies of exposure–effect and exposure–response relationships are usually of the follow-up kind. However, whenever exposure and effect are related in time, and provided validity aspects such as selection can be controlled, cross-sectional studies may also be useful. For example, Seppäläinen et al. (*22*) studied the relation between blood lead levels and

336

decreased nerve conduction velocities among workers in a factory making storage batteries. A statistically significant exposure–effect relationship was found between the maximum blood lead level ever measured and several average nerve conduction velocities, especially of the arm nerves. A dose–response relationship was also found in the sense that the prevalence of abnormal conduction velocities in one or more nerves increased from 27 % in the group with maximum blood lead levels between 40 and 49 μg/100 ml to 53 % in the group with maximum blood lead levels equal to or exceeding 70 μg/100 ml. In this study, however, the design was not completely cross-sectional, since earlier blood lead levels were also available and used, but the effect variables were measured cross-sectionally.

Identifying health-promoting factors

Not everything in our environment is harmful. On the contrary, there is little doubt that many environmental and lifestyle factors can actually promote health. Examples of these include good (but not "too" good) nutrition, enough sleep, sufficient physical activity, good housing, and availability of good health care. All these factors are associated with high social class, and it is known that morbidity and mortality are lower for this category of people than for other groups. Also, the very state of being employed in productive work promotes health, because the unemployed usually show a higher incidence of illness and death. However, when it comes to the study of exactly which circumstances exert the health promoting action, the problem becomes difficult because cause and effect are not easily separated methodologically.

As far as occupational health problems are concerned, it is self-evident that successful reduction of a harmful exposure must promote health and the qualitative aspect is even too trivial to justify a study. However, the quantitative aspects are more interesting, since they involve cost–benefit considerations, i.e., how much reduction of exposure will prevent how much disease or what is the cost of preventing a certain "amount" of disease? However, in practice this research is difficult to carry out, especially when the disease has a long latency period, e.g., silicosis or cancer, and when the preventive technology introduced at the same time promotes production economy.

It may be concluded, therefore, that the identification of health-promoting factors is a difficult exercise and it is therefore not surprising that so little research has been devoted to this field. It has become fashionable to emphasize the importance of such research, but this unfortunately does not provide the technical aid needed to overcome the methodological problems involved.

Evaluation of preventive measures (effect evaluation)

Intervention research is primarily concerned with the evaluation of a preventive programme, or in other words, with the effectiveness of the intervention. It can take the form of animal experiments, clinical trials, or epidemiological research, in which case it is community-based. In occupational health, the community may be a department, a plant, or an entire industry. The question is, what are the possibilities for epidemiological research in the evaluation of a preventive programme? The answer is very complex. First, epidemiological research in general is occupied with *unintended* (deleterious) effects, while evaluation of preventive efficiency is focused at *intended* (positive) effects. The study of the latter differs from that of the former, not least because "health" is more difficult to diagnose than "disease". Furthermore, a programme that acts preventively on one disease may have opposite effects on another. Second, there is a difficult problem in securing comparable groups. Self-selection may cause some individuals to collaborate more actively than others in a preventive programme and they may differ in many fundamental properties. For example, those who are willing to give up smoking probably belong to quite a different behavioural type from those who refuse to do so. Third, there may be difficulties in defining the outcome variables used as indicators of success. For example, when evaluating the effect of an antihypertensive programme, many difficulties arise. Is the benefit to be judged by a fall in blood pressure, a reduction in cerebrovascular accidents, or some other indicator? What about side-effects? How can comparability in diagnostic criteria be ensured if several physicians and hospitals participate, and how does the unavoidable knowledge of to which group the subject belongs influence judgement?

Fourth, the quality of intervention *practice* must inevitably influence the results. The preventive measures, although theoretically adequate, may be too ineffective in practice to bring about any changes, or their quality may be uneven. Also, feasibility problems disturb the procedure in the sense that it may be difficult to achieve collaboration (e.g., in changing the diet or in the use of personal protective equipment). Finally, the procedure of evaluation must be done *in retrospect* to avoid the Hawthorne effect, which means that the awareness of being subject to study influences the process. In other words, the evaluation must come as a surprise; otherwise it changes the practice. Therefore, the evaluation has to rely on routine records, which usually are too superfical and of poor quality.

The conclusion is, therefore, that the use of epidemiology in health care evaluation in general does not work, although there are some exceptions. These are mainly concerned with reducing harmful exposures and observing the decrease or disappearance of adverse effects that occurs as a result, as discussed on p. 318. In some instances, also, the effect of a screening programme can be evaluated epidemiologically by, for example, registering a possible decrease in the frequency of morbidity and mortality or, alternatively, a possible increase in survival times (*23*).

However, in most instances a blind, randomized, clinical trial is more valid, more informative, more effective and less expensive for evaluating the effects of health care and therefore such an approach should be preferred to interventive epidemiology whenever possible.

Evaluating health services in workplaces (programme evaluation)

The aim of evaluating health services in workplaces is to assess how the health care programme functions in relation to some normative concept of how the process should be. In other words, the results of a health service programme in terms of reduced morbidity are *not* the subject of the study, as in the effect evaluation discussed in the preceding section, but rather the procedure itself. A recent WHO publication (*24*) discusses the goals of programme evaluation and describes how such an evaluation is carried out. Epidemiological methods can well be used for this procedure. The greatest problem is to achieve a consensus from experts on how the ideal or an acceptable health service programme should function. If the health service is considered as a whole, recommendations for generally applicable health care programmes may be impossible to define, since the variety of work conditions, exposures, different composition of the work forces, etc., must leave room for flexibility. However, in the particular case of surveillance of certain exposures, e.g., lead, it may be quite possible to reach agreement on what is the ideal procedure. For example, one may agree that the surveillance should be based mainly on blood lead monitoring, supplemented by erythrocyte protoporphyrin measurements and/or determination of urinary concentrations of ALA if the blood lead concentration exceeds a certain level, say 40 or 50 μg/100 ml. It may also be possible to agree on how frequent these measurements should be under different exposure conditions, and at what blood lead level the worker should be removed from exposure and for how long. The programme evaluation then studies how these recommendations are followed in different industries or geographical areas and what reasons determine good or poor compliance; it also studies how accurate the measurements are, how conclusions are drawn, and to what extent hygienic improvements in the work conditions are implemented if the results of the surveillance indicate risks. Finally, recommendations for improvements of the practice should be given.

Another goal of programme evaluation is to survey what is going on in the *absence* of any normative recommendations, and to try to judge the need for some uniformity. Programme evaluation becomes more and more important as health care demands an ever increasing share of the public services, and here epidemiology will have a much greater role than in the evaluation of the effect of health care, which should rather be based on clinical trials.

References

1. WHO Technical Report Series, No. 647 1980, (*Recommended health-based limits in occupational exposure to heavy metals*. Report of a WHO Study Group).
2. **Blumberg, W. E. et al**. Zinc protoporphyrin level in blood determined by a portable hematofluorometer: A screening device for lead poisoning. *Journal of laboratory and clinical medicine*, **89**: 712–723 (1977).

339

3. **Kärävä, R. et al.** Prevalence of pneumoconiosis and chronic bronchitis in foundry workers. *Scandinavian journal of work, environment and health,* **2** (Suppl. 1): 64–72 (1976).

4. **Fox, A. J. & Adelstein, A. M.** Occupational mortality: work or way of life? *Journal of epidemiology and community health,* **32**: 73–78 (1978).

5. **Tola, S. et al.** Occupational lead exposure in Finland. VI. Final report. *Scandinavian journal of work, environment and health,* **2**: 115–127 (1976).

6. **Nordman, C-H.** *Environmental lead exposure in Finland. A study on selected population groups.* Thesis, Helsinki, 1975.

7. **Holmberg, P. C.** Central-nervous-system defects in children born to mothers exposed to organic solvents during pregnancy. *Lancet,* **2**: 177–180 (1979).

8. **Holmberg, P. C. & Hernberg, S.** Congenital defects and occupational factors. A comparison of different methodological approaches. *Scandinavian journal of work, environment and health,* **5**: 328–332 (1979).

9. **Miettinen, O. S.** *Notes for the WHO course in scientific research methods of the epidemiology of industrial intoxications, Helsinki, October 1975.* Copenhagen, WHO Regional Office for Europe, 1975 (Unpublished document).

10. **McMichael, A. J. et al.** An epidemiologic study of mortality within a cohort of rubber workers, 1964–72. *Journal of occupational medicine,* **16**: 458–464 (1974).

11. **Holmberg, P. C. & Nurminen, M.** Congenital defects of the central nervous system and occupational factors during pregnancy. A case-referent study. *American journal of industrial medicine,* **1**: 167–176 (1980).

12. **Saxén, L. et al.** A matched-pair register for studies of selected congenital defects. *American journal of epidemiology,* **100**: 297–306 (1974).

13. **Miettinen, O. S.** Proportion of disease caused or prevented by a given exposure, trait or intervention. *American journal of epidemiology,* **99**: 325–332 (1974).

14. **Cole, P. & Merletti, F.** Chemical agents and occupational cancer. *Journal of environmental and pathological toxicology,* **3**: 399–417 (1980).

15. **Morris, J. N.** *Uses of epidemiology.* Edinburgh, Livingstone, 1970.

16. **Hänninen, H. et al.** Psychological tests as indicators of excessive exposure to carbon disulfide. *Scandinavian journal of psychology,* **19**: 163–174 (1978).

17. **Seppäläinen, A. M. & Tolonen, M.** Neurotoxicity of long-term exposure to carbon disulfide in the viscose rayon industry: a neurophysiological study. *Work, environment, health,* **11**: 145 (1974).

18. **Tolonen, M.** Chronic subclinical carbon disulfide poisoning. *Work-environment-health,* **11**: 154–161 (1974).

19. **Hernberg, S.** Incidence of cancer in populations with exceptional exposure to metals. *In:* Hiatt, H. H. et al., ed. *Origins of human cancer. Book A: Incidence of cancers in humans,* vol. 4. New York, Cold Spring Harbor, 1977, pp. 147–157.

20. **Axelson, O. et al.** Arsenic exposure and mortality: A case-referent study from a Swedish copper smelter. *British journal of industrial medicine,* **35**: 8–15 (1978).

21. **Nordberg, G. F., ed.** *Effects of dose-response relationships of toxic metals. Proceedings of an International Meeting on the Toxicology of Metals, Tokyo 1974.* Amsterdam, Elsevier, 1976.

22. **Seppäläinen, A. M. et al.** Relationship between blood lead levels and nerve conduction velocities. *Neurotoxicology,* **1**: 313–332 (1979).

23. **Schneiderman, M. A.** The numerate sciences—Epidemiology and biometry. *Journal of the National Cancer Institute,* **59**: 633–644 (1977).

24. **WHO Regional Office for Europe.** *Evaluation of occupational health and industrial hygiene services.* Report on a WHO Working Group. Copenhagen, 1982 (EURO Reports and Studies, No. 56).

18

Reappraisal of an
epidemiological study

M. Nurminen[a]

The etiology of ischaemic (coronary) heart disease[b] (IHD) and its manifestations are characterized by their multifactorial nature. Current medical knowledge regarding the fundamental pathogenesis of this disease is still inadequate. The prevailing opinion is that IHD is either caused or prevented by many determinants. It has also been postulated that the simultaneous presence of several high-risk factors for IHD increases the liability to develop the disease.

Chemical exposures potentially predictive of IHD and perhaps even conducive to IHD are likely to be found not only in a person's microenvironment but also in the occupational environment. In the literature dealing with the health hazards of carbon disulfide (CS_2), circumstantial evidence published in the late 1940s, in the 1950s, and in the 1960s suggested that exposure to this compound could promote atherosclerosis and, therefore, also IHD. The source of the circumstantial evidence was, for the most part, clinical impressions of works physicians based on incidental cases of occupational disease or reports on small exposed groups without any adequate reference experience.

A theoretical impetus for reviewing the early work was that if the mechanism of CS_2-induced atherosclerosis and IHD in particular could be explained, it might promote the study of the etiology of atherosclerosis in general.

A practical motive was also evident: the discovery or confirmation of the operation of a new risk factor would compel action to be taken to improve the working conditions of some two hundred thousand workers employed in the viscose rayon industry in various industrialized

---

[a] Institute of Occupational Health, Helsinki, Finland.

[b] Since the project described here was first started, the term "coronary heart disease" has come to be used in English-language publications when referring to "ischaemic heart disease". However, the latter term is more appropriate since the coronary arteries have not, in fact, been examined (e.g., by means of angiography).

countries. In Finland, for instance, over 700 employees were occupationally exposed to CS_2. The corresponding number in Japan was some 20 000.

In the viscose rayon industry in the United States, no cases of any kind of occupational disease had ever been officially reported. This is possibly due partly to consideration of the legal consequences. In Finland, on the other hand, where the viscose rayon method has been used for the production of artificial fibres since 1938, compensation had been paid for a total of about 250 cases of occupational disease (acute and chronic) due to CS_2 at the time the study was initiated. During the period 1964 to 1967 alone, 76 cases had been reported to the Institute of Occupational Health in Helsinki, where a nationwide occupational disease register is kept. No occupational diseases resulting in IHD had been registered before the commencement of the project to be described.

Formulation of the problem

The specific scientific problem considered in the entire study project was: does prolonged exposure to CS_2 cause IHD? Clearly, therefore, it was a question of etiological or rather hazard research. From the point of view of medicine, the concern was to focus on the particular illness (IHD) and to try to trace the various pathways of its pathogenesis. In contrast, from the occupational health viewpoint, the focus was on a particular determinant of morbidity (CS_2) and the broad concern was with the entirety of the health effects of this chemical risk factor. It had been long known that overexposure to CS_2 may cause central and peripheral neurological disturbances, optic neuritis, atherosclerosis of the cerebral arteries and the arteries of the lower limbs, and psychiatric and psychological disorders. The possible effect of CS_2 on the coronary arteries had received less attention, at least in patients with overriding manifestations. In spite of the semispecific, exposure-centred nature of the general problem, the project concentrated more specifically on the health effects: on the spectrum of coronary mortality and morbidity, and on some factors known to be coronary risk indicators.

The particular nonscientific problem that had first to be solved (or "tested") was whether exposure to CS_2 at the concentrations measured $(30-120 \, mg/m^3)$ in a Finnish viscose rayon plant was a cause of overrepresentation of IHD in the working community. If this were found to be true, the next problem was to quantify ("to estimate") the strength of the association. The pragmatic problem was deemed important since, if a real relationship existed, it would be possible to consider specific interventive measures to prevent the effects of CS_2 exposure.

The scope of the study in terms of exposure was basically unrestricted, in the sense that the entire period since 1942 (when the plant was established) to the commencement of the study was included. An English mortality study (1) had indicated that excess cardiovascular mortality was strongly dependent on the intensity of exposure to CS_2, but there was no evidence that hydrogen sulfide (H_2S) would also increase the mortality. H_2S is a by-product of the viscose rayon process; H_2S and CS_2 vapours are released concomitantly into the ambient air. Furthermore, the concentrations of H_2S in the Finnish viscose rayon plant were reported to be usually about one-tenth of those of CS_2. On these grounds, H_2S was not regarded as an etiological factor for IHD; in any case, the role of H_2S remained subordinate. The English study referred to above had concentrated on men exposed for 10 years or more. The Finnish study was restricted to those men working in the contaminated departments—primarily the spinning and spinning-bath departments—for a minimum of five years. However, the decision was taken to study certain categories separately, e.g., the spinners, or men with at least 15 years of exposure, or men with the highest dosage, which meant considering both the intensity and the duration of exposure. Because the exposure had begun at various ages, had lasted for varying periods of time with interruptions, had involved highly fluctuating concentrations, and was considered to have a cumulative effect, the concepts of dose and time lag from onset of exposure to subsequent outcome were not of immediate practical use for operational purposes.

An important characteristic of the people studied was their employment status, which had implications not only for selection and for the validity of comparisons but also for the definition of the actual issue under investigation. In other words, should attention be concentrated on the mild forms of "coronary" electrocardiographic (ECG) findings among workers currently gainfully employed and with no previous history of coronary artery disease, or should the emphasis be placed on severe outcomes, in which case former employees on disability pensions ought to be included in order to ensure that the coverage extended to a sufficient number of cases? In the event, it was decided to opt for the second alternative, i.e., comprehensive coverage, as being the best way of establishing the role of the occupational factor in the development of IHD.

Study design

Study protocol
The protocol was drafted as completely as possible before the start of the study and is being updated as the study proceeds and modifications become necessary.

At the time when the idea of the CS_2-IHD study first began to take shape as a research project, it was not the prevailing custom in Finnish medical research practice to draft a study protocol as an indispensable first step. As far as a formal study plan is concerned, the reported study was no exception. In all fairness, however, it has to be remembered that for the first phase of the project in 1966/67 (see Table 1) no petition for a grant was submitted (this would have required an annexed study plan), nor did there exist, at that time, a research committee at the Institute of

Table 1. Time frame of IHD mortality and morbidity studies together with the reporting dates

Year of study		Type of study	Reporting year
1st phase	1942	Retrospective follow-up of IHD mortality and morbidity	1970
	1967/1968	Cross-sectional health examination	1970, 1971
2nd phase	1972	5-year prospective follow-up of IHD mortality and morbidity	1973, 1975
		Cross-sectional health examination	1975
3rd phase	1975	8-year follow-up of mortality	1976, 1976
	1977	10-year follow-up of mortality	1979, 1982
	1980	13-year follow-up of mortality	1981

Occupational Health, which would have demanded a rigid plan. However, in an attempt to achieve this goal, a meticulous record of progress and of the examination findings was kept by the principal investigator, who masterminded the project.

By the time the second phase of the project was being planned in 1972, the approach to research organization had matured, so that a proper study plan was formally presented to and accepted by the newly founded Research Committee of the Institute of Occupational Health in Helsinki.

Time frame

Table 1 shows schematically the overall time frame of the project, incorporating several different types of study. The first phase involved: (a) a retrospective follow-up of total and coronary mortality and an inquiry into previous coronary and other morbidity and the presence in the past of relevant coronary risk factors; (b) a cross-sectional health examination, which concentrated on the state of the cardiovascular system, including an exercise electrocardiogram, chest radiography, measurements of blood pressure, and determinations of blood lipids, glucose tolerance, and plasma creatinine, together with a questionnaire concerning the prevalence of current diseases, medication, and the presence of relevant coronary risk factors.

The second phase comprised a prospective follow-up of total and coronary mortality and incidence of coronary morbidity; this ended five years later, in 1972, with a cross-sectional medical re-examination.

The third and final phase of the project involved: (a) the ongoing follow-up of total mortality, and mortality from coronary heart diseases, cardiovascular diseases, and cancer; (b) a combination of retrospective and prospective studies, where information on coronary risk indicators was ascertained either retrospectively, i.e., prior to 1967, or at the examination in 1967/68, while the coronary end-points were determined prospectively.

A single (exposed) cohort was formed retrospectively, while two cohorts (one exposed and the other nonexposed) were followed up prospectively; the former approach might be termed a nonselective follow-up study, the latter an exposure-selective follow-up study. Similarly, the two cross-sectional studies may be defined as exposure-selective approaches, since subjects were drawn by taking different sampling fractions from the different exposure categories.

In the analysis of the second phase of the project, as will be shown later in the section on data analysis (p. 359), yet another approach was tried. Albeit not initially so designed, it bore a certain resemblance to a case–referent study nested in a cohort that is being followed up. Coronary death cases and noncases yielded by the 5-year follow-up of the exposed cases were classified according to their risk indicator status in 1967 or earlier. The design aimed at an evaluation of a possible dose–response relationship. But, as is almost always the case with *ex tempore* presentations, so here, too, the analysis was doomed to remain elementary.

Exposure category and exposure data

The exposed workers who constituted the study population came from an industrial environment that was almost unique at that time in Finland. The only other viscose plant, situated in the southernmost part of Finland (Hanko), was smaller and was founded in the 1950s; no cases of occupational disease were detected there until the early 1970s.

When considering the possibility of extrapolating from the sample selected to the population at large, it is perhaps instructive to examine the stepwise reduction process that led to the selection of the groups studied. First, we had the general population, which here means the occupationally active Finnish male population. Second, there was the target population exhibiting the particular condition to be studied, that is the workers exposed to CS_2. Third, we selected from these exposed workers those who could most readily be observed, namely, the workforce at the viscose plant in Valkeakoski, a small industrial town in the southern mainland of the country. Fourth, those meeting certain inclusion criteria were defined as the sampled population. Finally, we chose from the cohort population those who were available for follow-up. The above selection process underscores the nonstatistical nature of the approach commonly used in nonexperimental research when attempts are made to arrive at a suitable exposure category.

A serious problem in "industrial epidemiology" is the difficulty of obtaining relevant exposure data. Usually, only aggregate data or secular statistics are available and individual measurements of any sort can rarely be obtained. Different levels of measurement may be listed (from low to high): separation into exposed and nonexposed categories; classification by work area, type of work, or occupation; overall samples from stationary predetermined sites (and predetermined points in time,

345

usually once a month) taken under normal operational conditions; air samples taken from workers' breathing zones; biological monitoring.

Clearly, quantified dose or exposure indicators are needed for estimating the exposure–response relationships and for finding a "safe" exposure or a non-effect level. Some important dimensions of exposure that must be considered for this purpose are: quality; intensity; duration; fluctuation; calendar time; concurrent exposure(s).

In the CS_2 study, the mean concentrations of CS_2 and H_2S at the plant were measured and graphically recorded, first from 1945 to 1967 and later up to 1975. Approximately 3000 stationary measurements were available in 1967, based on 5-minute and 10-minute air samples taken 1–36 times yearly at 10–40 different sites. Roughly, the highly fluctuating concentrations of CS_2 and H_2S were higher than 40 ppm before 1950, 20–40 ppm between 1950 and 1960, 10–30 ppm in the 1960s, and not until the 1970s did they fall below 10 ppm[a] in nearly all the departments (see Fig. 1).

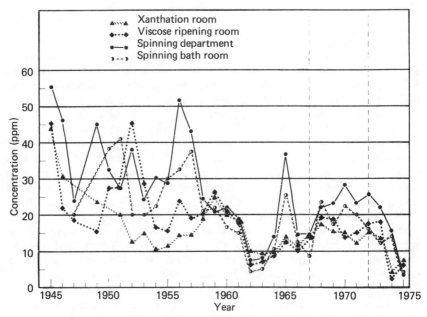

Fig. 1. Concentration of $CS_2 + H_2S$ in the rayon staple fibre factory

---

[a] Concentrations in ppm are used here to conform with the values given in the original publications. Conversion to mg/m³ is given by

$$\frac{\text{ppm} \times \text{molecular weight}}{24.45}$$

Note the decrease between 1972 and 1975.

346

An approximate measure of the personal history of exposure was developed for each subject using an index of exposure dosage. The index was a duration-weighted average of the intensity of exposure, the formulation of which was Sum(i) Sum(j) $n(ij)$ $m(ij)$, where $n(ij)$ = number of months worked at department i during year j and $m(ij)$ = annual arithmetic mean of the carbon disulfide and hydrogen sulfide concentrations (ppm) in the air of department i for year j.

Despite obvious limitations, including sparse work histories, the possible unrepresentative character of the air sampling with respect to the true individual exposure, and considerable intrafactory mobility, it was assumed that the index would give a reasonable measure of each subjects' total exposure to CS_2 and H_2S vapours.

Exposed workers were compared with the referent group at four levels: (1) total exposed group; (2) spinners exposed for at least two years; (3) workers with 15 years or more experience in polluted departments; and (4) workers whose exposure–dose index exceeded the mean for the exposed group.

The distribution of the 347 men examined in terms of the nature of their work is given in Table 2. Of the workers, 62% were engaged in exposed work at the time of medical check-up; the rest (38%) had a history of exposure. At that time, in 1967, no measurements of exposure were made using personal samplers.

Table 2. Distribution of exposed group according to type of work

Type of work	Number	
Management and laboratory personnel	18	
Rayon stable fibre factory	147	
Xanthation room		7
Viscose ripening room		23
Spinning department		53
Spinning-bath room		19
Salt factory		6
Repair, maintenance		39
Rayon filament factory	127	
Viscose ripening room		26
Spinning department		68
Spinning-bath room		10
Repair, maintenance		23
Viscose film factory	25	
Xanthation room		1
Viscose ripening room		8
Film-machine room		15
Repair, maintenance		1
Carbon disulfide factory	26	
Filtering		2
Retort operating room		8
Carbon room		12
Repair, maintenance		4
Total	343	

Criteria of admissibility

All men fulfilling the following eligibility criteria were included: (1) age 25 to 64 years at the time of the first examination (1967/68); (2) five years or more of continuous or interrupted exposure to CS_2 in the most heavily polluted departments (see below) during any period between 1942 and 1967. Also included were workers who were no longer employed by the plant or who were working in "clean" jobs at the time of examination; (3) in addition, those who had died before reaching the age of 65 years were included, provided they met criteria (1) and (2).

The factory records from 1942 onwards were used for the selection of subjects fulfilling the above criteria. The records were said to have been satisfactory, and so the group could be considered a complete tally of the relevant workers. The total exposed group (including 45 subjects who had died) comprised 410 men. Of these, 212 (211 examined, one unexamined) were working in the polluted departments at the time of study, but 81 had moved to "clean" jobs within the plant. There were another 72 men who no longer worked at the plant; most of them were traced and 51 examined. Of the remaining 21 men in this group, 12 returned a letter saying that they were not willing to undergo the examination. Thus, only 9 men were not traced at all, not even in registers, because the information regarding the person number (birth date) was incomplete. In all, 343 (93.3%) of the 365 men who were known not to have died, were examined.

When the periods of exposure, originally recorded from the employee rolls, were checked against the subject's own (and in few instances against his foreman's) reports, 21 subjects were found who had been exposed for less than five but at least one year. Nevertheless, these men were included in the study. The structure of the exposed group is given in Table 3.

The health status of the exposed group could have been affected by the various kinds of selection on entering employment at the plant, or by therapeutic measures undertaken for or by those employed. A possible bias may also have been introduced by untraced cases and those who refused to participate. Hypertensive applicants for jobs had usually been excluded at the pre-employment examination; they consequently did not appear in the sample. Exclusion because of cardiac conditions had been very rare. No subjects receiving drug therapy or prescribed diet were excluded from the sample on either of these grounds. Subjects receiving sickness insurance compensation (11 men), and accordingly not working, were excluded since appropriate referents could not be found. Five men who were receiving compensation for CS_2 poisoning were, however, included. The evidence available on the health status of the 12 untraced men who were known to be alive was not usable.

Choice of referents

In a comparative study, the choice of the reference entity is crucial to the attainment of a valid basis for comparison. In a follow-up study, an ideal reference group should share all the characteristics of the study

348

Table 3. Structure and coverage of the exposed sample

(a) *Alive*

| | Examined | | Not examined | | Total |
	No.	%	No.	%	
Employed at plant	292	99.7	1	0.3	293
Left the plant	51	70.8	21	19.2	72
Total	343	93.3	22	6.7	365

(b) *Dead*

| | Death certificate obtained | | Death certificate not obtained | | Total |
	No.	%	No.	%	
At the time of first examination	43	95.6	2	4.4	45
By 2 years after start of the follow-up	5	100.0	–	0.0	5

group relevant to the problem at issue, save the property that defines the groups, namely, the exposure. In the CS_2 study, the referents were selected from the employee rolls of a paper mill located in the same town, without any knowledge of their health status. The assumption was that selection for this work, if any, would be much the same as for the viscose workers, and also that the two types of occupation within these industrial branches had the same effects on the occurrence of IHD apart from the effect of CS_2.

The 343 exposed subjects who were alive and had been examined in 1967/68 were paired off or matched individually with the same number of referent subjects who had no histories of exposure or whose exposure had not been significant (less than six months' exposure to CS_2 or other industrial poisons). The matching was based on the criteria of (1) maximum intrapair age difference of three years at most; (2) approximately the same birth district; and (3) approximate equality in the physical requirements of the work. The matching by birth district was done since considerable interdistrict differences in cardiovascular mortality had been reported in Finland (the so-called East-West Study). This was of importance since people (personnel of a fibre company) from the ceded Karelia, formerly the easternmost province of Finland, had moved some 400 km westwards to work at this viscose rayon plant. Some difference had also been observed in cardiovascular status between referents born in the west and those born in the east of Finland. The two groups were roughly similar in occupational status (production work, repair and maintenance, technical management, and office work). A minimum period of five years' continuous employment in the papermill was also required. As the referents lived in the same macroenvironment as the exposed persons, systematic dissimilarities in factors such as softness of their drinking water seemed unlikely. Pre-employment examinations made at the mill had criteria not too dissimilar to those applied in the viscose rayon plant. Referent subjects under drug therapy or on a diet were not disqualified in the matching. All referents were also employed at the mill at the time of the investigation. The management were not allowed to influence the selection in any way. About 20 of the referent candidates selected did not show up for the examination. In place of these persons, new referents were selected as described above. The possible implications of this replacement procedure from the validity point of view were not investigated, however.

Handling of confounding variables

The two properties that a *confounder* must satisfy have generally been defined as:

1. It is predictive of the health outcome in the studied population-experience.

2. It is associated with the determinant under study.

The more demanding problem is the verification of the first condition of this definition. The recognition of health predictors is a matter of substantive judgement, which requires the theoretical arguments or the empirical evidence of earlier studies rather than a decision-making process based on statistical criteria.

The dilemma posed by a conceptual identification of every important confounder is that, in addition to those factors taken into account in this study, one may have been omitted that, together with the factors included, constitutes a confounder. There is a further problem: even when all the plausible confounders have been found to be identically distributed between the compared groups when taken separately, their joint effect may still cause confounding.

Naturally, there is a practical limit to the number of variables that can be admitted to the study, or at least simultaneously subjected to statistical adjustment. The problem is compounded by the search for operationally accurate variables and by the need to devise adequate indices for the confounders. This often underrated task is aggravated by the fact that the controlling of extraneous variables actually involves two goals. Primarily, bias should be prevented. Secondly, even if there seems to be no danger of bias, the presence of the disturbing variable may inflate the variance of the estimator of the outcome parameter between the compared groups to an extent that makes the comparison imprecise; and such imprecision should also be avoided.

When dealing with the problem of confounding, the first step in planning and executing a nonexperimental study is to list all the known and plausible confounders in conceptual terms. The confounders are generally arranged in three classes. (1) Major variables for which some kind of control is considered essential. The number of these variables is

350

preferably kept small in view of the complexities involved in controlling many variables at the same time. (2) Variables that, ideally, one would like to control, but instead one has to content oneself with some verification that their effects produce little or no bias of practical importance. (3) Variables whose direct or mediating effects are thought to be minor and can therefore be disregarded. Variables for which no data are available in the study may be included in this category. The next step is to design the study so as to record the confounders. The recordings are then translated into statistical variables. Further, those variables are selected that are to be adjusted, either at the design stage or at the stage of analysis. Finally, an attempt is made to find some method of removing or reducing the biases that such variables may cause.

The selection of the important confounding variables from the list of potential confounders does not involve a process of classical statistical inference, but rests mainly on substantive considerations. It could be argued that the notion of confounding inherently belongs to the realm of Bayesian inference in that prior knowledge of various causal links is necessary for distinguishing variables that are logical candidates for confounders.

Two main strategies are employed in designing a valid and precise study. In restriction and matching, the samples are selected from the populations in such a way that the distributions of the confounding variables are similar in some respects in the two sets of samples. At the stage of analysis, an attempt may be made to condition the effect of exposure by stratification analysis. When this is not feasible because the number of strata is too large, an alternative procedure is to estimate the magnitude of the effect that the bias and/or the imprecision will have on the outcome or effect parameter, in order to see whether a material improvement can be achieved through adjustment by modelling. Table 4 summarizes the procedure followed in the study under review.

In the above commentary, three types of covariable were considered in deciding on the necessity for matching. Using the following notation: X = suspected causal factor defining the compared groups, i.e., combined CS_2 and H_2S exposure; Y = outcome variable, i.e., occurrence of IHD; and Z = confounding variable, i.e., an extraneous covariable complicating the relation between X and Y, it is possible to construct simple causal models to describe the relation of the confounding variable to the cause–effect relationship being studied (see Fig. 2).

Age is often a good example of a confounder. This study was no exception. The incidence of IHD increased naturally with age in both the exposed and nonexposed populations, so that age was a genuine indicator of the risk of IHD. Age was also associated with occupational exposure to CS_2 because older people had usually been exposed longer to CS_2 than younger persons. Hence, aging complied with both the basic requirements for confounders; i.e., (1) it was predictive of the disease outcome (acute ischaemic event), and (2) it was associated with the determinant (exposure to CS_2).

Table 4. The attainment of comparability between the exposed and the referent group

Characteristic	Similarity	Methods
Sex	Same (only men)	Restriction
Age	Similar (range 25–64, pairs within ±3 years)	Restriction, pair-matching, modelling
Birth district	Similar (East, West, North Finland)	Pair-matching
Physical demands of work	Similar	Pair-matching
Occupational status	Similar (min. duration of employment 5 years)	Restriction, consequence of matching
Residential areas (drinking water, etc.)	Systematic differences unlikely	Questionnaire
Smoking habits	No major differences in frequency distribution	Questionnaire, modelling
Physical exercise during leisure time	No major differences in frequency distributions	Questionnaire
Previous or present disease (e.g., diabetes or myocardial infarction)	Minor differences tending to minimize the effect	Questionnaire, stratification
Drug therapy	No major differences in frequency distributions	Questionnaire
Relative body weight	No major differences in mean value and dispersion	Measurement
Physical work capacity	No major differences	Ergometry
Diet	No major differences	Deduction, estimation

Attention may also be drawn to the fact that, as already indicated, the birth district had to be taken into account. This is a specific Finnish problem. Eastern Finland has a considerably higher rate of coronary mortality than western Finland, and a preliminary screening had shown that 40 % of the viscose rayon workers were born in eastern Finland against 25 % of the papermill workers. This illustrates well the importance of knowing the background to the problem under study when the reference group is defined.

The groups were intentionally not equalized with respect to blood lipids, blood pressure, and blood glucose concentration because the literature reported that these indicators of risk may represent the mechanisms through which CS_2 causes IHD. It is important to avoid matching on the suspected intermediate steps in the etiological sequence from exposure to disease, since such a procedure is conducive to masking a true effect.

The question also arose whether cigarette smoking should be considered a confounder in the study. It is well known that smoking is a risk factor for IHD. Smoking had also proved to be a risk factor in this

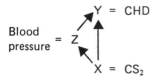

Model	Role of covariable
Age = Z (Y = CHD, X = CS₂)	A confounding variable Z causing systematic bias; it needs to be matched or otherwise controlled.
Birth district = Z (Y = CHD, X = CS₂)	It increases the variation of Y through the correlation between Y and Z; it may be matched for the sake of security should there be a connection between Z and X in these data.
Blood pressure = Z (Y = CHD, X = CS₂)	A link in the causal path; it should not be matched. In addition X has a direct effect on Y.

Fig. 2. Simple causal models depicting the role of a covariable Z in the study of the relationship between the outcome variable Y and the determinant X

Key: → = causation
/ = association

study by the five-year follow-up analysis; hence it fulfilled the first requirement for a confounder. For the other requirement to be fulfilled, the viscose rayon workers should have smoked either more or less than the referents. As smoking habits were not known in advance, the smoking histories had to be obtained during the examinations.

From Table 5 it can be seen that 55 % of the exposed and 49 % of the nonexposed subjects were smokers. Since the percentage of smokers differed in the compared groups by only 6 %, the comparability must be regarded as reasonably good.

Table 5. Frequency of cigarette smoking among the two cohorts compared

No. of cigarettes per day	Exposed[a] (%)	Nonexposed[a] (%)
None	45.4	51.3
1–9	9.9	9.0
10–19	24.2	24.2
20–29	17.2	12.8
30 +	1.8	1.2
Not known	1.5	1.5
Total	100.0	100.0

[a] N = 343

353

Although checks on the distribution of confounders are sometimes made by tests of significance, it is not at all clear what kind of assurance is provided by the finding of a nonsignificant result, or by a significant result for that matter. A recommendable approach would be to use regression modelling techniques to examine the relative increase in the variance of the outcome variable due to variations in the confounder. An alternative method of assessing the effects of a confounder is mentioned in the section on data analysis.

Selection processes

Among the factors that proved impossible to control was the unknown extent of selection at the time of entering employment. A certain degree of self-selection was probable, but it is hard to tell in which direction it might have influenced the results. Prior to 1967, the chemical hazards of CS_2 exposure were not well known to the public; the job seekers were inclined to avoid physically arduous posts. A selection made on the basis of the pre-employment examination at least excluded persons with manifest hypertension or a severe cardiac condition. It may be assumed that the disqualification was more strict in the viscose rayon plant. Owing to the low prevalence of such disorders, the bias was probably small.

More serious bias may have been occasioned by self-selection during employment. Although only nine out of the 51 examined who had quit employment at the viscose plant reported that health problems were the reason for leaving, the number with less than five years' exposure who had done the same was unknown. In addition, an unknown number of men had been moved to "clean" departments within the viscose plant after periods of less than five years of work in the polluted departments, health problems having been the most common reason. (This policy was not actively favoured by the plant management until the early 1970s, when the first results of this project became known.) As a result of these uncontrollable factors it may be assumed that "healthy" men could best stand the exposure and that such men are overrepresented in this series.

The exclusion of patients receiving disability pensions for causes other than CS_2 poisoning was most unfortunate, but was considered necessary because of the practical impossibility of forming an individually matched group of disabled persons from the papermill when the retrospective selection was started as early as 1942. A selection on any other basis seemed to be haphazard. Reliable comparisons between retrospectively compiled mortality rates in the two factories were also said to have been impossible. Finally, the fate of those 12 exposed persons who refused to participate in the examination was not discussed.

In all, there were approximately 9500 employee cards (men and women) in the employer's file at the papermill. It was estimated by the investigators that 2700 men would have complied with the inclusion criteria. Of these, about 700 (26 %) had left the mill for one reason or another. It must be admitted that a retrospective mortality study of the former papermill workers would have been a most onerous task.

354

Perhaps the validity of such a study would, in any case, have been questionable. But, it remains a cosmetic error in the study design that matched referents were not obtained for all those old age and disability pensioners from the viscose plant who were pensioned off for a reason other than the effects of CS_2.

Project organization
The study was planned in 1966 by a team consisting of three members, an assistant professor in occupational medicine, a graduate in biometry and science, and a final year medical student. In addition, the team included as a standing member a research assistant capable of computer programming. The design was submitted to an outside consultant experienced in epidemiological research methodology, and various specialists from the staff of the Institute of Occupational Health in Helsinki were also consulted.

The project was put into effect with the cooperation of the management, employees, and health personnel at the viscose plant and at the papermill.

In 1972, the 5-year IHD follow-up was expanded to include psychological, neurophysiological and neuro-ophthalmological investigations. Since then, the 8-, 10-, and 13-year follow-ups of IHD mortality data have been collected, analysed and published (see Table 1).

Efficiency aspects of the project design

The sampling procedure
The employee rolls kept by the two employers formed the basis for the selection of the sample or, as it is called in statistics, the sampling frame. The restrictions imposed on eligibility for the study have already been described. For those of the exposed population considered eligible, the study aimed at complete coverage, that is a 100 % sampling fraction, in the terminology of probability sampling theory. As noted earlier, this resulted in a sample of 365 men, a sample size that was judged to be statistically sound (see p. 356).

For the referent sample, the selection process was nonrepresentative in terms of statistical sampling theory, its sole purpose being to achieve comparability with the exposed group. Perhaps the method could best be described as a kind of quota sampling where latitude is allowed with regard to the referents to be included. The enumerator was instructed to continue the unsystematic inclusion of workers until the necessary "quota" had been obtained in each stratum defined by the three matching variables (age, birth district, and physical demands of work). If the investigators had wanted to obtain a statistically representative comparison sample of workers from the paper industry, they would have had to choose the referents at random, with assigned sampling fractions within the matching strata. A considerable amount of field work would have been required to fill all the quotas, however, since in the later stages most of the persons enumerated would have fallen into quotas already filled. The approach would have necessitated a much coarser age grouping, too. In fact, it would have meant a shift from individual matching to frequency or distribution matching, and the latter method is generally not sufficient for the elimination of bias.

Feasibility of matching

At this point, it is appropriate to consider the question: When should pair-matching be used to control bias? Roughly speaking, this popular technique could be employed in the following instances: (a) when there are only a few powerful confounders; (b) when the confounder possesses a complex nominal scale, i.e., when there are many possible combinations not having a natural grouping (e.g., birth district); (c) when the number of subjects studied is small; (d) when there is an ample reservoir of referent candidates, for whom the information on matching variables is readily retrievable (e.g., on a magnetic tape).

In this CS_2 project, individual matching was performed for three covariables, thought to be the most powerful confounders, namely, age, birth district, and type of work. Of these, the last two were matched for the reason given in (b) above. Had they been handled by stratification in the analysis, this would have resulted in unbalanced occupation of the cells in the stratification table, thus leading to low size efficiency, i.e., low information yield per subject.

The numbers of exposed and nonexposed subjects involved in the study were not very large, but because the matching process was carried out manually, the task was necessarily tedious. Perhaps the overriding argument in favour of matching was the greater size of the referent candidate pool. It was estimated that the number of men with at least five years' work experience and currently employed at the mill was about 2000, or four times as many as at the viscose rayon plant. This had the advantage of securing the finding of a suitable "match" for nearly every exposed subject. Even in the few instances where it turned out later that the chosen referent had previously been working at the viscose rayon plant, it was possible to find a replacement for him.

However, it was decided not to include more than one referent per index subject in the final series in view of the costly and time-consuming examinations required.

Determination of sample size

Another question that needs to be considered is: When is a sample statistically and otherwise of the proper size? The more formal issues involved in answering this question include: (a) the power of significance tests to detect differences and associations in the data; (b) the frequency/rarity of the outcome events *in the study population*; (c) the demand for controlling extraneous variables in the analysis stage by multiway stratification or modelling techniques, and the resultant search for adequacy of unconstrained observations ("degrees of freedom"); and (d) the roughness of approximation resulting from the use of large sample (asymptotic) methods versus the availability of small sample

356

(exact) algorithms, which were often cumbersome from the computational point of view. In addition to these formal issues, however, it was necessary to take into account a number of informal considerations in reaching a decision about the amount of information that the study should be designed to yield.

The retrospective follow-up of the exposed persons covered the period from the establishment of the viscose rayon plant in 1942 until the end of 1966. In this respect the follow-up was of maximum duration. The prospective follow-up was initiated with no decision as to its ultimate duration. This was, of course, understandable inasmuch as, in 1967, the investigators had only a vague idea of the existence or magnitude of the effect of CS_2 on the incidence of IHD. The relatively small cohort sizes of 343 men obliged the team to carry on with the follow-up for five years before the first statistically "significant" results could be made public at the end of 1972.

It is also conceivable that the statistical criterion of "significance" should be relaxed in small-scale studies, especially when such severe consequences as mortality are being considered. The results of studies of this kind can sound a warning of the need for preventive action if the estimated rate in the exposed group markedly exceeds the estimated rate in a comparable reference group, even when the customary significance test does not reach the critical P value of 0.05, mainly because of the small number of cases that accrue in a limited follow-up experience. It ought to be remembered that significance tests are statistical instruments for decision making. If they are intended for use as warning devices in industrial epidemiological surveys, they would need to sound the alarm before the "guilt" of the occupational exposures concerned has been proved beyond any reasonable doubt.

The results of the prospective mortality follow-up reflected, to a large extent, the effects of past exposure. After eight years of follow-up, less than 20% of the original cohort members were still working in conditions where they were subject to exposure. The number of deaths from IHD in both cohorts combined was 29, i.e., only 4.2% of the initial populations. Thus, the calculated rates still involved small numbers and tended to fluctuate over successive periods of time. The person-years totalled 5251 in 1975, and it was anticipated that the follow-up would be continued for at least another 8 years, which would bring the number of person-years to 10 000.

Experiences in data collection and data management

As far as the investigators recall, the feasibility of the initial study protocol was not jeopardized by any notable unexpected surprises in the data collection. It was recognized, however, that many of the laboratory data were collected mainly for the sake of creating a reliable "clinical" impression. For example, the exercise ECG programme, with one recording lasting about 45 minutes, was an extremely arduous

357

undertaking for both the subjects studied and the investigators. The enforcement of the schedule for the glucose tolerance testing was also complex and time-consuming.

The miniature pilot study was primarily intended to ensure the faultless installation and functioning of the ECG equipment. In addition, experience was gained in coding the ECGs and reading the radiographs. The pilot study also served the purpose of familiarizing the representatives of the workers and the management at the plant with the true nature of the examinations.

The exposure measurement data were collected by the laboratory staff at the factory, who also decided on the positioning of the stationary samplers. The compilation of the statistics was performed by the members of the research team. The sampling, which had been occasional from 1945 to 1949, thereafter became more regular. The values for the years 1942–44 were obtained by a kind of information "maximum likelihood" extrapolation, that is, by the so-called "best guess method". Because great emphasis was placed on validity considerations when both the exposure and the outcome indicators were measured, no compelling arguments arose during the data acquisition stage that would have led to a reformulation of the problem posed.

Data acquisition
A questionnaire covering birth place, migration history, smoking habits, physical activity, previous and current diseases, and present medication was prepared especially for the study and was mailed to each subject included in the study. The history of IHD was elicited by means of the Rose questionnaire (2), which allows for the classification of possible symptoms according to their severity and the probability of IHD etiology.

In order to ensure comparability, all the methods used in measuring blood pressure, in recording and coding ECGs, in calculating heart volumes, etc., were exactly the same in the two cross-sectional examinations of 1967 and 1972. Most of the instruments employed were also the same on the two occasions.

Death certificates were obtained from the Central Statistical Office of Finland. At that office the forms (originally filled out by the practising physician or, in some cases of obviously accidental or suicidal deaths, by the municipal police authorities) are checked for correctness and to achieve uniformity in coding. The coding employed was that given in the seventh revision of the International Classification of Diseases (ICD) (3), according to which the code numbers 420–422 were assigned to IHD. When the eighth revision of the ICD (4) became effective in Finland from the beginning of 1969 the new codes for IHD were adopted. The official codings were re-examined for correctness by the physicians of the research team. In particular, they checked that essential hypertension as the underlying cause of death had not been grouped together with coronary artery disease, but included amongst the cardiovascular diseases.

Qualitative monitoring and control were supported, *inter alia*, by the ECG recordings.Using the so-called Minnesota code, the recordings were coded independently by two trained technicians. Any differing codes were reanalysed jointly. The interobserver variability in coding was assessed in 1967 and in 1972 by the use of the earlier ECGs. The results did not indicate the existence of any bias (i.e., any systematic differences between the coders).

After coding, the primary data were transferred on to punched cards. The final data file on condensed coronary variables comprised about 60 items from the first-phase examination; the follow-up doubled the number of statistical variables. Data retrievability was planned jointly with the related CS_2 studies on other outcomes (neurophysiological, psychological, ophthalmological parameters, etc.). From the medical point of view, the IHD study may be considered sizable, if the total number of parameters is taken as the criterion. In computational terms, however, it may be rated as a small or medium-sized study. During the first and second phases of the study the investigators did not have direct access to the stored data but had to rely on the good offices of the data-processing staff.

The matched design presented some problems in compiling the data files, as no data management system capable of identifying the pairs existed. An easy solution was to treat the variables pertaining to a pair as one record, and to form difference variables.

Data analysis

Broadly speaking, both graphical and tabular displays were used for presentation of the results. Graphical reporting was chosen for the mean levels of exposure concentrations (see Fig. 1) (5–7). This was a natural choice considering that there were four factories and about 15 departments for which the mean levels of CS_2 and H_2S were reported annually from 1950 on. The graphs convey the general trend more vividly than a series of bulky tables would do. Another fitting application of graphical techniques was for illustrating the course of the multivariate risk (rate) function curves (8), the death risk curves (see Fig. 5) (9), and the survival curves (see Fig. 4) (10).

Thanks to the matched design, stratification by age was usually unnecessary. An exception, where the effect of aging (among other factors) was considered, was provided by the thorough analysis of a multidimensional $2 \times 2 \times 3 \times 4$ contingency table (11).

Estimation of effect parameters

The quantification of the effect of CS_2 on IHD was done by computing the incidence or prevalence rates for the exposed (R_1) and the nonexposed (R_0), and two types of relative measure. The rate ratio, $RR = R_1/R_0$, reflects the strength of the association, while the rate difference, $RD = R_1 - R_0$, provides a measure of the breadth of the problem. Various methods of constructing confidence limits for these

359

measures were tried. Refinements in the analysis were attempted, e.g., by the life-table technique. It should be noted, however, that this method provides compensation only for unequal follow-up and cannot correct for bias from loss of cohort members. Fortunately, no members of the original cohorts were lost in the prospective follow-up. The life-table permits the consideration of competing risks of death, that is the estimation of the sole effect of CS_2 on IHD in the hypothetical situation where IHD was the only factor responsible for a fatal outcome.

The 1967 study (5)

Life-table analysis for all causes of death, performed as a comparison between survival rates for the 30- to 54-year-old subgroup of the exposed sample and for the total urban male population in Finland, was uninformative. No tables for the occupationally active members of the population were available. The survival rate for the exposed group, calculated from retrospective data, was consistently higher than that for the general population. This result was interpreted as being due to the selection that operates in any industrial population, and was not taken as indicative of any hazard attributable to the exposure.

In an assessment of the proportionate mortality, twenty-five (52 %) of the 48 deaths occurring in the exposed cohort of 401 men (9 untraced) between the end of the five-year initial exposure period (from 1942 on) and two years after the start of the examination (June 1967) were attributed to IHD. The expected number of deaths from IHD that would have occurred without exposure, adjusted for age and year of death and based on official statistics for the general male population, was calculated to be 15.2. This gave an observed-to-expected ratio of 1.64.

The 1972 study (8)

During the five-year prospective follow-up period, fatal infarctions numbered 14 in the exposed and three in the reference cohort. The expected number of IDH deaths based on the total Finnish male population was about seven, showing how the "healthy worker effect" can indeed dilute the results because individuals incapable of working contribute heavily to the expected numbers in persons of working age. Other causes of death were evenly distributed, seven and six, respectively. The percentage distribution of "IHD events" ("coronary" ECGs, angina pectoris, nonfatal infarctions, and fatal infarctions) showed a disproportionally high fatality rate among the CS_2-exposed workers (Fig. 3).

It was also reported, as in the 1972 study, that blood pressures were significantly higher in the exposed group: there was a maximum difference of 17 % between the cumulative frequency distribution curves for systolic blood pressure of exposed workers and those of referents ($P < 0.001$). The difference for diastolic pressure was 14 % ($P < 0.01$). However, the authors did not comment on the clinical importance they attached to such differences (8 mmHg for systolic and 4 mmHg for

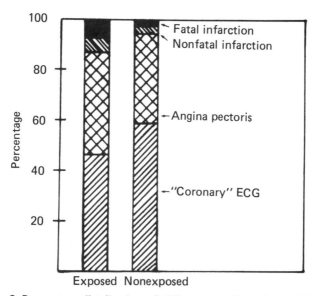

Fig. 3. Percentage distribution of different manifestations of IHD
in CS$_2$-exposed and nonexposed cohorts after five-year follow-up in 1972

Note: The total number of IHD events on which the percentage figures are based was 193
among the exposed and 122 among the nonexposed.

diastolic pressure): a small difference in the mean level of a group could indicate big changes in blood pressure for individuals already in high-risk categories.

The 1975 study *(10)*
Differences in mortality
The accumulated IHD and total mortality data for the two cohorts, along with number of person-years of observation, are given in Table 6. The cumulative incidence rate of IHD mortality was 5.8 % for the exposed cohort and only 2.6 % for the reference group (Table 7). The difference in the incidence rates (CID = 3.2 %) was statistically "almost significant" ($P > 0.05$, two-sided test). The relative occurrence of IHD deaths in the CS$_2$-exposed cohort, in comparison with that of the nonexposed, had stabilized to show an excess mortality of slightly over two-fold.

IHD mortality trend
Survival analyses of the compared groups (cohort life table analysis) revealed that excess mortality from IHD among the CS$_2$-exposed workers was most marked during the fifth year of follow-up (Table 8). In the last period of observation only one case of IHD death occurred in

361

Table 6. Number of person-years for the CS_2-exposed and nonexposed populations in June 1975, together with deaths from all causes and deaths from IHD by age during the eight years of follow-up

Age interval (years)	Exposed			Nonexposed		
	Person-years	All deaths	IHD deaths	Person-years	All deaths	IHD deaths
25–44	924	-	-	948	3	-
45–49	528	2	1	478	2	1
50–54	406	6	4	439	4	2
55–59	382	9	7	393	4	1
60–64	253	9	6	284	5	2
65–72	96	9	2	111	5	3
Total	2589	35	20	2662	23	9

Table 7. IHD and total mortality among the CS_2-exposed and nonexposed cohorts and various derived estimates of death rates (data from the eight-year follow-up)

Cohort (number)	IHD deaths	Deaths from other causes	Total deaths
Exposed (343)	20	15	35
Nonexposed (343)	9	14	23

IHD mortality
Eight-year cumulative incidence rate (CI):
    exposed: $CI_e = 5.8\%$
    nonexposed (reference group): $CI_r = 2.6\%$
Cumulative incidence difference: $CID = CI_e - CI_r = 3.2\%$
    95% two-sided confidence interval (lower and upper bound):
    CID_l, $CID_u = 0.13\%$, 6.29%
Rate ratio: $RR = CI_e/CI_r = 2.22$
    95% two-sided confidence interval: RR_l, $RR_u = 1.03$, 4.77
Etiological fraction: $EF = (RR - 1)/RR = CID/CI_e = 55\%$

the exposed cohort; it was therefore the first year of follow-up with no excess mortality. In place of the cumulative incidence of death, one may use its complementary measure, the survival probability. The accrued survival data are convincingly displayed in Fig. 4. The empirical eight-year survival rate for IHD sufferers was 94.1 % for the exposed cohort and 97.3 % for the reference group. These calculations took into account the changing structure of the cohorts as they were based on person-years of follow-up.

Age-specific mortality
Estimates of IHD mortality in the different age categories have been computed in Table 9. The incidence rate increased sharply with age in the exposed cohort, and only in the oldest age group (65–74 years) did it decrease. The age-specific rates for the workers exposed to CS_2

362

Table 8. Risk of dying from IHD, in terms of incidence densities during each of the eight follow-up years, with estimates of excess mortality attributable to IHD

Interval (x) since start of follow-up (years)	Death rate (incidence densities)[a]		Excess mortality due to IHD	
	Exposed cohort	Nonexposed cohort	Annual excess	Cumulative excess
< 1	5.8	0.0	5.8	5.8
1–2	8.9	0.0	8.9	14.7
2–3	6.0	2.9	3.1	17.8
3–4	9.0	3.0	6.0	23.8
4–5	12.3	3.0	8.3	32.1
5–6	6.3	3.0	3.3	35.4
6–7	9.5	6.1	3.4	38.8
7–8	3.2	9.3	− 6.1	32.7

[a] The incidence densities (ID_x) are expressed in units per 1000 person-years and were computed according to the formula $ID_x = C_x/(N_x - 0.5 (C_x + D_x))$, where C_x is the number of IHD deaths occurring during the follow-up interval x, D_x is the number of deaths from causes other than IHD in the same interval, and N_x is the number of subjects alive at the start of the corresponding interval.

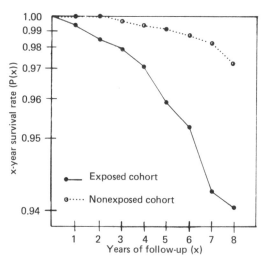

Fig. 4. Log survival curves of CS_2-exposed and nonexposed cohorts starting from the beginning of the follow-up in 1967, assuming that the risk of dying from causes other than IHD is statistically eliminated

Note: An estimate of the x-year survival rate P(x) was calculated according to the formula $P(x) = 1 - CI(x) = (1 - CI(0)) * (1 - CI(1)) * ... * (1 - CI(x-1))$, where CI(x) stands for the x-year cumulative incidence rate.

exceeded those for the comparison group from the age of 50 years on. The crude overall rates, with person-years as the denominators, were greater than the respective values obtained for the relative rates in Table 7, which were calculated using the initial cohort sizes as denominators.

Table 9. Age-specific death rates from IHD in the CS_2-exposed and nonexposed cohorts from 1967 to 1975, and the overall incidence (estimates of risk) of death.

Age (years) (a)	Age-specific death rates (cases per 100 person-years)	
	Exposed	Nonexposed
25–44	0.0	0.0
45–49	1.9	2.7
50–54	9.9	4.6
55–59	18.3	2.5
60–64	23.7	7.0
65–72	20.8	27.0
Overall death incidence rates:		
Crude[a]	7.7	3.4
Adjusted[b]	9.5	7.1
Cumulative[c]	35.3	25.9

[a] Crude incidence rate = (total number of IHD deaths)/(total number of person-years) = sum (a) D_a/sum (a) P_a = sum (a) P_a (ID_a)/sum (a) P_a. The crude rate thus involves implicit weighting of the age-specific rate ID_a to take account of the number of person-years P_a.

[b] In the adjusted incidence density rate (ID, force of mortality) the implicit weights in the crude rate have been replaced, explicitly, by weights proportional to the widths of the age categories.

[c] The cumulative incidence rate (CI_p) for an age period (a, a + p) depends on the adjusted incidence density over that period and is given by the equation: $CI_p = 1 - exp(-sum (i) (ID_i) p (i))$, where p(i) stands for the length of the ith age subinterval.

The adjusted rate of the exposed group corresponded approximately to the level of mortality in the age range 50–54 years. The 48-year cumulative incidence rate for the 25-year-olds, which is interpreted as the proportion of individuals who will die from this particular cause in the age range 25–72 years, was 9 % greater for the CS_2-exposed group.

Life table comparisons
In spite of the rigid age-matching procedures used to ensure the comparability of the two populations, the age of the subjects deserved to be considered as an additional factor. In all, 17 (85 %) of the 20 IHD deaths among the exposed individuals occurred in the 50- to 64-year age range (Table 6). A "current" life table analysis transformed the somewhat difficult-to-interpret statistical measure of incidence density and its derivatives into a meaningful concept, namely that of life expectancy. (For the methodology of the life table technique, the reader may consult Ref. *12*).

In a hypothetical situation in which IHD was the only risk acting, the difference in life expectancies of the two populations (Table 10) was two years for the age interval 25–54 years. For the nonexposed group, for example, this corresponds to 9 % of the remaining life-time (21 years) at the age of 50 years. Only in the age range 65 years and over did the difference in life expectancies show a better prospect for the exposed group when IHD was separated statistically as the sole cause of death. In all the other comparisons, the nonexposed individuals were expected to outlive the members of the exposed population.

Table 10. Life expectancy when IHD is separated statistically as the only risk factor with fatal outcome

Age interval (years)	Life expectancy (years)		Difference (years)
	Exposed	Nonexposed	
24–44	44.0	46.0	2.0
45–49	24.0	26.0	2.0
50–54	19.3	21.3	2.0
55–59	15.1	16.7	1.6
60–64	11.2	11.9	0.7
65–67	7.4	7.2	− 0.2

The 1979 study (9, 13)

The results of the eight-year follow-up made the investigators suspect that a shift in the trend might have taken place. But during the following two years, 9 exposed and 2 referents died from IHD. Hence, during the total ten-year period there were 29 IHD deaths among the exposed cohort and 11 among the referents, giving a RR of 2.6 ($P < 0.005$).

When the mortality data for IHD were divided into two successive five-year periods, the RR declined from the highly significant value of 4.7 during the first quinquennium to a nonsignificant 1.9 during the second one. This could well be an effect of aging study populations, since, as the base-line mortality increases, the RR tends to go down. Thus, the RD would be a more appropriate measure of comparison in this respect.

The follow-up experience of both cohorts has been summarized with regard to their estimated risks of death from IHD (Fig. 5). For the age span 40–69 years the risk was estimated to have been 32 % among the exposed cohort and 13 % among the nonexposed one.

The 1980 study (13)

The most recent data come from the eleventh, twelfth, and thirteenth years of the project (Table 11). The IHD mortality rate was almost twice as high in the exposed cohort as in the nonexposed one during the entire period from 1967 to 1980. Other cardiovascular deaths were more common among the exposed subjects, but this was counterbalanced by a lower cancer mortality in this cohort; in neither case was the difference significant, however. In the final 3-year period from 1977 to 1980, five IHD cases were detected in the exposed and eight in the nonexposed group (RR = 0.66). This, again, would suggest a levelling off of the risk, but in view of the previous experience with the tidal trend caution is necessary in interpreting these findings.

365

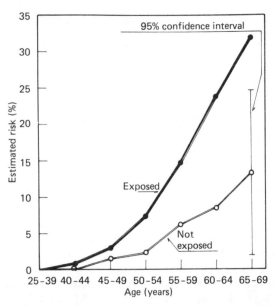

Fig. 5. Estimated risk of death from IHD, assuming that the distribution
of deaths is uniform or equivalent, i.e. that the survivor(ship) function
is linear within the five-year age intervals

Table 11. Mortality data from the 13-year follow-up

Cohort (number)	No. of deaths from:			
	IHD	other cardio-vascular causes	cancer	all causes
Exposed (343)	34	7	10	59
Nonexposed (343)	19	3	17	49
Mortality ratio	1.8	2.3	0.6	1.2
95% confidence interval	1.04 – 3.1	0.6 – 9.0	0.8 – 9.2	0.9 – 1.7
χ^2 test	4.25	1.60	1.81	0.93
P value	0.04	NS	NS	NS

NS = not significant

Multivariate approaches

The aim of the study was also to clarify the extent to which exposure to
CS_2 is a contributing factor to the development of IHD, allowing for the
action of covariables, whether as effect modifiers or confounders. In
addition, the study attempted to estimate the risk of an individual
exposed person dying from IHD.

An analysis was performed to differentiate among the exposed group between those who had died from IHD and those who had either survived or had died from another cause after a 5.5-year follow-up (8). The five risk indicators, determined at the start of the follow-up and included in the analysis, were: (1) age (in years); (2) cigarette consumption (coded in five ordinal scaled categories); (3) CS_2 exposure index; (4) diastolic blood pressure; and (5) serum cholesterol level. From these data, a so-called discriminant function was derived that described concisely the differences between the groups. Using this function, a score value was computed for each person characterizing his individual IHD risk. The 319 workers for whom data on all the five variables was available, were classified into six risk categories as shown in Table 12. The performance of the "retrospective screening test" may be described by means of several statistics. The sensitivity of the test was 75 % (= 12/16), the specificity of the test was 89 % (= 271/303), the predictive value of a negative test result was 99 % (= 271/275), and the predictive value of a positive test result was 27 % (= 12/44). A person belonging to the highest risk category had a 19-fold higher risk of dying from IHD than "normal" persons (in risk classes 1–5); risk ratio = 12 ×275/(4 ×44) = 19.

Table 12. Classification of exposed persons according to their IHD risk scores

IHD class	No. dying from IHD	No. of IHD survivors
6 Highest[a]	12	32
5	2	61
4	1	52
3	1	52
2	-	53
1 Lowest	-	53
Total	16	303

[a] Score exceeded 610; mean in total material approx. 500, SD = 100 (for method of computing risk score, see text).

The analysis did not provide a direct quantitative answer as regards the relative magnitude of the CS_2 exposure as a risk factor for IHD. In the graphs, the risk ratios are shown as a function of age. However, owing to the correlation between age and duration of exposure, aging may have channelled some of its effect through (the coefficient of) the exposure variable.

After the eight-year follow-up results became known (in 1979), it was thought that with more data on IHD deaths and both cohorts included, the estimation of the effect of CS_2 would become more accurate. On the contrary, however, the estimated risk model function showed poor, albeit statistically highly significant discriminatory power, probably for the following reasons. (1) The exposure variable was a multiplicative index of time and intensity; this approximate measure of personal exposure did not account for the potential separate effects of these two variables. (2) It was impossible to update the index for the ten years of follow-up—it thus referred to past exposure only. (3) The two groups compared were not entirely free from IHD in 1967. (4) As early as 1967 the groups differed with respect to blood pressure. (5) Missing measurements (32 cholesterol values) were replaced by group mean values. (6) No interaction term could be included in the model function because of the limiting size of the smaller group ($N = 40$). An attempt was made to employ another model form (the complementary log-log transform of the risk parameter), but the computational difficulties involved proved to be too great. Nevertheless, aging, CS_2 exposure, and elevated blood pressure stood out as prominent risk factors in this analysis.

Using as an alternative approach the log-linear model fitting and testing technique (see, for example, Ref. *14*), the magnitude of the independent and interdependent effects of the three risk factors—aging, CS_2 exposure, and elevated diastolic blood pressure—on IHD mortality was re-examined in the 10-year follow-up data (in 1981). The analysis demonstrated that the effect of CS_2 was largely independent of the effects of the other two risk factors. The analysis also revealed the dual role of blood pressure in the potential mechanism causing IHD death.

The subtle issue of confounding was examined in the analysis stage by application of the change-in-estimate criterion, basing it on the observed number of exposed cases and its null value estimated from the referent series (*15*).The observed number of IHD deaths in the exposed cohort was 29. The crude estimate ($\hat{E}$) of the corresponding null-expected number was 11; i.e., (number of exposed) ×(rate of IHD among the referents) = 343 (11/343). The corresponding adjusted E as the sum of the age-specific diastolic blood pressure values from the multiway contingency table was found to be 13.2. Thus, the crude and the adjusted estimates of the rate ratio were 2.6 and 2.2, respectively, indicating only minor confounding according to the two cofactors. In the above analysis, the known risk factors, aging and elevated diastolic pressure, were treated operationally as extraneous variables when the main aim was to examine the effect of CS_2 exposure.

Inferences

The following conclusion was drawn in the report of the 1967 study (*5*): "Although no definite evidence was found, the excess of coronary deaths in combination with a small hint of a higher prevalence of pathological electrocardiograms among the exposed subjects suggests that CS_2 may be an aetiological factor in the pathogenesis of coronary heart disease." One possible explanation given at that time of the small differences found in the ECG variables was "that CS_2 makes the prognosis of coronary heart disease graver. This could result in a higher number of fatal attacks and would therefore, not increase prevalence rates." The

368

authors pointed out the possible "diluting effects" of pre-employment screening out of unhealthy workers, of subsequent employee transfers and resignations, and of the exclusion of patients receiving disability pensions for causes other than poisoning due to either CS_2 or H_2S. It was emphasized that these selection factors increased the importance of significant or nearly significant findings as indicators of real effects of CS_2 poisoning and H_2S poisoning. But, even so, the authors were careful to state: " . . . our data are too vague to allow any conclusions to be drawn regarding whether or not the present hygienic standard is safe with regard to coronary heart disease." It should be noted that the issue of the hygienic standard was not explicitly formulated in the original research problem.

One interesting result of the survey was "the demonstration of the difficulties and fallacies that are inherent in any study dealing with IHD as a result of toxic exposure in any occupational population." Valuable experience was gained from the application of advanced epidemiological research methods in the field of occupational health. Finally, the hope was expressed "that a follow-up of both groups will provide us with more definite results." The prospective follow-up of mortality was then inevitable, but could perhaps have been avoided had a retrospective cohort of the referent population been formed; this, of course, is an afterthought.

In the next report (16), attention was drawn to several circumstances favouring a causal association between CS_2 and IHD: (1) the conformity of the results of the Finnish study with those of studies from other countries (England, Norway); (2) the strength of the association (rate ratios up to five); (3) the existence of possible biochemical mechanisms that might explain the phenomenon. In consequence, the observed difference in blood pressure was interpreted as a result of exposure, rather than as evidence of an independent risk factor.

The authors concluded that the highest RRs were associated with the most severe attacks of myocardial infarction and the lowest RRs (although larger than 1.0) with the mildest forms of IHD (e.g., history of angina, "coronary" ECGs). Accordingly, they suggested that CS_2 might cause worsening prognosis in cases of IHD of a given severity, while perhaps also slightly increasing the incidence and prevalence of symptoms related to IHD. (Thus, the previous conjecture was not far off the mark, although it was inferred from data that had been insufficiently examined. The clinical data were related to workroom concentrations of CS_2 and to the hygienic standard for that compound. The concentrations of H_2S plus CS_2 during the 10 years prior to this study had generally been near or below 20 ppm. The exposure sampling, which was done several times monthly and under prevailing (regular or irregular) conditions, might still have failed to detect peak exposures and periods without exposure. More importantly by 1968, 38 % of the workers were no longer exposed to CS_2 and about 40 % were not exposed during the period 1968–1973. This, the authors concluded (8), indicated irreversible, deleterious effects of CS_2 on the development of IHD and suggested that the levels of 10–30 ppm measured then were too high to safeguard the health of employees. However, the research group, acting according to the expressed formulation of the problem, did not dwell on the possible role of H_2S in bringing about the changes in IHD development.

In 1975, 165 men (48 %) out of the 343 original members of the exposed cohort assembled in 1967 were still employed in the plant, but only 64 (19 %) were still exposed to CS_2 (7, 10) (cf. Table 13). Because the environmental concentrations had greatly decreased by 1975, the authors (6) stated that the decrease in IHD coincided with the changed conditions. Their conclusion at that time was that this apparent reversibility (cf. above) indicated that the previously reported excess mortality due to IHD might have been caused by direct toxic effect of CS_2 on the myocardium rather than by the acceleration of the atherosclerotic process; many of the deaths due to IHD had been reported as "sudden deaths". The authors, however, based their "reversibility" theory on only three years of observation following the intervention and six IHD deaths in each cohort during this period. Furthermore, this finding in no way seemed to contradict the earlier findings of excess IHD mortality, because improved work practices, the introduction of a policy of transferring exposed workers to "clean" departments, administrative controls, construction of ventilated surveillance rooms, and extensive use of personal protective equipment had markedly improved working conditions.

Table 13. Status of the exposed cohort in June 1975

Status	No.		%	
Still employed at the plant	165		48	
Still exposed		64		19
No longer exposed		101		29
No longer employed at the plant	143		42	
Retired or pensioned off		66		19
Moved		77		22
Died	35		10	
Total	343		100	

When the relation between CS_2 exposure and IHD mortality was evaluated, another aspect considered was the possibility "that CS_2 hastens death in those who are prone to IHD so that, at the eighth year, the survivors are those whose risk for IHD is lower than in the general (comparison) population of the same age structure".

The extension of the follow-up to ten years did not furnish much new information, nor did it detract from the credibility of the conclusions already reached. A deeper analysis confirmed, for example, that the excess mortality from IHD, which had by then declined to show a RR of 1.9, was not attributable to differential bias due to previous nonfatal myocardial infarction or diabetes mellitus. It was concluded that occupational exposure to CS_2 was an additional risk factor that might increase the already notoriously high risk of IHD mortality among Finnish men.

When the results of the final 13-year follow-up are considered, it must be remembered that many of the exposed workers had been transferred to much less polluted or uncontaminated departments, and about half of them had already changed employment earlier (see Table 13). The time-weighted exposure levels were also drastically reduced in most departments, being in 1980 about 1–10 ppm expressed as background exposure and 0.5–3 ppm expressed as personal exposure. Hence the exposure situation had changed dramatically for the better, and the only question that can still be studied is whether the effects of CS_2 on IHD mortality are irreversible or not. At least in principle, one could compare the fate of those who had been removed from exposure with the fate of workers who had continued working in exposure conditions, provided that the removal had not been dictated by the IHD outcome. This is, however, a moot point and would perhaps require autopsy material and larger numbers to resolve the question. The investigators should have adopted a sequential method of inference, preferably using modelling techniques.

Sources of bias

The validity of epidemiological or biostatistical data may be undermined or vitiated by several kinds of bias. When two cohorts are compared with respect to the effects of the manoeuvers to which they have been exposed, the investigators (or readers) must constantly beware of unrecognized bias in the procedures selected for assembling the members and collecting the data on the cohorts. The objective of the cohort research described above was pathogenetic, that is the manoeuver was exposure to the alleged disease agent, CS_2. The effect of the exposure was noted by comparison with the (ontogenetic) development of IHD in the nonexposed ("normal") persons during the selected period of time.

Bias in the numerator may be due to unequal detection of the target event. This type of bias has been largely neglected in epidemiological research and is responsible for a major flaw in the validity of cohort statistics dealing with causes of disease. For example, as revealed by autopsy, about half of patients with IHD had not received the correct diagnosis during life (17). It is also possible that in some cases IHD was not appropriately recorded on the death certificates. However, is there any reason to suppose that the rate of false negatives was different in the two series of deceased persons?

In the case of an unexpected sudden death, Finnish legislation requires that an autopsy be performed to ascertain the cause of death. However, an autopsy may have been judged unnecessary for patients who had been treated by a physician (e.g., if hospitalized for cardiac pains) and who had subsequently died. The figures for the 10-year material show that autopsies had been performed more frequently among the exposed (70%) than among the nonexposed (58%).

Moreover, before 1 July 1972, the corresponding rates were 57% (exposed) and 56% (nonexposed); but from 1 July 1972 until 30 June 1977, the rates were 81% (exposed) and 55% (nonexposed). From 1973 onwards, every cause of death attribed to IHD among the former viscose rayon workers was certified on the basis of an autopsy, whereas only about half the cases in which death was attributed to another cause were submitted to autopsy. Coronary angiography, a highly sensitive investigation for IHD, was not done at all. For the deceased papermill workers, autopsies were performed with equal frequency before and after 1973, regardless of whether the cause of death was considered of cardiac origin or not (Table 14).

Table 14. Frequency of performing autopsies in the two periods of follow-up among the exposed and nonexposed cohorts according to IHD and other causes of death

| Follow-up period | Certified cause of death | Autopsy performed | | | |
| | | Exposed cohort | | Referent cohort | |
		Yes	No	Yes	No
1.7.1967–	IHD	8	6	2	1
30.6.1972	Other	4	3	3	3
1.7.1972–	IHD	15	–	6	2
30.6.1975	Other	7	5	6	8

Is it therefore conceivable that the results of the 5-year mortality study, which became known at the end of 1972, somehow altered the frequency with which autopsies were performed on former viscose rayon workers? The bias, if it existed, would have had the consequence of making the effect of CS_2 on IHD mortality look unduly large in the last 5-year period because some IHD cases were overlooked in the reference series. In other words, the impression that the risk had diminished would not have been an illusion. As a counter-argument, it might be pointed out that in Finland, where IHD is very common among men, cases of sudden death tend to be attributed to IHD even if an autopsy has not been done. In cases of sudden death, an increased frequency of autopsies might, in fact, bring down the rate of IHD as a cause of death, because it might be found, for example, that cerebral accidents were the actual causes of some deaths.

A practical example of this problem was created by the special policy adopted at the viscose rayon plant in the 1950s. According to this policy, workers were assigned to alternating jobs in the polluted departments and outdoors for defined periods of time (five months "in", one month "out"). Even if the assignment to tasks in polluted rooms was done more or less randomly by the management, rather than self-selectively by the cohort, the workers who chose to stay in those jobs were a self-selected group, and the results may be biased by unrecognized distinctions that were involved in the selection process.

As an arbitrary example of how such a process might work, let us assume that people who had an IHD-prone behaviour pattern (a risk indicator for IHD) were much more likely to continue working in the often better paid exposed posts than the rest of the workers. The exposed cohort would then have been composed mainly of "tense" people with "type A behaviour", because the "permanent" (or unchanging) jobs were associated with the "clean" departments. The reference group, however, which did not have to cope with the special policy of alternating jobs, was composed of a mixture of "tense" and "relaxed" people. Even if the CS_2 exposure had had no effect on IHD mortality, the resultant death rate would nevertheless have been higher in the exposed than in the referent group, because the exposed cohort would have contained a majority of "tense" people with high death rates.

Impact of the study

A report was published in 1972 announcing the alarming results of the 5-year prospective mortality study. The viscose rayon company responded promptly to the urgent recommendations. In a declaration, the managing director of the plant made it known that as soon as the results of the retrospective phase became known serious attention had been paid to the problem, which until then had been largely unrecognized. During the five years (from 1967 to 1972) the ventilation system of the chemical departments had been overhauled. For example, one former spinning department had discontinued operation and the ventilation in the rayon filament department had been greatly improved. The labour safety committee of the factory drafted a detailed long-term programme for lowering the CS_2 below the allowable 10 ppm. This programme was approved by the board of directors of the factory and was being enforced. The company had invested over 3 million Fmk during the previous five years in improving working conditions. An additional 0.65 million marks were donated to the work safety board to be spent in occupational safety projects in places of work chosen by the board. The management also introduced a new policy of transferring to uncontaminated departments men with a long history of exposure to CS_2.

From the scientific point of view, the main question of whether prolonged exposure to concentrations of CS_2 above the then prevailing hygienic standard of 20 ppm causes IHD, could be answered in the affirmative as far as mortality is concerned, on the condition that the true exposure concentrations were reflected by the measurement values supplied by the viscose plant laboratory.

A causal relationship between exposure to CS_2 and the development of IHD having been demonstrated, epidemiological research should now be directed towards quantitative risk assessment. Outlined below are areas for future research in this field in which international collaboration would certainly be of great help.

(1) More information is needed on the exposure–effect relationship, which cannot be established adequately on the basis of the available epidemiological data (by reason of the vague documentation of individual exposures to CS_2).

(2) Consequently, the no-effect level in terms of IHD also remains to be determined.

(3) Well designed, prospective studies of workers exposed to low levels (less than 10 ppm) of CS_2 would be needed.

References

1. **Tiller, J. R. et al.** Occupational toxic factor in mortality from coronary heart disease. *British medical journal,* 4: 407–411 (1968).
2. **Rose, G. A. & Blackburn, H.** *Cardiovascular survey methods.* Geneva, WHO, 1968, pp. 172–175.
3. **World Health Organization.** *Manual of the international statistical classification of diseases, injuries, and causes of death,* 7th revision, vol. 1, Geneva, 1955.
4. **World Health Organization.** *Manual of the international statistical classification of diseases, injuries, and causes of death,* 8th revision, vol. 1, Geneva, 1967.
5. **Hernberg, S. et al.** Coronary heart disease among workers exposed to carbon disulphide. *British journal of industrial medicine,* 27: 313–325 (1970).
6. **Tolonen, M. et al.** A follow-up study of coronary heart disease in viscose rayon workers exposed to carbon disulphide. *British journal of industrial medicine,* 32: 1–10 (1975).
7. **Hernberg, S. et al.** Eight-year follow-up of viscose rayon workers exposed to carbon disulfide. *Scandinavian journal of work, environment and health,* 2: 27–30 (1976).
8. **Hernberg, S. et al.** Excess mortality from coronary heart disease in viscose rayon workers exposed to carbon disulfide. *Work-environment-health,* 10: 93–99 (1973).
9. **Tolonen, M. et al.** Ten-year coronary mortality of workers exposed to carbon disulfide. *Scandinavian journal of work, environment and health,* 5: 109–114 (1979).
10. **Nurminen, M.** Survival experience of a cohort of carbon disulphide exposed workers from an eight-year prospective follow-up period. *International journal of epidemiology,* 5: 179–185 (1976).
11. **Nurminen, M. et al.** Quantitated effects of carbon disulfide exposure, elevated blood pressure and aging on coronary mortality. *American journal of epidemiology,* 115: 107–118 (1982).
12. **Chiang, C. L.** *Introduction to stochastic processes in biostatistics.* New York, Wiley, 1968.
13. **Hernberg, S. & Tolonen, M.** Epidemiology of coronary heart disease among viscose rayon workers. *Giornale italiano di medicina dei lavoro,* 3: 49–52 (1981). (Fifth International Symposium on Occupational Health in the Production of Artificial Fibers, Belgirate (Novara), Italy, 16–20 September 1980).
14. **Haberman, S. J.** *Analysis of qualitative data. Vol. 1. Introductory topics.* New York, Academic Press, 1978.
15. **Miettinen, O. S. & Cook, E. F.** Confounding: Essence and detection. *American journal of epidemiology,* 114: 593–603 (1981).
16. **Hernberg, S. et al.** Blood lipids, glucose tolerance and plasma creatine in workers exposed to carbon disulphide. *Work-environment-health,* 8: 11–16 (1971).
17. **Beadenkopf, W. G. et al.** An assessment of certain medical aspects of death certificate data for epidemiologic study of arteriosclerotic heart disease. *Journal of chronic diseases,* 16: 249–262 (1963).

Glossary of terms[a]

accuracy The degree to which a measurement, or an estimate based on measurements, represents the true value of the attribute that is being measured.

age-specific rate A rate for a specified age group. The numerator and denominator refer to the same age group.

Example:

$$\text{Age-specific death rate (age 25–34)} = \frac{\text{Number of deaths among residents age 25–34 in an area in a year}}{\text{Average (or midyear) population age 25–34 in the area in that year}} \times 100\,000$$

The multiplier (usually 100 000 or 1 000 000) is chosen to produce a rate that can be expressed as a convenient number.

algorithm Any systematic process that consists of an ordered sequence of steps with each step depending on the outcome of the previous one. The term is commonly used to describe a structured process, for instance, relating to computer programming or to health planning.

analytic study A hypothesis-testing method of investigating the association between a given disease or health state or other dependent variable and possible causative factors. Three types of analytic study are cross-sectional (prevalence), cohort (prospective), and case control (retrospective).

association The degree of statistical dependence between two or more events or variables. Events are said to be associated when they occur more frequently together than one would expect by chance. Association does not necessarily imply a causal relationship. Statistical significance testing enables us to determine how unlikely it would be to observe the sample relationship by chance if in fact no association exists in the population that was sampled. If the use of the term "association" is

---

[a] A selection of terms and definitions from: **Last, J.M., ed.** *A dictionary of epidemiology.* Oxford, Oxford University Press, 1983.

confined to situations in which the relationship between two variables is statistically significant, the terms "statistical association" and "statistically significant association" become tautological. However, ordinary usage is seldom so precise as this. The terms "association" and "relationship" are often used interchangeably.

attributable risk   This term has been used by different authors to denote a number of different concepts, including the attributable fraction in the population, the attributable fraction among the exposed, the population excess rate, and the rate difference.

attributable risk percent   Attributable fraction expressed as a percentage rather than as a proportion.

bias   Any effect at any stage of investigation or inference tending to produce results that depart systematically from the true values (to be distinguished from *random error*).

  The term "bias" does not necessarily carry an imputation of prejudice or other subjective factor, such as the experimenter's desire for a particular outcome. This differs from conventional usage in which bias refers to a partisan point of view.

blind experiment (syn: blind trial)   Experiment in which either the experimenter or the subject(s) does not know the group to which the subject(s) have been allocated. The intent of blind procedures is to eliminate the biases and prejudices of the subject(s) or of the investigator.

  If the experimenter but not the subject(s) of the experiment know the allocation, the experiment is said to be "single blind". If neither experimenter nor subject(s) knows the allocation the experiment is called "double blind". If analysis is also done "blind" it may be called a "triple-blind" experiment.

causality   The relating of causes to the effects they produce. Most of epidemiology concerns causality and several types of causes can be distinguished.

  A cause is termed "necessary" when a particular variable must always precede an effect. This effect need not be the sole result of the one variable. A cause is termed "sufficient" when a particular variable inevitably initiates or produces an effect. Any given cause may be necessary, sufficient, neither, or both. These possibilities are explained below.

  Four conditions under which independent variable X may cause Y:

	variable X may cause Y	
	X is necessary	X is sufficient
1.	+	+
2.	+	−
3.	−	+
4.	−	−

1. *X* is necessary and sufficient to cause *Y*.

Both *X* and *Y* are always present together, and nothing but *X* is needed to cause *Y*; *X* → *Y*.

2. *X* is necessary but not sufficient to cause *Y*.

X must be present when *Y* is present, but *Y* is not always present when *X* is. Some additional factor(s) must also be present; *X* and *Z* → *Y*.

3. *X* is not necessary but is sufficient to cause *Y*.

Y is present when *X* is, but *X* may or may not be present when *Y* is present, because *Y* has other causes and can occur without *X*. For example, an enlarged spleen can have many separate causes that are unconnected with each other; *X* → *Y*; *Z* → *Y*.

4. *X* is neither necessary nor sufficient to cause *Y*.

Again, *X* may or may not be present when *Y* is present. Under these conditions, however, if *X* is present with *Y*, some additional factor must also be present. Here *X* is a contributory cause of *Y* in some causal sequences; *X* and *Z* → *Y*; *W* and *Z* → *Y*.

causation of disease, factors in   The following factors have been differentiated (but they are not mutually exclusive).

Predisposing factors are those that prepare, sensitize, condition, or otherwise create a situation such as a level of immunity or state of susceptibility so that the host tends to react in a specific fashion to a disease agent, personal interaction, environmental stimulus, or specific incentive. Examples include age, sex, marital status, family size, educational level, previous illness experience, presence of concurrent illness, dependency, working environment, and attitudes towards the use of health services. These factors may be "necessary" but are rarely "sufficient" to cause the phenomenon under study.

Enabling factors are those that facilitate the manifestation of disease, disability, ill health, or the use of services or conversely those that facilitate recovery from illness, maintenance of enhancement of health status, or more appropriate use of health services. Examples include income, health insurance coverage, nutrition, climate, housing, personal support systems, and availability of medical care. These factors may be "necessary" but are rarely "sufficient" to cause the phenomenon under study.

Precipitating factors are those associated with the definitive onset of a disease, illness, accident, behavioural response, or course of action. Usually one factor is more important or more obviously recognizable than others if several are involved and one may often be regarded as "necessary". Examples include exposure to specific disease, amount or level of an infectious organism, drug, noxious agent, physical trauma, personal interaction, occupational stimulus, or new awareness or knowledge.

Reinforcing factors are those tending to perpetuate or aggravate the presence of a disease, disability, impairment, attitude, pattern of behaviour, or course of action. They may tend to be repetitive, recurrent, or persistent and may or may not necessarily be the same or similar to those categorized as predisposing, enabling, or precipitating. Examples include repeated exposure to the same noxious stimulus (in the absence of an appropriate immune response) such as an infectious agent, work, household, or interpersonal environment, presence of financial incentive or disincentive, personal satisfaction, or deprivation.

cohort   The component of the population born during a particular period and identified by period of birth so that its characteristics (e.g. causes of death and numbers still living) can be ascertained as it enters successive time and age periods.

The term "cohort" has broadened to describe any designated group of persons who are followed or traced over a period of time as in cohort study (prospective study).

confounding [from the Latin *confundere*, to mix together]   A situation in which the effects of two processes are not separated. The distortion of the apparent effect of an exposure on risk brought about by the association with other factors that can influence the outcome.

confounding variable (syn: confounder)   A factor that distorts the apparent magnitude of the effect of a study factor on risk. Such a factor is a determinant of the outcome of interest and is unequally distributed among the exposed and the unexposed.

contingency table   A tabular cross-classification of data such that subcategories of one characteristic are indicated horizontally (in rows) and subcategories of another characteristic are indicated vertically (in columns). Tests of association between the characteristics in the columns and rows can be readily applied. The simplest contingency table is the fourfold, or 2×2, table. Contingency tables may be extended to include several dimensions of classification.

control (group)   A comparison group, identified as a rule before a study is done, comprising persons who have not been exposed to the disease, intervention, procedure, or other variable whose influence is being studied.

controls, matched   Controls who are selected so that they are similar to the study group, or cases, in specific characteristics. Some commonly used matching variables are age, sex, race, and socioeconomic status.

cumulative incidence rate   Proportion of an initially disease-free group developing the disease over a fixed time interval. Distinct from force of morbidity. Proper estimation of this rate requires the use of methods for cumulating observed proportions falling ill within short subintervals of the interval of interest, such as the *life table*. Sometimes referred to simply as "cumulative incidence".

cumulative incidence ratio   The ratio of the cumulative incidence rate in the exposed to the cumulative incidence rate in the unexposed.

cumulative mortality (rate)   The proportion of a defined population dying over a fixed time period.

data processing   Conversion (as by computer) of crude information into usable or storable form. Data generated by epidemiologic studies are usually transferred to punch cards or optical mark-sense forms and thence to a computer for storage and retrieval. The term is often loosely used to mean also the statistical analysis of data by a computer program.

dependent variable   1.   A variable the value of which is dependent on the effect of other variable(s) [independent variable(s)] in the relationship under study. A manifestation or outcome whose variation we seek to explain or account for by the influence of independent variables.

2.   In statistics, the dependent variable is the one predicted by a regression equation.

determinant   Any factor, whether event, characteristic, or other definable entity, that brings about change in a health condition, or other defined characteristic.

discrete data   Data that can be arranged into naturally occurring or arbitrarily selected groups or sets of values, as opposed to data in which there are no naturally occurring breaks in continuity, i.e. continuous data. An example is number of decayed, missing, and filled teeth (DMF).

disease   Literally, *dis-ease*, the opposite of *ease*, when something is wrong with a bodily function. The words "disease", "illness", and "sickness" are loosely interchangeable, but are better regarded as not wholly synonymous. M.W. Susser has suggested that they be used as follows.

Disease is a physiological/psychological dysfunction;

Illness is a subjective state of the person who feels aware of not being well;

Sickness is a state of social dysfunction, i.e. a role that the individual assumes when ill.

distribution   The complete summary of the frequencies of the values or categories of a measurement made on a group of persons. The distribution tells either how many or what proportion of the group were found to have each value (or each range of values) out of all the possible values that the quantitative measure can have.

distribution-free method (syn: nonparametric method)   A method of testing a hypothesis or of setting up a confidence interval that does not depend on the form of the underlying distribution; in particular, it does not depend upon the variable following a normal distribution.

dose–response relationship   A relationship in which a change in amount, intensity, or duration of exposure is associated with a change—either an increase or a decrease—in risk of a specified outcome.

double-blind trial   A procedure of blind assignment to study and control groups and blind assessment of outcome, designed to insure that ascertainment of outcome is not biased by knowledge of the group to which an individual was assigned. "Double" refers to both parties, i.e. the observer(s) in contact with the subjects, and the subjects in the study and control groups.

effect modifier, effect modification (syn: conditional variable, moderator variable)   A factor that modifies the effect of a causal factor under study. The action is termed "effect modification". For example, age is a modifier of the effect of measles infection on mortality risk.

Effect modification may be expressed by differences in a measure of association such as the *odds ratio, risk ratio,* or risk difference across levels of a third factor.

error 1. A false or mistaken result obtained in a study or experiment. Several kinds of error can occur in epidemiology, for example, due to *bias.*

2. Random error (sampling error) is that due to chance, when the result obtained in the sample differs from the result that would be obtained if the entire population ("universe") were studied. Two varieties of sampling error are type I, or alpha error, and type II, or beta error.

In an experiment, if the experimental procedure does not in reality have any effect, an apparent difference between experimental and control groups may nevertheless be observed by chance, a phenomenon known as type I error. Another possibility is that the treatment is effective but by chance the difference is not detected on statistical analysis—type II error.

In the theory of testing hypotheses, rejecting a null hypothesis when it is actually true is called "type I error". Accepting a null hypothesis when it is incorrect is called "type II error".

3. Systematic error is that due to factors other than chance, such as faulty measuring instruments. It is further considered in *bias.*

estimate A quantitative statement of a measurement or value. We commonly use the term when referring to measurements that are believed to be approximate rather than accurate or precise.

exposure ratio The ratio of rates at which persons in the case and control groups of a case control study are exposed to the *risk factor* (or to the protective factor) of interest.

false negative Negative test result in a subject who possesses the attribute for which the test is conducted. The labelling of a diseased person as healthy when screening in the detection of disease.

false positive Positive test result in a subject who does not possess the attribute for which the test is conducted. The labelling of a healthy person as diseased when screening in the detection of disease.

fourfold table See *contingency table.*

frequency distribution See *distribution.*

health risk appraisal (HRA) (syn: health hazard appraisal [HHA]) A generic term applied to methods for describing an individual's chances of becoming ill or dying from selected causes. The many versions now available share several common features. Starting from the average risk of death for the individual's age and sex, a consideration of various lifestyle and physical factors indicates whether the individual is at greater or lesser than average risk of death from the commonest causes of death for his age and sex. All methods also indicate what reduction in risk could be achieved by altering any of the causal factors (such as cigarette smoking) that the individual could modify.

The premise underlying such methods is that information on the extent to which an individual's characteristics, habits, and health practices are influencing his future risk of dying will assist health care workers in counselling their patients.

health status index   A set of measurements designed to detect short-term fluctuations in the health of members of a population; these measurements generally include physical function, emotional wellbeing, activities of daily living, feelings, etc. Most indexes require the use of carefully composed questions designed with reference to matters of fact rather than shades of opinion. The results are usually expressed by a numerical score that gives a profile of the wellbeing of the individual.

health survey   A survey designed to provide information on the health status of a population. It may be descriptive, exploratory, or explanatory.

healthy worker effect   A phenomenon observed initially in studies of occupational diseases. Workers usually exhibit lower overall death rates than the general population, due to the fact that the severely ill and disabled are ordinarily excluded from employment. Death rates in the general population may be inappropriate for comparison if this effect is not taken into account.

historical cohort study (syn: historical prospective study, nonconcurrent prospective study, prospective study in retrospect)   A cohort study conducted by reconstructing data about persons at a time or times in the past. This method uses existing records about the health or other relevant aspects of a population as it was at some time in the past and determines the current (or subsequent) status of members of this population with respect to the condition of interest. Different levels of past exposure to risk factor(s) of interest must be identifiable for subsets of the population.

incidence rate   A measure of the rate at which new events occur in the population. The number of new events, e.g. new cases of a specified disease diagnosed or reported during a defined period of time, is the numerator, and the number of persons in the stated population in which the cases occurred is the denominator.

These events may be acute short-term episodes, e.g. of acute upper respiratory infection, or the initial (diagnosed) onset of a long-term, chronic condition such as tuberculosis or cancer. When the *incidence rate* is calculated for a period of a year, the number of the former type of episodes may exceed the number of persons at risk; with chronic or long-term conditions, the annual incidence rate is usually lower than the *prevalence rate*.

Mathematically, two types of incidence rate may be distinguished. These are force of morbidity, in which the number of new cases is the numerator, and person-time units of experience is the denominator; and *cumulative incidence rate*, in which the denominator is the number of persons initially at risk.

independent variable 1. The characteristic being observed or measured that is hypothesized to influence an event or manifestation (the dependent variable) within the defined area of relationships under study; that is, the independent variable is not influenced by the event or manifestation but may cause it or contribute to its variation.

2. In statistics, an independent variable is one of (perhaps) several variables that appear as arguments in a regression equation.

index group (syn: index series) 1. In an experiment, the group receiving the experimental regimen.

2. In a case control study, the cases.

3. In a cohort study, the exposed group.

individual variation Two types are distinguished.

1. *Intraindividual variation.* The variation of biological variables within the same individual, depending upon circumstances such as the phase of certain body rhythms and the presence or absence of emotional stress. These variables do not have a precise value, but rather a range. Examples include diurnal variation in body temperature, fluctuation of blood pressure, blood sugar, etc.

2. *Interindividual variation.* As used by Darwin, the term means variation *between* individuals. This is the prefered usage; the first usage is better described as personal variation.

inference The act of passing from observations to generalizations. In statistics, the development of generalization from sample data, usually with calculated degrees of uncertainty.

instrumental error Error due to faults arising in any or in all aspects of a measuring instrument, i.e. calibration, accuracy, precision, etc. Also applied to error arising from impure reagents, wrong dilutions, etc.

interaction 1. The interdependent operation of two or more causes to produce an effect. *Biological interaction* means the interdependent operation of two or more causes to produce, prevent, or control disease.

2. Differences in the effects of one or more factors according to the level of the remaining factors.

3. In statistics, the necessity for a product term in a linear regression model.

intervention study An epidemiologic investigation designed to test a hypothesized cause–effect relationship by modifying a supposed causal factor in a population.

latent period (syn: latency) Delay between exposure to a disease-causing agent and the appearance of manifestations of the disease. After exposure to ionizing radiation, for instance, there is a latent period of five years, on average, before development of leukaemia, and more than 20 years before development of certain other malignant conditions. The term "latent period" is often used synonymously with "induction period", that is, the period between exposure to a disease-causing agent and the appearance of manifestations of the disease. It has also been defined as the period from disease initiation to disease detection.

382

life table A summarizing technique used to describe the pattern of mortality and survival in populations. The survival data are time-specific and cumulative probabilities of survival of a group of individuals subject, throughout life, to the age-specific death rates in question. The life table method can be applied to the study not only of death, but also of any defined endpoint such as the onset of disease or the occurrence of specific complication(s) of disease.

The first rudimentary life tables were published in 1693 by the astronomer Halley. These made use of records of the funerals in the city of Breslau. In 1815 in England, the first actuarially correct life table was published, based on both population and death data classified by age.

logistic model A statistical model of an individual's risk (probability of disease y) as a function of a risk factor x:

$$P(y|x) = \frac{1}{1 + e^{-\alpha - \beta x}}$$

where e is the (natural) exponential function. This model has a desirable range, 0 to 1, and other attractive statistical features. In the multiple logistic model, the term βx is replaced by a linear term involving several factors, e.g. $\beta_1 x_1 + \beta_2 x_2$ if there are two factors x_1 and x_2.

Mantel-Haenszel estimate, Mantel-Haenszel odds ratio Mantel and Haenszel[a] provided an adjusted *odds ratio* as an estimate of relative risk that may be derived from grouped and matched sets of data. It is now known as the Mantel-Haenszel estimate, one of the few eponymous terms of modern epidemiology.

The statistic may be regarded as a type of weighted average of the individual odds ratios, derived from stratifying a sample into a series of strata that are internally homogeneous with respect to confounding factors.

The Mantel-Haenszel summarization method can also be extended to the summarization of rate ratios and rate differences from follow-up studies.

Mantel-Haenszel test A summary chi-squared test developed by Mantel and Haenszel for stratified data and used when controlling for *confounding*.

matching The process of making a study group and a comparison group comparable with respect to extraneous factors. Several kinds of matching can be distinguished.

Caliper matching is the process of matching comparison group subjects to study group subjects within a specified distance for a continuous variable (e.g. matching age to within two years).

---

[a] **Mantel, N. & Haenszel, W.** Statistical aspects of the analysis of data from retrospective studies of disease. *Journal of the National Cancer Institute,* **22**: 719–748 (1959).

Frequency matching requires that the frequency distributions of the matched variable(s) be similar in study and comparison groups.

Category matching is the process of matching study and control group subjects in broad classes such as relatively wide age ranges or occupational groups.

Individual matching relies on identifying individual subjects for comparison, each resembling a study subject on the matched variable(s).

Pair matching is individual matching in which study and comparison subjects are paired.

mathematical model A representation of a system, process, or relationship in mathematical form in which equations are used to simulate the behaviour of the system or process under study. The model usually consists of two parts: the mathematical structure itself, e.g. Newton's inverse square law or Gauss's "normal" law, and the particular constants or parameters associated with them, such as Newton's gravitational constant or the Gaussian standard deviation.

A mathematical model is deterministic if the relations between the variables involved take on values not allowing for any play of chance. A model is said to be statistical, stochastic, or random, if random variation is allowed to enter the picture.

measurement The procedure of applying a standard scale to a variable or to a set of values.

misclassification The erroneous classification of an individual, a value, or an attribute into a category other than that to which it should be assigned. The probability of misclassification may be the same in all study groups (nondifferential misclassification) or may vary between groups (differential misclassification).

monitoring 1. The performance and analysis of routine measurements, aimed at detecting changes in the environment or health status of populations. Not to be confused with surveillance. To some, monitoring also implies intervention in the light of observed measurements.

2. Ongoing measurement of performance of a health service or a health professional, or of the extent to which patients comply with or adhere to advice from health professionals.

3. In management, the continuous oversight of the implementation of an activity that seeks to ensure that input deliveries, work schedules, targeted outputs and other required actions are proceeding according to plan.

morbidity Any departure, subjective or objective, from a state of physiological or psychological wellbeing. In this sense, *sickness*, *illness*, and *morbid condition* are similarly defined and synonymous (but see *disease*).

The WHO Expert Committee on Health Statistics noted in its Sixth report (1959) that morbidity could be measured in terms of three units: (1) persons who were ill; (2) the illnesses (periods or spells of illness) that these persons experience; and (3) the duration (days, weeks, etc.) of these illnesses.

morbidity rate A term, preferably avoided, used indiscriminately to refer to incidence or prevalence rates of disease.

morbidity survey A method for the estimation of the prevalence and/or incidence of disease or diseases in a population. A morbidity survey is usually designed simply to ascertain the facts as to disease distribution, and not to test a hypothesis.

multivariate analysis A set of techniques used when the variation in several variables has to be studied simultaneously. In statistics, any analytic method that allows the simultaneous study of two or more *dependent variables*.

nonparticipants (syn: nonresponders) Members of a study sample or population who do not take part in the study for whatever reason, or members of a target population who do not participate in an activity. Differences between participants and nonparticipants have been demonstrated repeatedly in studies of many kinds, and this is often a source of *bias*.

norm This term has two quite distinct meanings.

    1. The first is "what is usual", e.g. the range into which blood pressure values usually fall in a population group, the dietary or infant feeding practices that are usual in a given culture, or the way that a given illness is usually treated in a given health care system.

    2. The second sense is "what is desirable", e.g. the range of blood pressures that a given authority regards as being indicative of present good health or as predisposing to future good health, the dietary or infant feeding practices that are valued in a given culture, or the health care procedures or facilities for health care that a given authority regards as desirable.

    In the latter sense, norms may be used as criteria when evaluating health care, in order to determine the degree of conformity with what is desirable, the average length of stay of patients in hospital, etc. A distinction is sometimes made between norms, defined as quantitative indexes based on research, and standards, which are fixed arbitrarily.

normal This term has three distinct meanings. Conceptual difficulties may arise if these different meanings are not specified, or if the area of their overlap is not clearly understood.

    1. Within the usual range of variation in a given population or population group; or frequently occurring in a given population or group. In this sense, "normal" is frequently defined as, "within a range extending from two standard deviations below the mean to two standard deviations above the mean", or "between specified (e.g. the 10th and 90th) percentiles of the distribution".

    2. In good health, indicative or predictive of good health, or conducive to good health. For a diagnostic or screening test, a "normal" result is one in a range within which the probability of a specific disease is low.

    3. (Of a distribution) Gaussian.

normal limits The limits of the "normal" range of a test or measurement, in the sense of being indicative of or conducive to good health. One way to determine normal limits is to compare the values obtained when the measurements are made in two groups, one that is healthy and has been found to remain healthy, the other ill, or subsequently found to become ill. The result may be two overlapping distributions. Outside the area where the distributions overlap, a given value clearly identifies the presence or absence of disease or some other manifestation of poor health. If a value falls into the area of overlap, the individual may belong to either the normal or the abnormal group. The choice of the normal limits depends upon the relative importance attached to the identification of individuals as healthy or unhealthy. See also *false negative, false positive, sensitivity and specificity.*

null hypothesis In simplest terms, the null hypothesis states that the results observed in a study, experiment, or test are no different from what might have occurred as a result of the operation of chance alone. The hypothesis that the independent variable(s) under study have no association with or effect upon the dependent or outcome variable(s).

observational study (syn: nonexperimental study, survey) Epidemiologic study in situations where nature is allowed to take its course; changes or differences in one characteristic are studied in relation to changes or differences in other(s), without the intervention of the investigator.

occurrence (syn: frequency) In epidemiology, a general term describing the frequency of a disease or other attribute or event in a population without distinguishing between incidence and *prevalence.*

odds The ratio of the probability of occurrence of an event to that of nonoccurrence, or the ratio of the probability that something is so, to the probability that it is not so.

odds ratio (syn: cross-product ratio, relative odds) The ratio of two odds.

outliers Observations differing so widely from the rest of the data as to lead one to suspect that a gross error may have been committed, or suggesting that these values come from a different population.

overmatching A situation that may arise when groups are matched. Several varieties can be distinguished.

1. The matching procedure partially or completely obscures evidence of a true causal association between the independent and dependent variables. Overmatching may occur if the matching variable is involved in, or is closely connected with, the mechanism whereby the independent variable affects the dependent variable. The matching variable may be an intermediate cause in the causal chain or it may be strongly affected by, or a consequence of, such an intermediate cause.

2. The matching procedure uses one or more unnecessary matching variables, e.g. variables that have no causal effect or influence on the dependent variable, and hence cannot confound the relationship between the independent and dependent variables.

3. The matching process is unduly elaborate, involving the use of numerous matching variables and/or insisting on very close similarity with respect to specific matching variables. This leads to difficulty in finding suitable controls. See also *matching.*

parameter A constant in a statistical formula or model. The word is often used loosely as a synonym for "factor" or "variable".

parametric method A method of hypothesis testing that requires the assumptions of a particular model concerning the distribution of data (normal, Poisson, etc.).

pathogenesis The postulated mechanisms by which the etiologic agent produces disease. The difference between *etiology* and *pathogenesis* should be noted. The etiology of a disease or disability consists of the postulated causes that initiate the pathogenetic mechanisms; control of these causes might lead to prevention of the disease.

pathogenicity The property of an organism that determines the extent to which overt disease is produced in an infected population, or the power of an organism to produce disease. Also used to describe comparable properties of toxic chemicals, etc. Pathogenicity of infectious agents is measured by the ratio of the number of persons developing clinical illness to the number exposed to infection.

permissible exposure limit (PEL) An occupational health standard to safeguard employees against dangerous chemicals or contaminants in the workplace.

person-time A unit of measurement combining persons and time, used as denominator in instantaneous incidence rates. It is the sum of individual units of time that the persons in the study population have been exposed to the condition of interest.

pilot investigation (syn: pilot study) An inquiry, survey, or study of limited scope, employed to test the feasibility of more extensive investigation.

pollution Any undesirable modification of air, water, or food by substance(s) that are toxic or may have adverse effects on health or that are offensive though not necessarily harmful to health.

precision 1. The quality of being sharply defined or stated. One measure of precision is the number of distinguishable alternatives from which a measurement was selected, sometimes indicated by the number of significant digits in the measurement. Another measure of precision is the standard error of measurement, the standard deviation of a series of replicate determinations of the same quantity.

2. In statistics, precision is defined as the inverse of the variance of a measurement or estimate.

prevalence The number of instances of a given disease or other condition in a given population at a designated time; sometimes used to mean *prevalence rate*. When used without qualification the term usually refers to the situation at a specified point in time (point prevalence).

prevalence, annual (an occasionally used index) The total number of persons with the disease or attribute at any time during a year. It includes cases of the disease arising before but extending into or through the year as well as those having their inception during the year.

prevalence rate (ratio) The total number of all individuals who have an attribute or disease at a particular time (or during a particular period) divided by the population at risk of having the attribute or disease at this point in time or midway through the period. A problem may arise with calculating period prevalence rates because of the difficulty of defining the most appropriate denominator.

prevention   The goals of medicine are to promote health, to preserve health, to restore health when it is impaired, and to minimize suffering and distress. These goals are embodied in the word "prevention", which is easiest to define in the context of levels, customarily called primary, secondary, and tertiary prevention. Authorities on preventive medicine do not agree on the precise boundaries between these levels, nor on how many levels can be distinguished, but the differences of opinion are semantic rather than substantive.

An epidemiologic interpretation of the distinction between primary and secondary prevention is that primary prevention is aimed at reducing incidence of disease and other departures from good health, secondary prevention aims to reduce prevalence by shortening the duration, and tertiary prevention is aimed at reducing complications.

Primary prevention can be defined as the promotion of health by personal and community-wide efforts, e.g. improving nutritional status, physical fitness, and emotional wellbeing, immunizing against infectious diseases, and making the environment safe.

Secondary prevention can be defined as the measures available to individuals and populations for the early detection and prompt and effective intervention to correct departures from good health.

Tertiary prevention consists of the measures available to reduce impairments and disabilities, minimize suffering caused by existing departures from good health, and to promote the patient's adjustment to irremediable conditions. This extends the concept of prevention into the field of rehabilitation.

protocol   The plan, or set of steps, to be followed in a study or investigation, or in an intervention programme.

public health   Public health is one of the efforts organized by society to protect, promote, and restore the people's health. It is the combination of sciences, skills, and beliefs that are directed to the maintenance and improvement of the health of all the people through collective or social actions.

quantitative data   Data in numerical quantities such as continuous measurements or counts.

quasi-experiment   An experiment in which the investigator lacks full control over the allocation and/or the timing of the intervention.

questionnaire   A predetermined set of questions used to collect data— clinical data, social status, occupational group, etc. This term is often applied to a self-completed survey instrument, as contrasted with an interview schedule.

randomization   Allocation of individuals to groups, e.g. for experimental and control regimens, by chance. Within the limits of chance variation, randomization should make the control and experimental groups similar at the start of an investigation and ensure that personal judgement and prejudices of the investigator do not influence allocation.

random sample   A sample that is arrived at by selecting sample units such that each possible unit has a fixed and determinate probability of selection. See also *sample.*

range of distribution   The difference between the largest and smallest values in a distribution.

ranking scale (ordinal scale)   A scale that arrays the members of a group from high to low according to the magnitude of the observations, assigns numbers to the ranks, and neglects distances between members of the array.

rate   This term is used inconsistently in different contexts. We offer a "strict" definition and then identify some of its less strict applications.

Definition   A rate is a ratio whose essential characteristic is that time (per minute, per hour, etc.) is an element of the denominator and in which there is a distinct relationship between numerator and denominator. The numerator may be a measured quantity (e.g. litres per day: centimetres per year) or a counted valued (e.g. hospital admissions per year). Additional terms may be included in either numerator or denominator and usually are in the latter (e.g. calories per kilogram per day: attacks per 1000 population per annum).

Less strict application   While a number of commonly used demographic and epidemiologic "rates" are true rates in the above sense, others do not meet this definition. An attack rate is not a true rate, but a birth rate (births per 1000 population per year) is. An infant mortality rate (deaths under one year, per 1000 live births) is not a true rate, but rather a proportion. Perinatal mortality rate, neonatal mortality rate, and case fatality rate (unless very carefully specified) are usually dimensionless in time, and thus not true rates. Force of mortality is a true rate but the *prevalence rate* and the *cumulative incidence rate* are not. See also *ratio.*

ratio   The value obtained by dividing one quantity by another. A general term of which rate, proportion, percentage, prevalence, etc. are subsets.

record linkage   A method for assembling the information contained in two or more records, e.g. in different sets of medical charts, and in vital records such as birth and death certificates, and a procedure to ensure that the same individual is counted only once.

register, registry   In epidemiology the term "register" is applied to the file of data concerning all cases of a particular disease or other health-relevant condition in a defined population such that the cases can be related to a population base. With this information incidence rates can be calculated. If the cases are regularly followed up, information on remission, exacerbation, prevalence, and survival can also be obtained. The *register* is the actual document, and the *registry* is the system of ongoing registration.

relative risk   1.   The ratio of the *risk* of disease or death among the exposed to the risk among the unexposed; this usage is synonymous with *risk ratio.*

2. Alternatively, the ratio of the cumulative incidence rate in the exposed to the cumulative incidence rate in the unexposed, i.e. the cumulative incidence ratio.

reliability The degree of stability exhibited when a measurement is repeated under identical conditions.

representative sample The term "representative" as it is commonly used is undefined in the statistical or mathematical sense; it means simply that the sample resembles the population in some way.

The use of probability sampling will not ensure that any single sample will be "representative" of the population in all possible respects. If, for example, it is found that the sample age distribution is quite different from that of the population, it is possible to make corrections for the known differences. A common fallacy lies in the unwarranted assumption that, if the sample resembles the population closely on those factors that have been checked, it is "totally representative" and that no difference exists between the sample and the universe or reference population.

research design The procedures and methods, predetermined by an investigator, to be adhered to in conducting a research project.

risk A probability that an event will occur, e.g. that an individual will become ill or die within a stated period of time or age. Also, a nontechnical term encompassing a variety of measures of the probability of a (generally) unfavourable outcome.

risk factor This term is used by different authors with at least three different meanings.

1. An attribute or exposure that is associated with an increased probability of a specified outcome, such as the occurrence of a disease. Not necessarily a causal factor. A risk marker.

2. An attribute or exposure that increases the probability of occurrence of disease or other specified outcome. A *determinant*.

3. A determinant that can be modified by intervention, thereby reducing the probability of occurrence of disease or other specified outcomes. To avoid confusion, may be referred to as a modifiable risk factor.

risk ratio The ratio of two risks.

sample A selected subset of a population. A sample may be random or nonrandom and may be representative or nonrepresentative. Several types of sample can be distinguished, including the following.

Cluster sample Each unit selected is a group of persons (all persons in a city block, a family, etc.) rather than an individual.

Grab sample (syn: sample of convenience) These ill defined terms describe samples selected by easily employed but basically nonprobabilistic methods. "Man-in-the-street" surveys and a survey of blood pressure among volunteers who drop in at an examination booth are in this category. It is improper to generalize from the results of a survey based upon such a sample for there is no way of knowing what sorts of bias may have been operating. See also *bias*.

sampling   The process of selecting a number of subjects from all the subjects in a particular group or "universe". Conclusions based on sample results may be attributed only to the population sampled. Any extrapolation to a larger or different population is a judgement or a guess and is not part of statistical inference.

sampling variation   Since the inclusion of individuals in a sample is determined by chance, the results of analysis in two or more samples will differ, purely by chance. This is known as "sampling variation".

sensitivity and **specificity** (of a screening test)   *Sensitivity* is the proportion of truly diseased persons in the screened population who are identified as diseased by the screening test. Sensitivity is a measure of the probability of correctly diagnosing a case, or the probability that any given case will be identified by the test (syn: true positive rate).

Specificity is the proportion of truly nondiseased persons who are so identified by the screening test. It is a measure of the probability of correctly identifying a nondiseased person with a screening test (syn: true negative rate).

The relationships are shown in the following fourfold table, in which the letters a, b, c, and d represent the quantities specified below the table.

Screening test results	True status		Total
	Diseased	Not diseased	
Positive	a	b	$a + b$
Negative	c	d	$c + d$
Total	$a + c$	$b + d$	$a + b + c + d$

a = Diseased individuals detected by the test (true positives).

b = Nondiseased individuals positive by the test (false positives).

c = Diseased individuals not detectable by the test (false negatives).

d = Nondiseased individuals negative by the test (true negatives).

$$\text{Sensitivity} = \frac{a}{a + c}$$

$$\text{Specificity} = \frac{d}{b + d}$$

$$\text{Predictive value (positive test result)} = \frac{a}{a + b}$$

$$\text{Predictive value (negative test result)} = \frac{d}{c + d}$$

391

sequential analysis   A statistical method that allows an experiment to be ended as soon as an answer of the desired precision is obtained. Study and control subjects are randomly allocated in pairs or blocks. The result of the comparison of each pair of subjects, one treated and one control, is examined as soon as it becomes available and is added to all previous results.

standardized rate ratio (SRR)   A rate ratio in which the numerator and denominator rates have been standardized to the same (standard) population distribution.

stratification   The process of or result of separating a sample into several subsamples according to specified criteria such as age groups, socioeconomic status, etc. The effect of confounding variables may be controlled by stratifying the analysis of results. For example, lung cancer is known to be associated with smoking. To examine the possible association between urban atmospheric pollution and lung cancer, controlling for smoking, the population may be divided into strata according to smoking status. The association between air pollution and cancer can then be appraised separately within each stratum. Stratification is used not only to control for confounding effects but also as a way of detecting modifying effects. In this example, stratification makes it possible to examine the effect of smoking on the association between atmospheric pollution and lung cancer.

survey   An investigation in which information is systematically collected but in which the experimental method is not used. A population survey may be conducted by face-to-face inquiry, by self-completed questionnaires, by telephone, postal service, or in some other way. Each method has its advantages and disadvantages. For instance, a face-to-face (interview) survey may be a better way than a self-completed questionnaire to collect information on attitudes or feelings, but it is more costly. Existing medical or other records may contain accurate information, but not about a representative sample of the population.

survey instrument   The interview schedule, questionnaire, medical examination record form, etc., used in a survey.

variable   Any quantity that varies. Any attribute, phenomenon, or event that can have different values.

vital records   Vital records are certificates of birth, death, marriage, and divorce required for legal and demographic purposes.

vital statistics   Systematically tabulated information concerning births, marriages, divorces, separations, and deaths based on registrations of these vital events.